Frontiers of Metallurgy and Materials Technology

Frontiers of Metallurgy and Materials Technology

Editor

V. V. Kutumba Rao

BSP **BS Publications**

An unit of **BSP Books Pvt., Ltd.**

4-4-309, Giriraj Lane, Sultan Bazar,
Hyderabad - 500 095
Phone : 040 - 23445605, 23445688

Published by :

BSP **BS Publications**

An unit of **BSP Books Pvt., Ltd.**

4-4-309, Giriraj Lane, Sultan Bazar,
Hyderabad - 500 095
Phone : 040 - 23445605, 23445688
e-mail : info@bspbooks.net

ISBN: 978-93-85433-32-0 (HB)

Dedicated to the Memory of

Late Dr. Patcha Ramachandra Rao (1942 – 2010)
Chairman, Organizing Committee, fMMT 2009

Dr. Ramachandra Rao guided the organisation of fMMT 2009 from the time the idea of the Conference was conceived till his untimely passing away in January, 2010. His exemplary dedication, deep involvement and boundless enthusiasm were a major source of inspiration for the other workers which led them to achieve the highly acclaimed success of the Conference.

Dr Ramachandra Rao was one of the most outstanding materials scientists of our country. He achieved international recognition through lasting contributions in numerous areas of metallurgy and materials science. A highly innovative experimentalist, he pioneered Rapid Solidification studies in India. He developed intriguingly original models for the formation and structure of several meta-stable and quasi-crystalline materials. Towards the latter years of his highly fruitful research career he was deeply involved with Biomimetic materials and perfected several innovative experiments with them. At the time of his sudden passing away, he was on the verge of completing a novel treatise on Thermodynamics that no doubt would have made a great impact on the world literature in this area. Like all true masters, he spawned a huge band of talented students and scientists who have made their own mark in research. Recognitions and honours galore, too numerous to list here, came naturally his way.

An able administrator that Dr Ramachandra Rao was, he guided the destinies of the prestigious National Metallurgical Laboratory, Jamshedpur as its Director for over a decade during which the laboratory rose to great heights of excellence in metallurgical research. Later he provided remarkable leadership to bring stability and growth to two major institutions namely, the Banaras Hindu University, Varanasi and the Defence Institute of Armament Technology, Pune as their Vice Chancellor.

Above all, Dr Ramachandra Rao possessed an extraordinarily charismatic personality that drew a large number of friends who enjoyed his company and shared a good hearty laugh with him. The picture that would linger in their minds for a long time is that of a gifted and unforgettable friend who lighted up their world for the all-too-brief period of his mortal existence.

– V V Kutumbarao

Preface

The world today is witness to lightning-rapid strides in technology that are helping people to live easier and better lives. Many of these developments have become possible because of the matching advances in materials technology that are making available an increasing variety of better and cheaper materials. For students of metallurgy and materials science as also the practising engineers and scientists, it is difficult to keep pace with these advances on a routine basis owing to the much slower evolution of University syllabi or the compulsions of day-to-day work life. Thus it is essential to hold periodic gatherings of experts to take stock of the developments in their respective specialisations and pass on their knowledge to others.

One such attempt is the present Conference organised jointly by the Department of Metallurgy and Materials Technology of the Mahatma Gandhi Institute of Technology (MGIT), Hyderabad and the local Student Affiliate Chapter of the Indian Institute of Metals during January 29-31, 2009. Venue of the Conference was the Taramati Baradari, one of the most beautiful and spacious auditoriums in Hyderabd equipped with most modern amenities and located amidst picturesque surroundings dotted with historical structures. It attracted a large participation with over — delegates from India and — from other countries. Particularly heartening was the enthusiastic response from the huge body of students of several engineering colleges who contributed greatly to the success of the Conference.

The proceedings over three days of the Conference served a veritable knowledge feast to the students and other delegates with 38 papers of very high calibre on the latest developments in Metallurgy and Materials Technology the world over. It is important to record that nearly a third of these papers (29%) have come from countries other than India represented by top-notch academic institutions as well as the industry thus making the Conference truly international. It is satisfying to note that over 80% of the papers presented were concerned with results of original cutting-edge research or reviews thereof thus helping to achieve the objective of the Conference to touch the frontiers in Materials Technologies. Large participation by the Industry which accounted for nearly a quarter of the papers presented is another notable feature of the Conference.

Materials covered in these presentations encompass a wide range from traditional ones like Al alloys, Ti alloys and steels to new classes of materials like superalloys, aluminides, metal foams, nuclear fuels, carbon nanotubes and composites. Processing operations dealt with here range from electro-slag refining, powder metallurgy processing and welding to bio-processing, cold spraying, laser processing, nano-coating, non-equilibrium processing and even processing in Space. Papers on modelling and statistical analysis eloquently bring out the importance of these aspects of research. The paper by an expert from the celebrated British Museum on a detailed scientific analysis of the popular Indian art form of Bidri as also the paper dealing with the results of outstanding original research on the enigmatic Delhi iron Pillar help to highlight the proud Indian metallurgical heritage. Imperatives in modernising metallurgical education to serve the emerging requirements of the industry have been brought out by a highly respected veteran of the Indian steel industry.

Proceedings of the Conference are presented in this Volume consisting of the full text of all the papers with figures and tables. It is hoped that the earnest readers would find that the material presented here provides adequate information on recent developments in Materials Technology that would enable them to pursue these studies further or to initiate new lines of research.

– Organizing Committee

Contents

COMPOSITE AND CELLULAR MATERIALS

MANUFACTURING PROCESSES AND INDUSTRIAL PRACTICES

POWDER TECHNOLOGY

SURFACE ENGINEERING

MICROSTRUCTURAL ENGINEERING AND CHARACTERISATION

Non-Equilibrium Processing of Advanced Materials

C. Suryanarayana

Department of Mechanical, Materials and Aerospace Engineering
University of Central Florida, Orlando, USA
e-mail: csuryana@mail.ucf.edu

Abstract: *Mechanical alloying (MA) is a powder metallurgy processing technique that involves repeated cold welding, fracturing and rewelding of powder particles in a high-energy ball mill. Started in 1960 to produce oxide dispersion strengthened (ODS) nickel-based superalloys to obtain high strength both at room and elevated temperatures, this technique is now widely used to synthesize a variety of non-equilibrium alloys including solid solutions, intermetallics, and glassy alloys. Apart from the ODS alloys, the products of MA are now finding applications in newer and innovative areas. The present article briefly summarizes the recent contributions of the author and his colleagues and collaborators in the field of MA to synthesize commercially useful and advanced materials. Both the scientific and technological aspects of MSAZ have been reviewed. Four specific topics dealing with advanced materials have been covered – nanocomposites, Fe-based metallic glasses, photovoltaic materials, and use of MA powder particles in metal combustion. The article concludes with an indication of the topics that need special attention in the near future.*

1. Introduction

Advanced materials are defined as those where first consideration is given to the systematic synthesis and control of the crystal structure and microstructure of materials to provide precisely tailored properties for demanding applications [1]. With the rapid advances taking place in the technology sectors, there has been a great demand for materials that are stronger, stiffer, lighter, usable at higher temperatures, and less expensive than the existing and available materials. Thus, the high-tech industries have given an immense fillip to efforts in developing novel materials that will perform better for demanding applications. Some of the advanced materials developed during the last few decades include metallic glasses [2,3] (including the relatively recently developed bulk metallic glasses [4]), quasicrystals [5,6], high-temperature superconductors, superhard carbo-nitrides, thin-film diamond synthesis, and nanostructured materials [7-9]. A number of new techniques for processing and characterizing these advanced materials have also been developed towards achieving this goal. But, a common underlying theme in obtaining advanced materials has been to process the materials under far-from-equilibrium conditions, i.e., the materials developed are mostly in a metastable condition. Turnbull [10] has coined the phrase "energize and quench" to describe these processes, and sometimes these are also referred to as "driven" materials [11]. It is through such processes that the constitution and microstructure of the materials can be drastically altered leading to achieving the desired properties. Some of these "non-equilibrium processing" techniques include rapid solidification from the melt, mechanical alloying, laser processing, plasma processing, spray forming, physical and chemical vapor deposition techniques, ion mixing, among others. The relative advantages and disadvantages of these techniques have been discussed in a recent publication [12]. However, we will focus on the technique of mechanical alloying in the following paragraphs and describe some of the advanced materials that have been developed using the processing method.

2. The Process of Mechanical Alloying

Mechanical Alloying (MA) is a process that was developed by John Benjamin of INCO International in the late 1960's to produce oxide dispersion-strengthened (ODS) materials [13]. The process can be described as a high-energy milling process in which powder particles are subjected to repeated cold welding, fracturing, and rewelding [14,15]. The generally accepted explanation for alloying to occur from blended elemental powders and formation of different types of phases is that a very fine and intimate mixture of the components (often lamellar if the constituent elements are sufficiently ductile) is formed after milling, if not the final product. The crystalline defects (grain boundaries, interfaces, dislocations, stacking faults, vacancies, and others), introduced into the material due tot the intense cold working operation, which act as fast diffusion paths, and a slight rise in the powder temperature during milling as a result of frictional forces and impact of the grinding balls against other balls and the surfaces of the container were found to facilitate alloy formation. If the final desired phase had not formed directly by MA, then a short annealing treatment at an elevated temperature was found to promote diffusion and consequently alloy formation. Accordingly, it was shown that by a proper choice of the process parameters during MA and choosing an appropriate alloy composition, it is possible to produce solid solutions (both equilibrium and supersaturated), intermediate phases (crystalline and quasicrystalline) and amorphous alloys. The thermodynamic criterion that the phase with the lowest free energy, under the continuous deformation conditions, would be the most "stable" phase has been found to be valid. A significant attribute of MA has been the ability to even alloy metals with positive heats of mixing that are difficult to alloy otherwise.

The effects of MA could be categorized under two groups:

1. Constitutional changes: Milling of the blended elemental powders could lead to the formation of solid solutions (both equilibrium and supersaturated), intermetallic phases (equilibrium, metastable and quasicrystalline), and amorphous phases in alloy systems.

4

The actual type of phase formed could be different depending on the alloy system and the milling conditions employed.

2. Microstructural changes: As a result of MA, the processed materials could develop ultrafine-grained and nanostructured phases. In addition, uniform dispersion of a large volume fraction of very fine oxide or other ceramic particles can be achieved in different matrices; not possible or easy by other processes.

In many cases, chemical reactions have also been found to take place leading to the formation of novel materials and microstructures. Such a process has been referred to as mechanochemical processing [16]. All these effects have been well documented in the literature [14,15,17].

As has been frequently pointed out in the literature, the technique of MA was developed out of an industrial necessity in 1966 to produce nickel-based superalloys that combined the room temperature strength obtained by precipitation hardening with the high-temperature strength achieved through dispersion hardening [13,18]. The historical development of MA during the last 40 years or so can be divided into three major periods. The first period, covering the first twenty years (from 1966 up to about 1985), was mostly concerned with the development and production of oxide dispersion strengthened (ODS) superalloys for applications in the aerospace industry. Several alloys, with improved properties, based on Ni and Fe were developed that found useful applications. These included the MA754, MA760, MA956, MA957, MA6000, and others [19]. The second period of about fifteen years, covering from about 1986 up to about 2000, witnessed lot of advances in the fundamental understanding of the processes that take place during MA. Along with this, a large number of dedicated conferences were held and there was a burgeoning of publication activity [17]. Comparisons have also been frequently made between MA and other non-equilibrium processing techniques, notably rapid solidification processing (RSP). It was shown that the metastable effects achieved by these two non-equilibrium processing techniques are similar. Revival of the mechanochemical processing (MCP) took place and a variety of novel substances were synthesized. Several modeling studies were also conducted to enable prediction of the phases produced or microstructures obtained, although with limited success. The third period, starting from about 2001, saw a reemergence of the quest for new applications of the MA materials, not only as structural materials, but also for chemical applications such as catalysis and hydrogen storage, with the realization that contamination of the milled powders is the limiting factor in the widespread applications of the MA materials. Innovative techniques to consolidate the MA powders to full density while retaining the metastable phases (including glassy phases and/or nanostructures) in them were also developed. All these are continuing with the ongoing investigations to enhance the scientific understanding of the MA process. Fig. 1 shows these developments schematically.

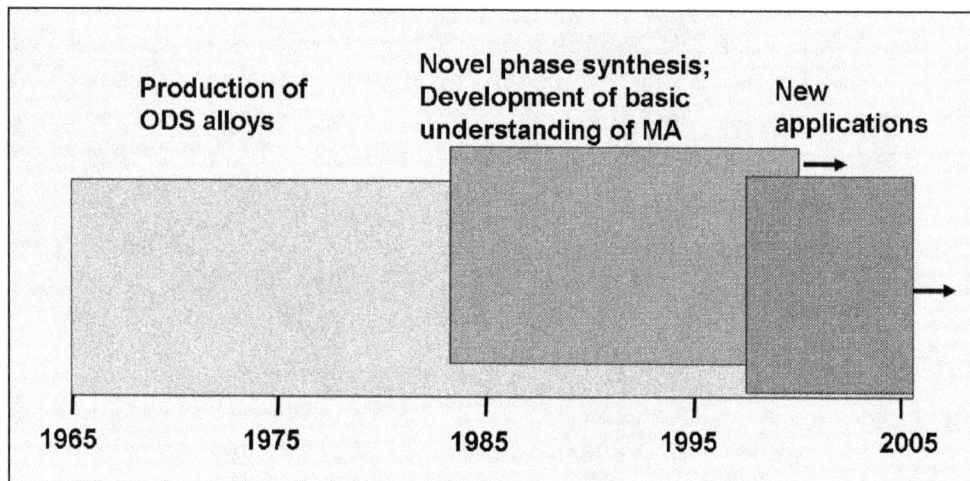

Fig. 1 Schematic showing the different periods of development in mechanical alloying (MA) since its inception in 1966.

It should also be mentioned in this context that, at present, activities relating to both the science and technology of MA have been going on simultaneously. While some groups have been focusing on developing a better understanding of the process and optimizing the process variables to achieve a certain phase or phase combination in the alloy system, others have been concentrating on exploiting this technique for different applications. The bottom line is that newer materials are being developed for novel applications. We will try to give a bird's eye view of both these aspects, with emphasis on the work conducted in the author's laboratories.

3. Science of Mechanical Alloying

A knowledge of the reasons for the formation of metastable phases and the mechanism of their formation will greatly aid in achieving these constitutional effects in a number of alloy systems. Even though solid solutions, intermetallic phases, and metallic glasses have been produced by MA, we will discuss only briefly discuss the results obtained on Solid Solutions and Metallic Glasses.

3.1. *Solid Solutions*

Solid solution alloys show many desirable properties such as increase in strength and modulus of elasticity, changes in density, and suppression or elimination of undesirable

6

second phases. Alloy strengthening through precipitation of a second phase from a supersaturated solid solutions during aging has been known to materials scientists for about a hundred years. Further, since the strengthening effect is related to the size and volume fraction of the second phase particles, and both these could be controlled by the extent of supersaturation and the aging parameters, achievement of extended solid solubility limits in alloy systems is very desirable.

Empirical rules for determining the limits of solid solubility limits in alloy systems under equilibrium conditions, especially those based on the noble metals Cu. Ag. And Au, were elegantly described Hume-Rothery [20]. These include favorable atomic sizes (< about 15% difference), not much of a difference in the electronegativities, similar valences, and same crystal structures. A similar set of rules seem to be generally applicable to supersaturated solutions obtained by MA, even though exceptions have been noted in quite a few cases [14,15]. Fig. 2 shows the Darken-Gurry plot for solid solubility limits achieved by MA in Cu-based alloy systems [15]. In this plot, electronegativity is plotted against the atomic radius for all the elements. An ellipse with ± 15% atomic radius and ± 0.3 units in electronegativity is drawn around the solvent element (Cu in this case). All elements inside this ellipse are expected to exhibit extensive solid solubility limits, while those outside the ellipse are expected to have limited solid solubilities. It is interesting to note from this figure that even though complete solid solubility of Cu in Ag and that of Ag in Cu were achieved by MA [21], the element Ag lies well outside the ellipse representing extensive solid solubility. Further, Cr and Fe, which do not show extensive formation of solid solutions lie inside the ellipse. This could be due to the fact that the crystal structures and valences are not considered in this plot. But, as a general rule of thumb, it has been noted that solute elements which exhibit limited solubility under equilibrium conditions exhibit high solubility under the non-equilibrium conditions of MA.

3.2. *Metallic Glasses*

Metallic glasses are novel materials first obtained in a Au-Si eutectic alloy by Duwez et al. using the RSP technique in 1960 [22]. Since then a large number of metallic glasses have been obtained in different alloy systems and their mechanical, physical, and chemical properties have been evaluated. A variety of different applications have been found for these novel materials, the most important of them being as core laminations in distribution transformers [2,3]. Metallic glasses (or amorphous alloys to be more precise, since traditionally the tem glass is

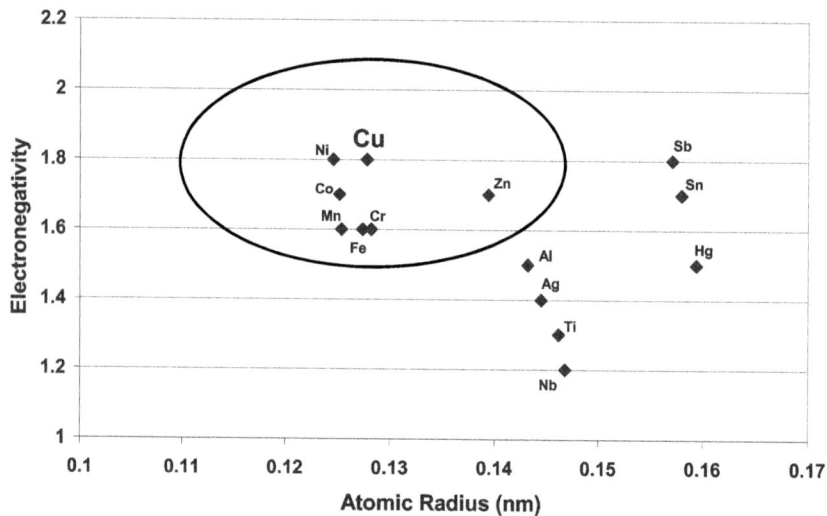

Fig. 2 Darken-Gurry (electronegativity vs. atomic size) plot for mechanically alloyed copper-based powder mixtures. Elements lying inside the ellipse drawn with ± 15% atomic size ands ± 0.3 units of electronegativity are expected to show high solid solubilities.

reserved for solid non-crystalline alloys obtained by continuous cooling from the liquid state) were also synthesized by MA first in an intermetallic compound [23] and later in a blended powder mixture [24]. Amorphous phases have now been synthesized in a very large number of alloy systems starting from either blended elemental powder mixtures, intermetallic compounds, or mixtures of them [14,15,25]. Comparisons have been frequently drawn between the amorphous alloys produced by MA and RSP techniques, and both similarities and differences have been noted [15]. Further, the mechanism of glass formation appears to be significantly different by MA and RSP methods.

A number of criteria were developed to predict glass formation in alloy systems by the RSP technique. These include the reduced glass transition temperature, defined as $T_{rg} = T_g/T_l$, where T_g is the glass transition temperature and T_l is the liquidus temperature, presence of deep eutectics in the phase diagrams, significant differences in the atomic sizes of the constituent elements, and several others. With the development of bulk metallic glasses since 1988, additional criteria were also proposed such as the width of the supercooled liquid region defined as $\Delta T_x = T_x - T_g$, where T_x is the first crystallization temperature of the amorphous alloy, and a host of other parameters based on the glass transformation temperatures [26]. The three most important criteria that need to be satisfied for obtaining a bulk metallic glass are (i) the alloy system should contain at least three elements, (ii) a significant difference in atomic sizes should exist among the constituent elements, and (iii) the constituent elements should have a negative heat of mixing among them. Alloys with bulk metallic glass compositions have also been synthesized by MA.

8

There are some significant differences in the formation of metallic glasses by the MA and RSP methods. For example, metallic glasses are obtained by MA in almost every alloy system provided that sufficient energy has been stored in the powder. But, metallic glasses are obtained by RSP methods only in certain composition ranges – about 20 at.% metalloid content in metal-metalloid systems and different compositions in metal-metal systems, mostly at compositions corresponding to deep eutectic compositions. Additionally, glass formation by MA occurs by the accumulation of crystal defects, which raises the free energy of the crystalline phase above that of the hypothetical glassy phase, under which conditions the crystalline phase becomes destabilized. On the other hand, for glass formation by the RSP methods, the critical cooling rate for glass formation needs to be exceeded so that formation of the crystalline nuclei is completely suppressed. Since MA is a completely solid-state powder processing technique, we wished to determine if the criteria used for RSP method would be applicable to the solid-state processed amorphous alloys. For this purpose, we had undertaken a comprehensive investigation on the glass formation behavior and the stability of several Fe-based glassy alloys synthesized by MA [27-32].

Systematic and detailed investigations were conducted on Fe-based alloy composition such as $Fe_{42}X_{28}Zr_{10}B_{20}$ with X = Al, Co, Ge, Mn, Ni, or Sn. The subscripts represent the composition of the alloy in atomic percent. The element X was chosen on the basis of the number of intermetallics it forms with Zr under equilibrium conditions at room temperature. This number increases from 1 with Mn to 8 with Al [Ref. 33], which provides a basis to analyze the results systematically. Further, the negative heat of formation of the intermetallics with Zr is much higher than that with either Fe or B [Ref. 34]. Thus, the probability of forming an intermetallic with Zr during milling is higher than with Fe or B.

It was noted from the results that amorphous phases were obtained only in alloy systems containing X – Al, Ge, and Ni and not in the other systems. The respective binary phase diagrams show that the total number of intermetallics is more than 10 in systems that formed the amorphous phase and less in others. Thus, it could be concluded that the ease of glass formation, as measured by the time required for glass formation, increases with the number of intermetallics in the phase diagrams [27].

The presence of intermetallics in an alloy system easily contributes to amorphization. This is due to two important effects. Firstly, disordering of intermetallics contributes an energy of about 15 kJ/mole of atoms to the system. For example, in strongly ordered intermetallics such as NiAl and γ-TiAl that continue to be in the ordered state till melting, the disordering energy has been estimated to be about 17.5 kJ/mol. Secondly, a slight change in the stoichiometry of the intermetallic increases the free energy of the system drastically. Additionally, grain size reduction contributes about 5 kJ/mole. Since MA reduces the grain size to nanometer levels and also disorders the usually ordered intermetallics, the energy of the milled powders is significantly raised. In fact, it is raised to a level above that of the hypothetical amorphous phase. This condition leads to a situation where the preferential formation of the amorphous phase is favored over the crystalline phase.

On the other hand, it is noted that in alloy systems which did not amorphize on milling, a solid solution phase only had formed. Since the only way a solid solution could contribute to an increase in the energy of the system is by grain size refinement, the total increase in the energy is not sufficiently high to amorphize the system. Thus, these observations provide a simple and visual means of identifying alloy systems which are likely to become amorphous on subjecting them to milling.

It has also been observed that continued milling of the powders beyond the formation of the amorphous phase resulted in crystallization of the amorphous phase, and this has been termed mechanical crystallization (Fig. 3) [Ref. 28,29]. The main reason for the mechanical crystallization to occur during mechanical milling is that the relative stabilities of the different phases change due to the incorporation of a variety of defects.

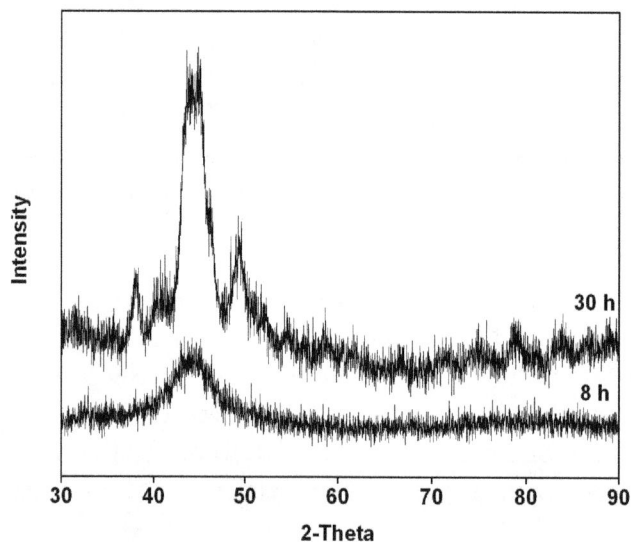

Fig. 3 X-ray diffraction patterns of the $Fe_{42}Ni_{28}Zr_{10}C_{10}B_{10}$ powder blend milled for 8 and 30 h. Note that the powder milled for 8 h shows the presence of an amorphous phase. On continued milling of this powder containing the amorphous phase to 30 h, mechanical crystallization had occurred resulting in the formation of several crystalline phases.

The influence of additional alloying elements with different atomic sizes and heats of mixing were also investigated with useful attributes for consolidating the milled powders to full density. It was noted, for example, that addition of C hastens the formation of the amorphous phase [30], and the addition of Nb, which has a positive heat of mixing with Zr also enhances the glass-forming ability. However, there appears to be a critical concentration of Nb to achieve the best glass-forming ability [31]. Addition of 2 at. % Nb shows the best glass-forming ability; deviation from this value reduces the ability to form the glass (Fig. 4). This requirement of the critical composition for achieving glass formation most easily and the associated lattice contraction during amorphization have been explained using the concept of lattice strain present due to atomic size mismatch and the different co-ordination number as a result of the atomic sizes of the individual components in the alloy system [32].

From the above paragraphs, succinctly describing the results on the formation of solid solutions and metallic glasses obtained by MA, it becomes amply clear that very important basic information can be obtained on the mechanism of formation of metastable phases. Such knowledge could aid in the design and development of new materials to explore their applications in advanced technologies.

4. Technology of Mechanical Alloying

As mentioned earlier, the technique of MA was developed as an industrial necessity to produce oxide-dispersion strengthened (ODS) superalloys for use in the aerospace industry. This has been the major application of MA products, in addition to other small-scale applications that have been developed later. But, in recent years, there has been a resurgence of search for novel applications of MA products and efforts are being continuously made to exploit the mechanically alloyed materials for different applications. We will summarize our recent activities in this direction and we will discuss these under three specific topics:

(1) Development of homogeneously-dispersed nanocomposites,

(2) Synthesis of photo-voltaic materials, and

(3) Use of energetic metal particles in combustion.

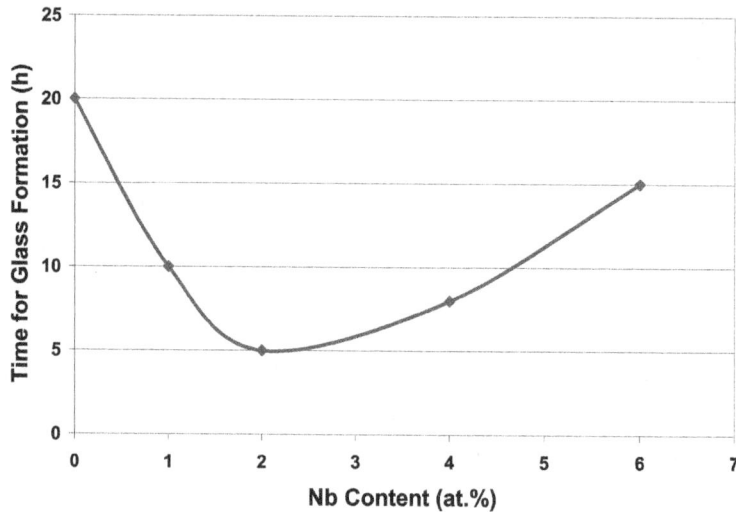

Fig. 4 Time required for formation of the glassy phase in mechanically alloyed $Fe_{42}Ni_{28}Zr_{10-x}Nb_xB_{20}$ alloys as a function of the Nb content. Note that the time required decreases with increasing Nb content up to certain level and then it increases again.

4.1 Nanocomposites

Achievement of a uniform distribution of the reinforcement in a matrix is essential to achieving good mechanical properties of the composites. Further, the mechanical properties of the composite tend to improve with increasing volume fraction and decreasing particle size of the reinforcement [35]. Traditionally, a reasonably large volume fraction of the reinforcement could be added, if the size of the reinforcement is large (on a micrometer scale). But, if the reinforcement size is very fine (of nanometer dimensions), then the volume fraction added is restricted to about 2 to 4 %. This is because the fine powders tend to float to the top of the melt during processing of the composites through solidification processing. However, if we are able to introduce a large volume fraction of nanometer-sized reinforcement, the mechanical properties of the composite are likely to be vastly improved. In our investigations, during the last few years, we have successfully achieved a very uniform distribution of the reinforcement phases in different types of matrices through the solid-state powder processing technique of MA [36-45]. These include homogeneous dispersion of graphite in an Al6061 alloy matrix [36], effect of clustering of the reinforcement on the mechanical properties of the composites [37], dispersion of a high volume fraction of Al_2O_3 in Al [41], and of Ti_5Si_3 in γ-TiAl [38,44,45], synthesis of $MoSi_2+Si_3N_4$ composites for high-temperature applications [39,40,43], and synthesis of amorphous+intermetallic composites in Al-Mg alloys [42]. However, we will describe only two typical examples.

12

A. Al-Al$_2$O$_3$ Composites

Aluminum-based metal matrix composites are ideal materials for structural applications in the aircraft and automotive industries due to their lightweight and high strength-to-weight ratio. Reinforcement of the ductile aluminum matrix with stronger and stiffer second-phase reinforcements like oxides, carbides, borides, and nitrides provides a combination of properties of both the metallic matrix and the ceramic reinforcement. Uniform dispersion of the fine reinforcements and a fine grain size of the matrix contribute to improving the mechanical properties of the composite.

Most of the work carried out to-date on discontinuously reinforced aluminum was primarily concerned with dispersing reinforcements on the micrometer scale (a few to a couple of hundred micrometers). Even though studies were carried out on nanometer-sized reinforcements, the volume fraction of the reinforcements was very small, typically about 2-4% by volume. Thus, there is a gap in understanding the structure and properties of high volume fraction composites containing nanometer-sized reinforcements. Therefore, we have conducted investigations to synthesize and characterize Al-Al$_2$O$_3$ composites with the Al$_2$O$_3$ reinforcement varying in size from 50 nm to 5 µm, and the volume fraction from 5 to 50% by volume. The reasons behind the choice of this combination of particles sizes and volume fractions were to check (i) whether there is a maximum volume fraction of Al$_2$O$_3$ beyond which it will be difficult to achieve a uniform distribution and also (ii) to see if there is a minimum particle size, below which again it will be difficult to achieve a uniform distribution of the two components in the composite.

Mechanical alloying of the powder blends containing different amounts and size fractions of Al$_2$O$_3$ was found to reach a stable and uniform distribution of the reinforcement. A milling time of 20 h was found to be sufficient in all the combinations of the composites. Fig. 5 shows that a very uniform dispersion of the fine 50-nm Al$_2$O$_3$ particles was achieved even when the volume fraction of the reinforcement was as much as 50%. This corresponds to the finest particle size and largest volume fraction, the maximum value achieved so far in any nanocomposite. Similar uniform distributions were obtained in the larger size and smaller volume fractions of the reinforcement as well. The uniform distribution of the reinforcement phase was also confirmed through the X-ray elemental mapping technique. An interesting observation made during this study was that the stable α-Al$_2$O$_3$ transformed to γ-Al$_2$O$_3$ on milling, when the Al$_2$O$_3$ powder article size was small, e.g., 50 and 150 nm [46]. However, no such transformation occurred when the Al$_2$O$_3$ particle size was large, e.g., 5 µm.

These composites with a large volume fraction of the reinforcement of the Al_2O_3 powders were very hard and strong and consequently it was not easy to consolidate them to full density by any of the single different technique that is presently available. Therefore, to determine the effect of reinforcement particle size and volume fraction, nanocomposites with 50 nm and 150 nm particle size and 5 and 10 vol.% were consolidated to full density. Even at these small volume fractions, full (close to 100%) density was achieved only by a combination of vacuum hot pressing followed by hot isostatic pressing. Compression testing was done on the fully dense samples, and the strength properties and modulus values were determined. Table I lists the mechanical properties of these composites. Fig. 6 shows the variation of the modulus of

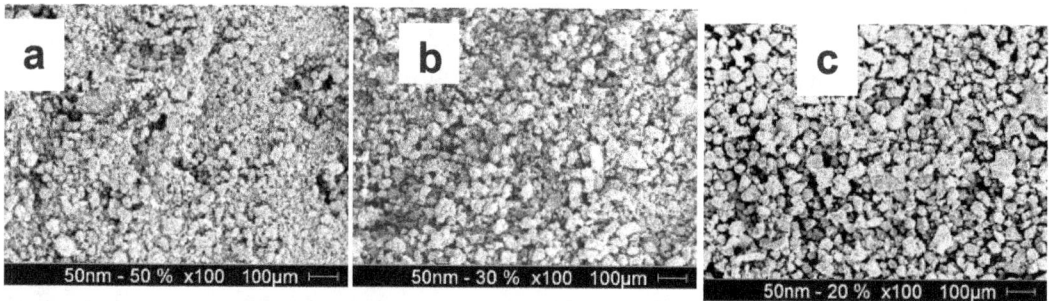

Fig. 5 SEM micrographs of Al-Al_2O_3 (50 nm) powders milled to the steady-state condition (20 h) showing a uniform distribution of Al_2O_3 in the Al matrix. (a) 20 vol.%, (b) 30 vol.%, and (c) 50 vol.% Al_2O_3.

Table 1 Mechanical properties of Al-Al_2O_3 nanocomposites obtained by milling and subsequent consolidation by vacuum hot pressing and hot isostatic pressing.

Al_2O_3		Yield Strength (MPa)	Compressive Strength (MPa)	Elastic Modulus (GPa)	Elastic Modulus calculated by the rule of mixtures* (GPa)
Particle size	Volume fraction				
50 nm	5%	488	605	78	83
50 nm	10%	515	628	90	95
150 nm	5%	409	544	75	83
150 nm	10%	461	600	77	95

*$E_c = V_mE_m+V_rE_r$ where E and V represent the elastic modulus and volume fraction, respectively and the subscripts c, m, and r represent the composite, matrix, and reinforcement, respectively.

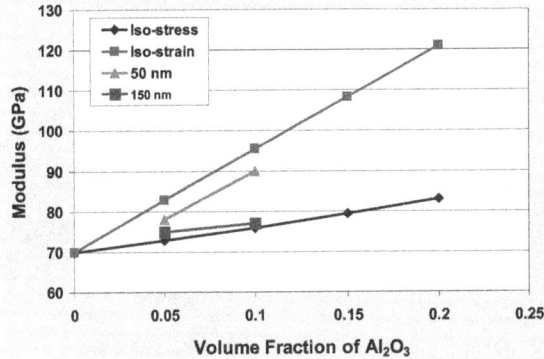

Fig. 6 Variation of modulus of elasticity with volume fraction of Al_2O_3 for two different particle sizes (50 and 150 nm). The solid lines were drawn on the basis of iso-stress or iso-strain conditions using the rule of mixtures. Note that the variation is close to the iso-strain condition for 50 nm-size Al_2O_3 particles and close to the iso-stress condition for the 150 nm-size Al_2O_3 particles.

elasticity as a function of the Al_2O_3 content. From this figure, it may be noted that the strength and modulus increased with (i) increasing volume fraction and (ii) decreasing size of the reinforcement. Comparison of these modulus values with those calculated using the iso-strain and iso-stress conditions suggested that composites with the smaller reinforcement particle size (50 nm) behaved closer to the iso-strain condition, while composites with the larger reinforcement size (150 nm) behaved closer to the iso-stress condition [47,48]. This observation clearly suggests that it is possible to tailor the modulus (and strength) of the nanocomposite by choosing the appropriate reinforcement size and volume fraction. Such a processing methodology should be equally applicable to other systems, even though the details of processing and consolidation would be different.

TiAl-ξTi$_5$Si$_3$ Composites

Lightweight intermetallic alloys based on γ-TiAl are promising materials for high-temperature structural applications, e.g., in aircraft engines or stationary turbines [49,50]. Even though they have many desirable properties such as high specific strength and modulus both at room and elevated temperatures, and good corrosion and oxidation resistance, they suffer from inadequate room temperature ductility and insufficient creep resistance at elevated temperatures, especially between 800 and 850 °C, an important requirement for elevated temperature applications of these materials.

Therefore, current research programs have been addressing the development of high-temperature materials with adequate room temperature ductility for easy formability and ability to increase the high-temperature strength by a suitable heat treatment or alloying additions to obtain sufficient creep resistance.

It has been shown that the compressive strength of binary γ-TiAl alloys with nanometer-sized grains is about 2600 MPa at room temperature and that, at temperatures higher than about 500 °C, the strength drops very rapidly to low values [51]. In fact, the strength was found to decrease at a faster rate for ultrafine-grained materials than for the coarse-grained counterparts. That is, the smaller the grain size of the specimen, the stronger and sharper is the drop in yield strength on increasing the temperature. This observation suggests that monolithic nanostructured materials may not be suitable for achieving the desired creep resistance.

The Ti-Al-Si alloy system was chosen because it is a model system to study the influence of phase distribution and microstructure on the high-temperature deformation behavior of ceramic-intermetallic composites. Earlier work has demonstrated that favorable deformation behavior could be obtained in nanostructured/submicron-sized TiAl-based intermetallics. Thus, it was decided to check whether the microstructure has a similar influence on the mechanical behavior of ceramic-based ceramic-intermetallic composites which could then lead to easy deformability and eventually the possibility of achieving superplasticity. It is possible that the creep properties of these composites will be poor at low temperatures, but, similar to that in TiAl-based alloys, it could be improved through grain coarsening after deformation at elevated temperatures.

Composites of γ-TiAl and ξ-Ti_5Si_3 phase, with the volume fractions of the ξ-Ti_5Si_3 phase varying from 0 to 60 vol.%, were produced by MA of a combination of the blended elemental and prealloyed intermetallic powders. Fully dense and porosity-free compacts were produced by hot isostatic pressing, with the resulting grain size of each of the phases being about 400 nm. Fig. 7 shows a scanning electron micrograph of the γ-TiAl + 60 vol.% ξ-Ti_5Si_3 composite showing that the two phases are very uniformly distributed throughout the microstructure. Such a microstructure is expected to be conducive to superplastic deformation behavior. To test this hypothesis, both compression and tensile testing of these composite specimens were conducted at different temperatures and strain rates. From the tensile stress-strain plots of the γ-TiAl + 60 vol.% ξ-Ti_5Si_3 composite specimen shown in Fig. 8(a), we can draw the following conclusions [44].

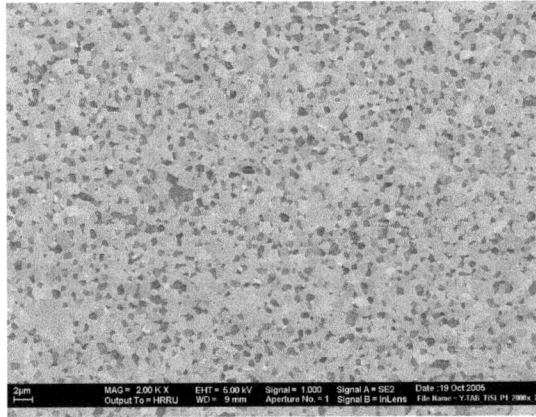

Fig. 7 Scanning electron micrograph of the γ-TiAl + 60 vol.% ξTi₅Si₃ composite specimen showing that the two phases are very uniformly distributed in the microstructure. Such a microstructure is conducive to observing superplastic deformation under appropriate conditions of testing.

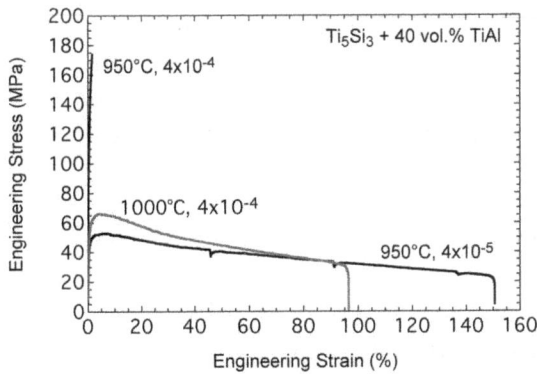

(a)

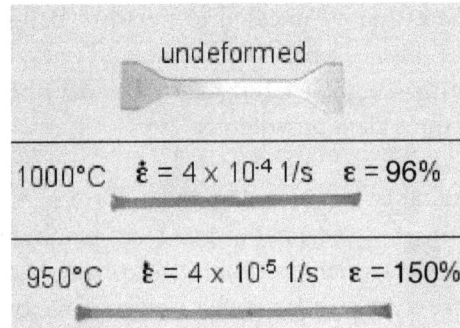

(b)

Fig. 8 (a) Stress-strain curve of γ-TiAl + 60 vol.% Ti₅Si₃ composites showing that superplastic deformation was achieved at 950 °C and a strain rate of $4x10^{-5}$ s^{-1} and 1000 °C and a strain rate of $4x10^{-4}$ s^{-1}. (b) Photographs of the actual tensile specimens before and after tensile testing.

Firstly, the strength decreased with increasing temperature and decreasing strain rate, both expected from normal mechanical behavior of metallic materials. Secondly, the specimens tested at 950 °C and a strain rate of $4x10^{-5}$ s^{-1} and 1000 °C and a strain rate of $4x10^{-4}$ s^{-1} exhibited large ductilities of nearly 150 and 100%, respectively. Considering that this composite is based on a ceramic material (Ti₅Si₃) this is a very high amount of deformation, suggestive of superplastic deformation. Final proof is provided by TEM investigations that confirm the continued stability of the microstructure after deformation. Thirdly, even though the strain rate employed is relatively low, it is

17

interesting that superplasticity was observed at 950 °C, corresponding to about 0.5 T_m, where T_m is the melting temperature of the alloy. This should be compared with the coarse-grained material which shows the superplastic behavior only at temperatures about 300-400 °C higher than the temperature observed in this investigation.[36] Fig. 8(b) shows photographs of the actual specimens before and after tensile testing, showing significant amounts of elongation in the samples.

4.2. Synthesis of Photovoltaic Materials

Copper indium diselenide ($CuInSe_2$)-based photo-voltaic thin film solar cells continue to receive attention because of their high conversion efficiency [52]. It has been shown that the band gap of $CuInSe_2$ can be modified continuously by substituting Ga for In; the band gap increases with increasing Ga content. For achieving the maximum efficiency of the cell, a Ga/(Ga+In) ratio of 0.3 and a slight deficiency in the copper content have been found to be beneficial.

The $CuInSe_2$-based alloy films are usually obtained either by multisource thermal evaporation and co-deposition or reactive annealing of precursor films. These processes are very involved, contain a number of deposition and annealing steps, pose difficulties in maintaining the desired film composition, and can be expensive. Therefore, we decided to perform band gap engineering through the technique of MA. The two simple steps involved will be synthesis of stoichiometric powder of the $CuIn_{0.7}Ga_{0.3}Se_2$ (CIGS) alloy by MA and obtain the bulk target by hot isostatic pressing of the milled powder.

Milling of the powders even for a short time resulted in the formation of the desired $CuIn_{0.7}Ga_{0.3}Se_2$ phase. For example, by milling at a BPR of 20:1 and a speed of 300 rpm, the tetragonal $CuIn_{0.7}Ga_{0.3}Se_2$ phase had formed. But, small amounts of other phases were also present at this stage. But, on continued milling of this powder for 20 min, only the tetragonal $CuIn_{0.7}Ga_{0.3}Se_2$ phase had formed homogeneously (Fig. 9). This phase could be indexed on the basis of a tetragonal structure with the lattice parameters $a = 0.5736$ nm, $c = 1.1448$, and $c/a = 1.9958$ having the space group $I\bar{4}2d$. Even though the quaternary Cu-In-Ga-Se tetragonal phase exists over a wide composition range, the present XRD pattern could be most satisfactorily indexed on the basis of the tetragonal phase with the stoichiometry $CuIn_{0.7}Ga_{0.3}Se_2$. Scherrer analysis of the peak widths confirmed that the crystallite size reached a saturation value of about 8 nm. The phase identification and crystallite size determination were confirmed by transmission electron microscopy (TEM) studies.

The chemical analysis of the milled powder under different milling conditions is shown in Table II. From the results it is clear that the composition of the milled powder is reasonably close to the targeted composition. But, because some copper has been picked up from the container and/or the grinding balls during milling, the copper content appears to be higher than what the stoichiometry requires. But, the powder milled at a BPR of 10:1 and a speed of 150 rpm for 40 min did not pick up much copper. Consequently, the chemical composition of the powder milled under these conditions is quite close to the starting powder mixture. By further optimizing the powder processing conditions (BPR, speed of rotation. and time of milling), it is possible to control the composition of the final powder to the exact desired value.

The milled powder containing the homogeneous $CuIn_{0.7}Ga_{0.3}Se_2$ phase was consolidated by hot isostatic pressing to full density at 750 °C and 100 MPa for 2 h. Examination in a scanning electron microscope indicated that the compact was fully dense and did not contain any porosity. XRD patterns of the hot isostatically pressed alloy suggested that the $CuIn_{0.7}Ga_{0.3}Se_2$ phase was retained with the exact stoichiometry and that the peaks are sharper than what they were in the milled powder. This is mainly due to grain growth, confirmed by TEM investigations. The TEM micrograph shows that the grain size now is about 50-60 nm.

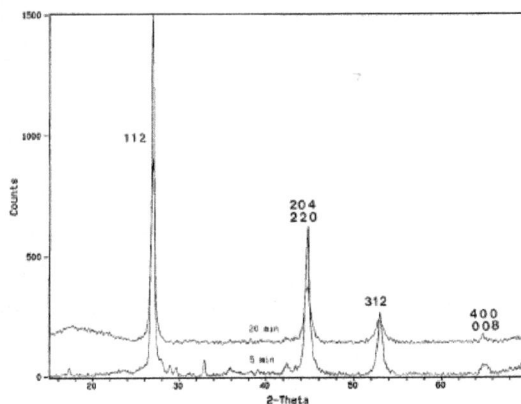

Fig. 9 X-ray diffraction patterns of the CIGS blended elemental powder mixture milled at a BPR of 20:1 and a speed of 300 rpm for 5 and 20 min. The powder milled for 5 min contained the tetragonal $CuIn_{0.7}Ga_{0.3}Se_2$ phase and small amounts of elemental Cu and In. When the powder was milled for 20 min, it contained only the tetragonal $CuIn_{0.7}Ga_{0.3}Se_2$ phase.

Table 2 Chemical analysis (at.%) of the Cu-In-Ga-Se powder mechanically alloyed under different processing conditions.

BPR	Speed (rpm)	Time (min)	Cu	In	Ga	Se
20:1	300	120	32.61	15.23	6.53	45.62
10:1	300	240	31.73	15.63	6.71	45.93
10:1	150	150	25.98	16.98	7.25	49.78
Targeted ($Cu_{0.97}In_{0.7}Ga_{0.3}Se_2$)			24.43	17.63	7.56	50.38

This investigation clearly proves that it is possible to obtain the CIGS targets of the required stoichiometry through a combination of MA and hot isostatic pressing or vacuum hot pressing. Further, it is possible to control the stoichiometry by optimizing the MA process parameters. Such an approach should be equally applicable to synthesize other target materials.

4.3. Combustion Studies of Metal Particles

One of the potential applications for materials with fine grain sizes is in improving the combustion characteristics of propellants [54]. It is known that Mg can be easily combusted, but it has a low heat content (14.9 MJ/kg). On the other hand, Al has a higher heat content (32.9 MJ/Kg), but is more difficult to combust. Therefore, it was decided to see if Al and Mg can be alloyed together to increase the net heat content and also be able to combust relatively easily. Accordingly, a series of Al-Mg powder blends of different compositions were mechanically alloyed in a SPEX mill. Since the milled powder normally contains a range of particle sizes, the milled powder was sieved to well-defined particle sizes in the ranges of 20-25 μm, 45-53 μm, and 90-106 μm (Fig. 10). These powder particles were then burned in a methane-air premixed flame on a Meeker burner, at a temperature of approximately 1,800 K. Burning of the powder particles was recorded using chopped images by long-exposure digital photography. Knowing the rotational speed of the chopper and the number of exposures per revolution, the burning time, τ_b and the ignition time, τ_{ig} could be determined. Fig. 11 shows the plot of variation of τ_{ig} and τ_b with particle size for two different alloy compositions (Al-20 at.% Mg and Al-90 at.% Mg) [Ref. 55]. It is clear from this plot that both the ignition and burning times decrease with (a) a decrease in particle size and (b) an increase in Mg content. Thus, by a judicious choice of particle size and alloy composition, it should be possible to control the ignition and burning times.

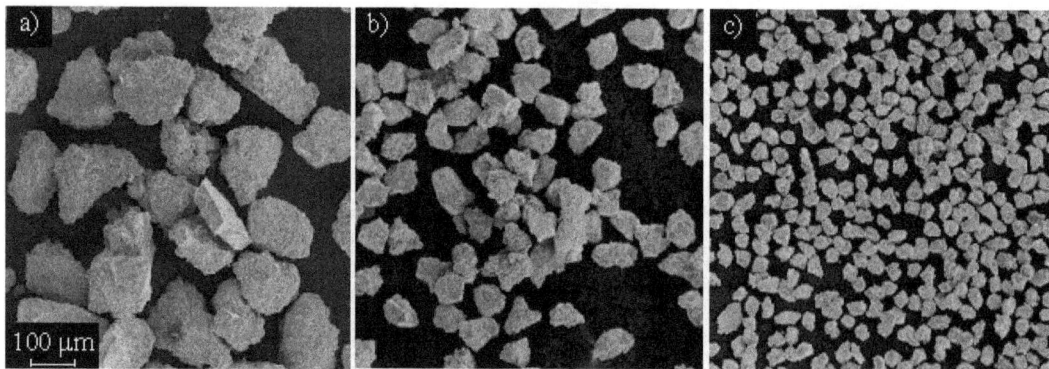

Fig. 10 Scanning electron micrographs of the sieved Al-Mg alloy particles after milling. (a) 90-106 μm, (b) 45-53 μm, and (c) 20-25 μm.

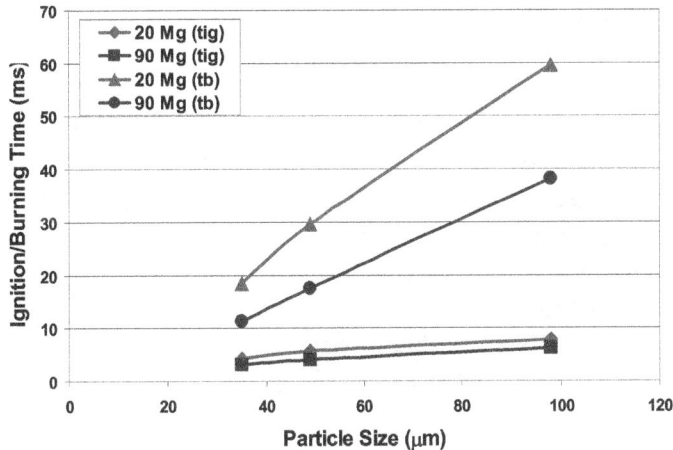

Fig. 11 Ignition time, τ_{ig} and burning time, τ_b as a function of particle size for two different alloy compositions (Al-20 at. % Mg and Al-90 at.% Mg).

5. Conclusions

From the above description of the type of activities, it becomes amply clear that the technique of MA is very efficient in synthesizing a variety of equilibrium and non-equilibrium advanced materials. Some of the present issues that need immediate attention are that we need homogeneous, pure, macro-defect (i.e., porosity and cracks)-free and fully dense materials for obtaining reproducible results. Further, large-scale applications require materials that could be produced in tonnage quantities, and materials that have been well characterized and exhibit reproducible properties. But, the bane of MA has been the three C's – Cost, Consolidation, and Contamination. Powder processing is expensive and therefore, unless one is able to identify niche markets for the MA products and the powders produced in large volumes, cost is likely to continue to be high. However, in some cases, e.g., hydrogen storage materials or magnetic alloys, the high cost may be acceptable in view of the improved properties and enhanced performance.

Consolidation of the milled powder continues to be a serious problem. There have not been many investigations to report successful consolidation of MA powders to achieve full density and simultaneously retain the metastable effects. Newer and improved methods or modification/combination of the existing methods may be necessary.

The last point is that contamination of the MA powder has been a serious issue in many cases.[13,14] Some solutions have also been suggested to minimize/avoid contamination completely. These include enclosing the mills inside chambers that have been evacuated and/or filled with an inert gas, or using high-purity gaseous atmosphere in which the powders could be milled. Both options could be expensive and/or impractical.

The MA powders could also be used in the as-produced powder condition without consolidation. Examples for such applications include use as catalysts, pigments, solder, hydrogen storage materials, materials for printed electronics, (CIGS powders are already in use for printed solar cells, Nanosolar), etc. In that case, some of the above problems mentioned above may not be relevant. Further, even if a bulk component made of MA powder needs to be used, then if that material could tolerate high gaseous impurities, then also there is no problem. With the intense research activity going on all over the world, there is great potential for MA products and bright future for research in this area.

Acknowledgements

The work reported here is supported by the US National Science Foundation under grants DMR-0314212 and DMR-0334544 and by the Office of Naval Research. The author also acknowledges fruitful collaboration and useful discussions with Drs. Eugene Ivanov of Tosoh SMD, Grove City, OH, Professors Thomas Klassen of Helmut-Schmidt University, Hamburg, Germany, Professor Rüdiger Bormann of the Hamburg-Harburg Technical University, Hamburg, Germany, Raj Vaidyanathan, Linan An, and Ruey-Hung Chen of the University of Central Florida, Orlando, USA, and Dr. S.J. Hong of Kongju National University, Kongju, South Korea. I am also thankful to the experimental input from his graduate students – Devender Singh, Umesh Patil, Pushkar Katiyar, Balaji Prabhu, Satyajeet Sharma, and Uma Maheswara Rao Seelam, all from the University of Central Florida.

References

1. K.H.J. Buschow, R.W. Cahn, M.C. Flemings, B. Ilschner, E.J. Kramer, and S. Mahajan (eds.), *Encyclopedia of Materials: Science and Technology*, Elsevier, 2001.

2. T.R. Anantharaman and C. Suryanarayana, *Rapidly Solidified Metals: A Technological Overview*, Trans Tech Pub., Zurich, Switzerland, 1987.

3. H.H. Liebermann, *Rapidly Solidified Alloys: Processes, Structures, Properties, Applications,* Marcel Dekker Inc., New York, NY, 1993.

4. A. Inoue, *Bulk Amorphous Alloys*, Trans Tech Publications, Zurich, Switzerland, Part I: Vol. 4 of Materials Science Forum, 1998 and Part II: Vol. 6 of Materials Science Forum, 1999.

5. C. Suryanarayana and H. Jones, Int. J. Rapid Solidification Vol. 3, 1988, p. 253.

6. H.R. Trebin (ed.), *Quasicrystals: Structure and Physical Properties,* Wiley-VCH, 2003.

7. H. Gleiter, Acta Mater. Vol. 48, No. 1, 2000

8. C. Suryanarayana, Int. Mater. Rev., Vol. 40, No. 41, 1995.

9. C.C. Koch (ed.), *Nanostructured Materials*, 2nd ed.,William Andrew Pub. Co., New York, 2007.

10. D. Turnbull, Metall. Trans. Vol. 12A, 1981, p. 695.

11. G. Martin and P. Bellon, Solid State Phys. Vol. 50, 1997, p. 189.

12. C. Suryanarayana (ed.), *Non-equilibrium Processing of Materials,* Pergamon, Oxford, 1999.

13. J.S. Benjamin, Metal Powder Report, Vol. 45, 1990, p.122.

14. C. Suryanarayana, Prog. Mater. Sci. Vol. 46, No. 1, 2001

15. C. Suryanarayana, *Mechanical Alloying and Milling,* Marcel Dekker, Inc., New York, NY, 2004.

16. L. Takacs, Prog. Mater. Sci. Vol. 47, 2002, p. 355.

17. C. Suryanarayana, *Bibliography on Mechanical Alloying and Milling*, Cambridge Int. Science Pub., Cambridge, UK, 1995.

18. J.S. Banjamin, Metall. Trans. Vol. 1, 1970, p. 29.

19. J.J. deBarbadillo and G.D. Smith, Mater. Sci. Forum, Vol. 88-90, 1992, p. 167.

20. W. Hume-Rothery, R.E. Smallman, and C. Haworth, *Structure of Metals and Alloys*, 5th ed., Institute of Metals, London, 1969.

21. K. Uenishi, K.F. Kobayashi, K.N. Ishihara, and P.H. Shingu, Mater. Sci. Eng. Vol. A 174, 1994, p. 119.

22. W. Klement, R.H. Willens, Jr., and P. Duwez, Nature, Vol. 187, 1960, p. 869.

23. A.E. Yermakov, Y.Y. Yurchikov, and V.A. Barinov, Phys. Met. Metallogr. Vol. 52 No. 6, 1981, p. 50.

24. C.C. Koch, O.B. Cavin, C.G. McKamey, and J.O. Scarbrough, Appl. Phys. Lett. Vol. 43, 1983, p. 10.

25. C. Suryanarayana, *Bibliography on Mechanical Alloying and Milling*, Cambridge Int. Science Pub., Cambridge, 1995.

26. C. Suryanarayana, I. Seki,a nd A. Inoue, J. Non-Cryst. Solids, **(in press)**, 2009.

27. *S. Sharma, R. Vaidyanathan, and C. Suryanarayana, Appl. Phys. Lett. Vol. 90, 2007, 111915.*

28. U. Patil, S.J. Hong, and C. Suryanarayana, J. Alloys & Compounds, Vol. 389, 2005, p. 121.

29. S. Sharma and C. Suryanarayana, J. Appl. Phys. Vol. 102, 083544. (2007).

30.   S. Sharma and C. Suryanarayana, J. Appl. Phys. 103, 013504 (2008).

31.   S. Sharma and C. Suryanarayana, Scripta Mater. 58, 508 (2008).

32.   C. Suryanarayana and S. Sharma, J. Appl. Phys. 104, 103503 (2008).

33.   T.B. Massalski (ed.), *Binary Alloy Phase Diagrams*, ASM Int., Materials Park, OH, 1986.

34.   F.R. de Boer, R. Boom, W.C.M. Mattens, A.R. Miedema, and A.K. Niessen, *Cohesion in Metals, Transition Metal Alloys*, North-Holland, Amsterdam, 1988

35.   T.W. Clyne and P.J. Withers, *An Introduction to Metal Matrix Composites*, Cambridge University Press, Cambridge, 1995.

36.   H.T. Son, T.S. Kim, C. Suryanarayana, and B.S. Chun, Mater. Sci. Eng. Vol. A, No. 348, 2003, p. 163.

37.   S.J. Hong, H.M. Kim, D. Huh, C. Suryanarayana, and B.S. Chun, Mater. Sci. & Eng. Vol. A, No. 347, 2003, p.198.

38.   T. Klassen, R. Bohn, C. Suryanarayana, G. Fanta, and R. Bormann, in *Processing and Properties of Structural Nanomaterials*, eds. L. Shaw, C. Suryanarayana, and R.S. Mishra, TMS, Warrendale, PA, 2003, p. 93.

39.   P. Mohan, C. Suryanarayana, and V. Desai, in *Nanomaterials: Synthesis, Characterisation, and Application*, eds. S. Bandyopadhyay et al., Tata McGraw-Hill Pub. Co. Ltd., New Delhi, India, 2004, p. 171.

40.   P. Mohan, C. Suryanarayana, T. Du, and V. Desai, in Proc. 206th ECS PV 2005-14, 2005, p. 461.

41.   B. Prabhu, C. Suryanarayana, L. An, and R. Vaidyanathan, Mater. Sci. & Eng. Vol. A, No. 425, 2006, p. 192.

42.   N. Al-Aqeeli, G. Mendoza-Suarez, C. Suryanarayana, and R.A.L. Drew, Mater. Sci. Eng. Vol. A, No.480, 2008, p. 392.

43.   C. Suryanarayana, Mater. Sci. Eng. Vol. A479, 2008, p. 23.

44.   T. Klassen, C. Suryanarayana, and R. Bormann, Scripta Mater. Vol. 59, 2008, p.455.

45.   C. Suryanarayana, R. Behn, T. Klassen, and R. Bormann, Acta Mater. (to be submitted).

46.   Y. Wang, C. Suryanarayana, and L. An, J. Amer. Ceramic Soc. Vol. 88, 2005, p.780.

47.   B. Prabhu, M.S. Thesis, University of Central Florida, Orlando, USA, 2005

48.   W.D. Callister, *Materials Science and Engineering: An Introduction*, 7th ed., John-Wiley & Sons Inc., New York, 2007.

49. F.H. Froes, C. Suryanarayana, and D. Eliezer, J. Mater. Sci. Vol. 27, 5113 (1992).

50. F. Appel and R. Wagner, Mater. Sci. Eng. Reports Vol. 5, No. 22, 1998, p. 187.

51. R. Bohn, T. Klassen, and R. Bormann, Intermetallics Vol. 9, 2001, p. 559.

52. R.W. Birkmire and E. Eser, Ann. Rev. Mater. Sci. Vol. 27, 1997, p. 625.

53. C. Suryanarayana, E. Ivanov, R. Noufi, M.A. Contreras, and J.J. Moore, J. Mater. Res. Vol. 14, 1999, p. 377.

54. E.L. Dreizin, Prog. Energy & Combust. Sci. Vol. 26, 2000, p. 57.

55. R.H. Chen, C. Suryanarayana, M. Chaos, Adv. Eng. Mater. Vol. 8, 2006, p. 563.

Microstructure Engineering by Liquid Phase Sintering in Tungsten Alloys

Bijoy Sarma and Ravi Kiran

Defence Metallurgical Research Laboratory, Hyderabad

Abstract: *Performance of any component is related to condition of the materials of construction. Establishing the structure property correlation of every material in its final useful configuration is the ultimate challenge for the process designer. With increasing demands of performance, the choice of materials has moved from the usual copper and iron to refractory metals like tungsten and molybdenum. The powder metallurgy technique of consolidating elemental powders in desired proportions and then net shaping through application of pressure and temperature seems to be the ultimate in component processing. High melting point refractory metals which are extremely difficult to process through the conventional melt route can be densified from their powder form by mixing suitable low melting elements to result in liquid phase sintering at considerably lower temperatures than required for solid state sintering. Nickel, iron and cobalt combinations have been found to be very effective while sintering tungsten. Processing temperatures can be brought down to around 1500°C against 2400°C for pure tungsten. This has led to a whole new system of dense alloys which have been investigated extensively. The advantages of liquid phase sintering and some of the limitations there of are described. The paper elucidates some of the factors which effect the liquid phase formation and its subsequent effects on bulk properties.*

Keywords: Liquid phase sintering, Tungsten heavy alloys

1. Introduction

Every development endeavor aims at maximizing performance attributes. Material behaviour invariably depends on its constituent elements, their disposition and orderly arrangement at the atomic level. While atoms may not be visible (size is too small) their arrays in crystallographic arrangements and clusters of these constituting phases can be resolved by microscopy (optical and electron). The effort of processing parameters is to suitably affect these arrangements of the constituent elements in a controlled fashion to ensure reproducibility, and predictability. This, in essence is engineering microstructures to obtain desired product performance.

The conventional method of processing most materials is to melt, cast and then work them with some heat treatment thrown in to obtain suitable microstructure and thus the properties for enabling performance. Sometimes an alternate route of product manufacture is adopted where the individual constituents are taken in particulate form. A typical powder metallurgy component starts with elemental or alloy powders in the size range of a fraction of a micron to few hundred microns. These are intimately mixed and brought together into a predetermined shape by a consolidation process known as compaction. The individual particles comprising the porous compact are made to react metallurgically by exposing the compacted mass to elevated temperatures for densification. In most cases sintering temperatures are selected somewhat below the melting point of any of the constituents and densification proceeds through diffusion of different constituents along the surface of the particles and the bulk. During this densification process a microstructure eventually develops such that each phase has its identity and role to play in the subsequent performance of the component. Some times deformation processes are also used to affect the microstructure. In specific cases constituent elements may have widely varying melting points and if there is mutual solubility then activation of the densification process occurs and such an event is termed as liquid phase sintering. Liquid phase sintering is evidenced in the following systems shown in table 1.

When selecting materials for specific application regimes we often refer to the Ashby diagrams (Figure.1a & 1b). Tungsten alloys enjoy an exclusive niche with a high strength, toughness and density combination.

Table 1 Liquid phase sintering systems and applications

System	Application
WC+Co	Cutting & Machining Tools
Cu+Sn	Oil less bearings
Al+Pb	Wear and Bearing surfaces
W+Ni+Fe	Radiation Shields, C.G. Adjusters, Ordnance
W+Ag	Electrical Contacts
Fe+Cu+C	Structural components and gears
Ag+Hg	Dental amalgam for fillings
Fe+P	Soft magnetic components
Al+Si+Cu	Light weight structural components
$BaTiO_3$+LiF	Electrical capacitors
Si_3N_4+Y_2O_3	High temperature turbines

1.1 The PM Approach

Powder Metallurgy (P/M) has been defined as the art and science of producing metal powders and making semi finished and finished objects from individual, mixed or alloyed powders with or without the addition of nonmetallic constituents.

Production of P/M parts involves mixing of elemental or alloy powders with additives and lubricants, compacting the mixture in a suitable die and heating the resulting green compacts in a controlled atmosphere furnace so as to bond the particles metallurgically. This process is known as sintering. In many cases subsequent thermo mechanical treatments are necessary to maximize mechanical properties through microstructure modification. The driving force for the entire densification in a PM approach is the reduction in surface energy from finely divided powder to a fully densed product.

One of the material system which is very difficult to process by conventional melt route is tungsten due to its extremely high melting point. However, the other attributes of the element make it very valuable. The PM option has been found commercially viable and is extensively used.

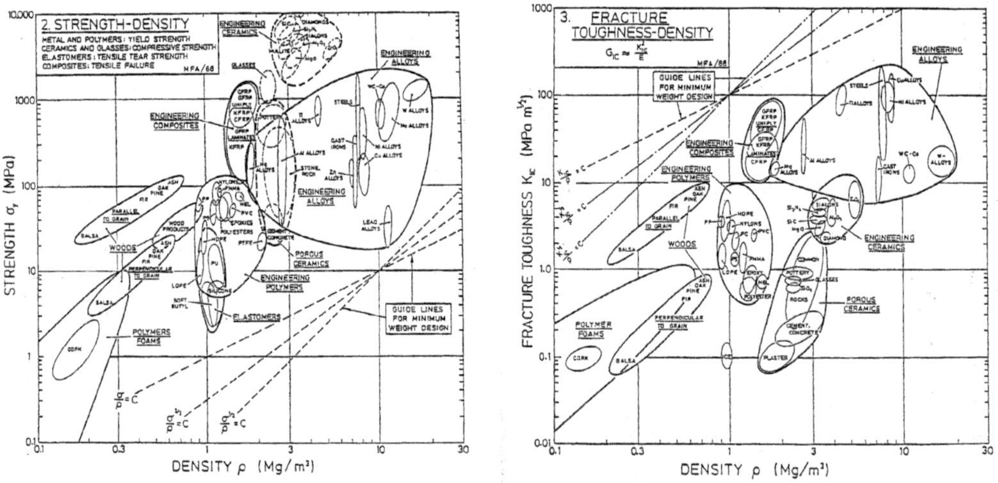

(a)   (b)

Figure 1 Ashby Material Selection Diagram.

2. Tungsten Heavy-Metal Alloys (WHAs)

The high melting point (3440° C) of tungsten makes it an obvious choice for structural applications for very high temperature regimes. Tungsten is used at lower temperatures for applications that can use its high elastic modulus, density, or shielding characteristics to advantage. The sintered and wrought tungsten is brittle and very difficult to machine (Figure 2). As such its applications get limited to sheets and wires.

Tungsten heavy metal alloys are a category of tungsten-base materials that typically contain 90 to 98 wt% W in combination with some mix of nickel, iron, copper, and/or cobalt with melting point in the range of 1400-1500° C. The bulk of WHA production falls into the 90 to 95% W range. The Ni, Fe, Cu and Co combination forms a liquid phase at sintering temperature (1400-1500° C) wets the tungsten and has reasonable solubility of tungsten in the liquid metal. Rapid sintering ensues with solution re-precipitation and a microstructure develops (Figure.3a) showing spheroidal tungsten dispersed in a strong matrix alloy. The matrix alloy forms a complete network as shown in Figure. 3b.

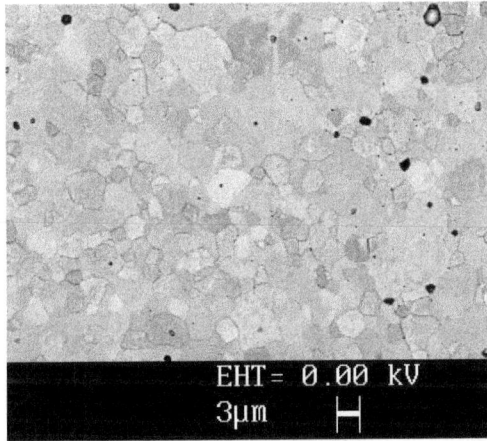

Figure 2 Sintered Pure Tungsten

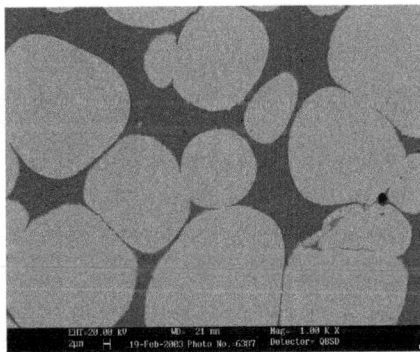

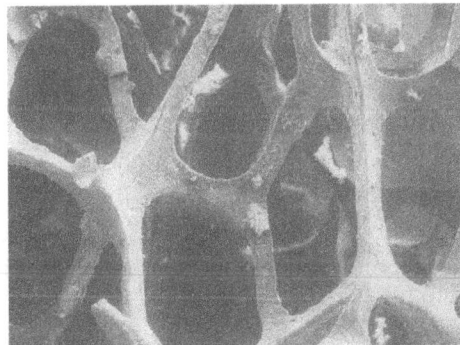

(a)   (b)

Figure 3 (a) Liquid phase sintered tungsten alloy and (b) matrix skeleton after removing spheroidal tungsten.

Most commercial WHAs are two-phase structures, the principal phase being nearly pure tungsten in association with a binder liquid phase containing the transition metals plus dissolved tungsten. As a consequence, WHAs derive their fundamental properties from those of the principal tungsten phase, which provides for both high density and high elastic stiffness. It is these two properties that give rise to most applications for this family of materials. It is evident that significant advantage is obtained by adopting the liquid phase sintering technique for modifying the tungsten sinter microstructure to facilitate commercial exploitation.

2.1 Conditions for Liquid Phase Sintering

- There should be at least two phases, one with high melting point and other with low melting point.

- Wide gap should exist between the melting points of these two phases.

- Sintering temperature should be well below the melting point of the high melting-point phase.

- A liquid phase (transient or persistent) should form during sintering.

- High melting-point phase should have good solubility in the molten phase.

- Molten phase should have poor solubility (in the solid phase) but good wettability.

2.2 Stages of liquid phase sintering

- Rearrangement (Stage I). – In the first stage of liquid phase sintering liquid will form and rearrangement of grains will takes place.

- Solution-Reprecipitation (Stage II). - Densification by rearrangement slows down, Solid grains get dissolved in the liquid phase and liquid phase get saturated with high melting point constituent. Grains become several times larger than the original particle size and grain shape changes from irregular to rounded. The high melting point constituent gets reprecipitated on existing solid grains. Material gets transported from small grains to large grains leading to grain coarsening (Ostwald ripening). Progressive growth of large grains with wider spacing occurs. Neighboring faces of large grains get flattened for tighter packing (grain shape accommodation) leading to complete pore elimination.

- Solid state Grain Growth (Stage III) - In the final stage a solid stable skeleton is formed.

In this condition, a typical 7: 3: 90 Ni Fe W after sintering to full density results in UTS of 650 MPa, tensile elongation 5% and impact strength of 30 J (Table 2). The material has a density of 17.1 g/cc and finds application in a variety of ballasts; C.G. Adjusters, radiation shields and nose plug (figure.4). The microstructure shows spheroidal tungsten floating in a matrix alloy of Ni-Fe-W which has excellent machinability (Figure.5a). The same material, finds extensive ordnance application after thermo- mechanical treatment (Figure.5b). The microstructure is modified to elongated tungsten in the matrix of Ni-Fe-W through extensive working and this raises the property level to UTS 1500 MPa, tensile elongation 10% and impact strength of 100 J (Figure 6).

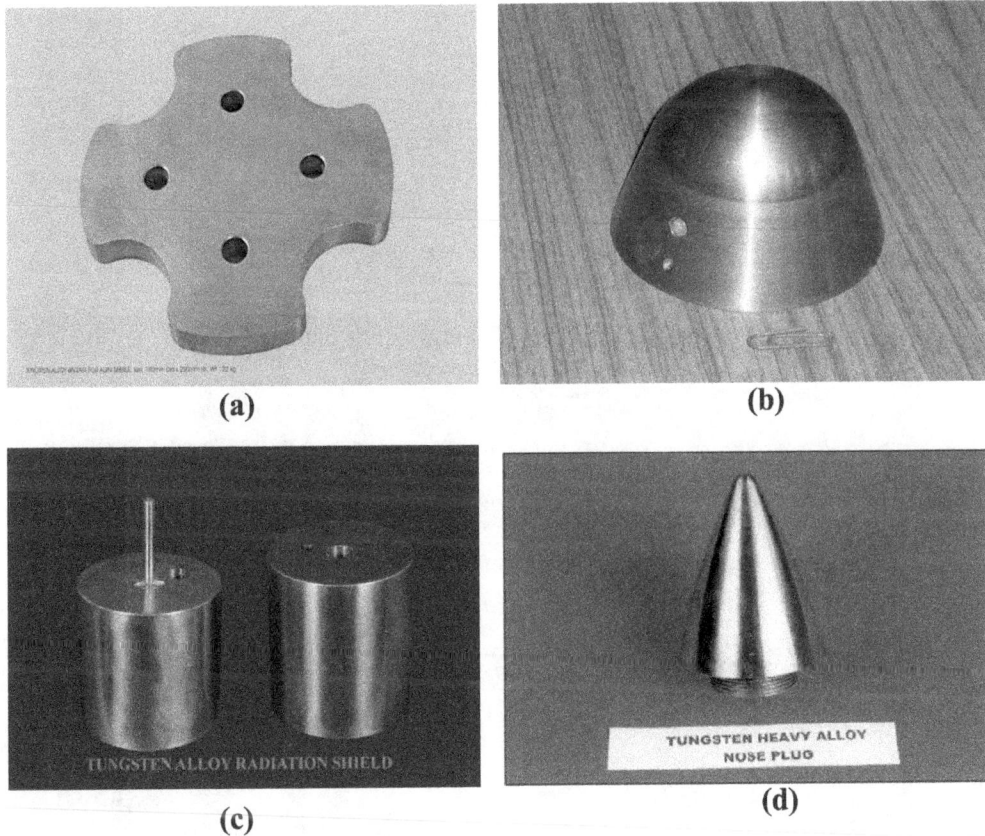

Figure 4 (a) ballast (b) C G adjuster (c) radiation shield and (d) nose plug.

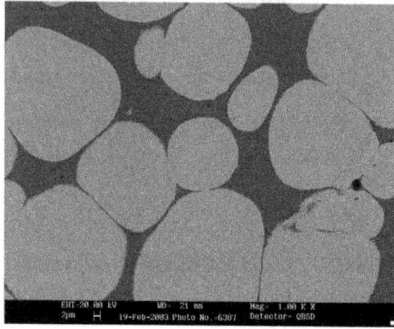
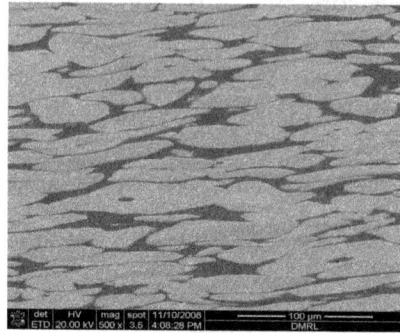

(a) **(b)**

Figure 5 (a) Microstructure on un-swaged heavy alloy (b) swaged heavy alloy.

Properties	As sintered	As swaged
Reduction in Area, %	0	56
UTS, MPa	650	1560
Elongation, %	5	8
Impact strength, J	30	100
Hardness, Hv	320	450

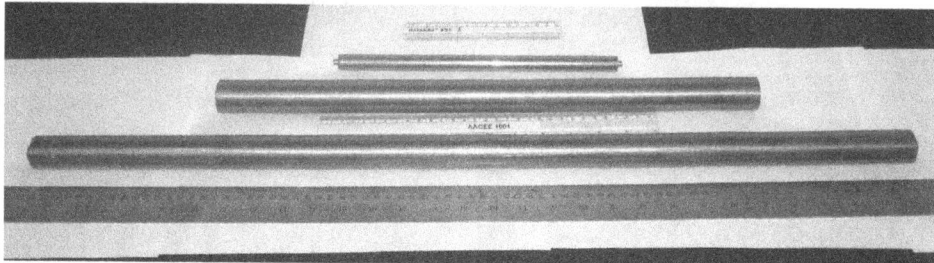

(a)

(b)

Figure 6 (a) High strength tungsten alloy swaged rod and (b) Penetrator for anti tank ammunition.

3. Conclusion

Microstructure engineering through liquid phase sintering of tungsten using transition metals combination and adopting a PM processing route results in a very favorable properties and thereby finds wide applications.

References

1. R. M. German, *Liquid Phase Sintering*, Plenum Press, New York, NY, 1985.

2. Hirschhorn, J. S., *Introduction to Powder Metallurgy*, American Powder Metallurgy Institute, New York, 1969.

3. R.M. German., *Sintering Theory and Practice*, John Wiley & Sons, Inc., New York, 1996.

4. Ashby, M. F., *Materials and process selection in mechanical design*, Butterworth Heinemann, Oxford, 1999.

The First Indian Materials Processing Experiments in Space Aboard ISRO's Orbital Micro-G Platform

S.C. Sharma, K. Chattopadhyay* and P.P. Sinha

Materials and Mechanical Entity, Vikram Sarabhai Space Centre, Thiruvananthapruam, India
* Materials Engg. Dept., IISc, Bangalore, India
sinha pp@vssc.gov.in

Abstract *One of the core objectives of Indian Space Research Organization's maiden mission of space capsule recovery experiment has been to conduct materials processing experiments under micro-g environment of space. The recoverable capsule provides an orbital platform for carrying out long duration space-based experiments. The platform has carried two microgravity payloads into orbital space: (i) Isothermal Heating Furnace, and (ii) Biomimetic Material Processing Reactor. The former is used for study of the growth of Ga- Mg-Zn based quasicrystals in space environment and the latter facilitates synthesis of hydroxyapatite nanoparticles based biomimetic material under the reduced gravity conditions of space.*

The platform has the provision of accommodating scientific payloads up to 40 kg of weight and 0.0125 m3 of volume (recoverable) and the mission enables maximum 90 days of orbital microgravity environment. To provide high quality onboard micro-g conditions, 3-axis stabilization of the platform is maintained through magnetic torquers which minimize onboard g-jitters.

The present paper highlights salient features of the micro-g platform and the two payloads and brings out the details of the in-orbit experiments performed by employing these scientific payloads.

1. Introduction

Indian Space Research Organization (ISRO) successfully conducted the first mission of Space capsule Recovery Experiment (SRE) in the year 2007. One of the core objectives of the mission has been to provide an orbital platform for carrying out long duration microg experiments. The recoverable capsule performed two materials processing experiments in space. The experiments belonged to remarkably varied range of scientific disciplines namely crystallography and biomimetic materials synthesis. The former experiment aimed at the growth of Ga-Mg-Zn based stable quasicrystal while the latter dealt with the synthesis of a hydroxy apatite based biomimetic material under the microgravity environment. Two payloads viz, Isothermal Heating Furnace (IHF) and Biomimetic Material Processing Reactor (BMPR) were specially designed and developed for the quasicrystal growth and biomimetic material synthesis experiments respectively. Thus, the SRE mission validated an Indian orbital micro-g platform which now offers exciting opportunities to the community of material scientists.

An Orbital Platform is one of the most effective experimental micro-g facilities as it provides very low micro-g levels of the order of 10-6 g for long durations extending up to several weeks/months. An orbital platform enables the entire range of micro-g material processing experiments including those which involve very slow physical phenomena such as diffusion or need high quality of micro-g environment such as crystal growth cell culture, electrophoresis, etc. Moreover, an unmanned orbital platform or Free Flyer, free from gjitters to a large extent, further adds to the quality of micro-g conditions available aboard the platform.

The present paper brings out the salient design features of the above Indian orbital micro-g platform. It provides an outline of the envelope in terms of volume, weight and power available aboard this platform for micro-g experiments. Various interfaces of the satellite with ground for tele-conduction of the experiment are also covered. The paper highlights important details of the two payloads used for conducting micro-g materials processing experiments and presents major scientific results of the experiments.

2. Recoverable Space Capsule as Orbital Micro-G Platform

As per the SRE mission objectives, the recoverable space capsule serves as a free flyer or micro-g platform during its orbital phase. Micro-g experiment set-ups or scientific payloads are accommodated on the uppermost deck or payload Bay of the capsule (Figs.1a and 1b).

(a)   (b

Fig. 1 (a) Uppermost deck or Payload Bay of the Capsule houses the micro-g experiment set-ups or payloads,
(b) Uppermost deck containing the micro-g experiment payloads IHF Payload

Details of the envelope allotted for micro-g experiments in the payload bay are given in Table-I. More than one experiment can be housed within available mass, volume, and power budgets.

TABLE - I : Details of Envelope for Micro-g Experiment

Payload Mass	40 kg
Payload Volume	0.0125 m³ (200 mm dia x 400 mm height)
Power	85 W (Normal), 145 W (Peak)
Micro-g Levels	10^{-5} g or better
Data Rate	32 kbps in S band
Experiment Duration	3 to 30 days

Features of recoverable space capsule as micro-g platform includes microgravity environment of the order of 10-5g or better, electric power of 85 W and 145 W in normal and peak modes respectively, thermal management, and internal and external communication with the micro-g experiment payloads during the orbital phase. To ensure high quality micro-g levels, the capsule is put into 3-axis stabilized mode with control through magnetometer / dual coil magnetic torquers (60 Am2). Onboard power is based on battery tied single bus system with bus voltage of 28-40 V. A 10 Ah Li-ion battery is incorporated to support the power requirements during eclipse and re-entry phases of mission, when solar arrays do not generate energy.

A centrally processing onboard computer manages the function of telemetry, telecommand, sensor processing and thermal control. To facilitate the teleconduction of micro-g experiments, computer executes real- time as well as time-tagged auto commands. Simultaneous storage of the experimental data during the eclipse period and their playback and transmission to the ground stations during immediately next visible period, are also enabled by the on-board computer.

3. Quasicrystal Growth Experiment (Ihf Payload)

3.1 Scientific Basis

Microgravity (µg) conditions of space provide an ideal environment for crystal growth since phenomena of convection and buoyancy driven sedimentation are greatly reduced in space. Motivated by these prospects, a large number of experiments under µg conditions have been conducted on the growth of single crystals of electronic materials and protein crystals mainly to increase their size and reduce the level of structural defects in them. As a result, the performance characteristics of these crystals is enhanced appreciably.

The experimental efforts to synthesize quasicrystals in the μg environment of space have mostly been confined so far to the non-equilibrium processing routes which yield metastable structures [1]. Consequently, there is lack of systematic efforts to grow stable quasicrystals in space. The need of such an effort assumes greater importance in light of the fact that operational performance of quasicrystals also can benefit a great deal from any gain in the size and quality of these crystals. For instance, size enlargement and structural improvement can further enhance the thermoelectric (TE) performance or figure-of-merit of a certain class of quasicrystals, which at present, is comparable to that of the best semiconducting TE materials developed so far [2].

In fact, stable quasicrystals synthesized on earth are reported to have poor quality, which set a limit for achievable structural resolution [3]. The slowly cooled crystals have more distorted diffraction patterns than the rapidly solidified ones. The monochromatic transmission Laue X- ray patterns of the single quasicrystals reveal a high density diffraction spots and also complex diffuse scattering. One of the causes of diffuse scattering and peak broadening can be local atomic disorders or defects such as phasons / dislocations or wrong connections [4].

Shape and size of the quasicrystal also are affected by the gravity related buoyancy forces. As nuclei grow in the melt, they eventually reach a size where buoyancy dominates their movement in the fluid rather than Brownian motion. Depending upon the relative density of the quasicrystal with respect to the surrounding liquid, movement of the quasicrystal in the melt takes place either in the same or in the opposite direction of the gravitational force acting on it.

After moving through solution, the crystal either settles at the bottom of the container or floats to the top surface, where it continues to grow. One or more of the crystal faces, particularly those adjacent to the container wall, may cease to grow. This may distort the final crystal shape. Similar distortion will be experienced by faces of the floating crystal, which will be protruding out of the melt and these faces will be devoid of any supply of solute atoms which make them grow. In a nutshell, quasicrystals that stay suspended in the melt, grow free of such boundary restrictions and hence, will inherit a more uniform shape and large size[5].

The first stable quasi-periodic crystal was found in Al-Li-Cu system which solidifies with triacontrahedral morphology. Since then, a number of compositions have been discovered which show the growth of stable quasicrystals. For our experiment, we selected Ga-Mg-Zn based ternary system, which is reported to form stable quasicrystal ico-Ga20.4Mg38.7Zn42.9 with pentagonal dodecahedral growth morphology [6]. The reasons for this selection include:

(i) Low peritectic reaction temperature of 690 K in Ga-Mg-Zn system matches with the power available for conducting μg experiment aboard Recoverable Space Capsule.

(ii) Buoyancy effects on the size of the growing quasicrystals in the Ga-Mg-Zn system can be studied clearly. These quasicrystals tend to float on the top surface [6] under 1g conditions as the force of the buoyancy are eliminated under the μg environment

3.2 Development of the Payload

For the experiment, a space-worthy μg payload comprising Isothermal Heating Furnace (IHF) was specially designed and developed (Fig. 2). The IHF is based on isothermal conditions for heating and re-radiative concept for the thermal insulation. Heater is resistor based and insulation is provided by the radial and axial shields/reflectors (Fig. 3).

Fig. 2 Isothermal Heating Furnace(IHF)   Fig. 3 Exploded view of the IHF

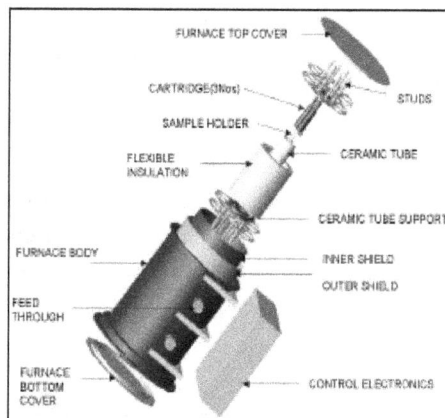

The isothermal heating furnace (IHF) can attain a maximum temperature of 800 K in its The isothermal heating furnace (IHF) can attain a maximum temperature of 800 K in its central hot (Isothermal) zone. The hot zone contains six Ga-Mg-Zn alloy samples placed in three twin-chamber graphite crucibles, which in turn, are housed in three sealed OHFC grade copper cartridges. Control electronics mounted on the IHF body and specially tailored for the experiment, executes the double soak temperature-time cycle through telemetry and telecommand (TTC) system. It has a provision to control /

maintain the isothermal temperature as well as to map the overall thermal health of the IHF during experimental phase in the orbit. Data retrieval occurs in the real-time through ground tracking stations. The P/L consumes 70W of power, weighs 6 kg and occupies an onboard envelope of 0.0035m3 .

The metallurgical characteristics of Ga-Mg-Zn alloy such as high vapour pressure and tendency for oxidation, phenomena like Marangoni convention associated with the microgravity and prevalence of high order of vacuum in space, necessitate the incorporation of special design features in sample containment / holding mechanism (Fig. 4) of the IHF for this experiment.

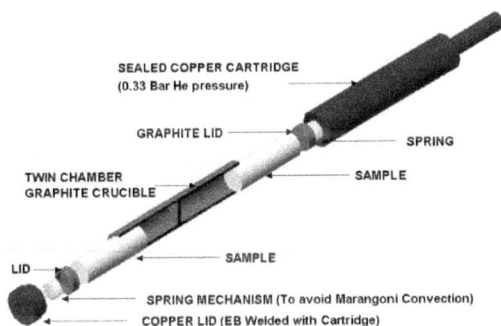

Fig. 4 Sample containment mechanism

The salient features of sample containment mechanism are:

(1) Presence of 0.30 bar inert atmosphere around sample in the crucible in order:

    (a) to provide positive pressure to prevent the loss of constituent elements through vaporization

    (b) to ensure protective atmosphere against the likely oxidation of elements at high temperature

(2) Use of graphite as crucible material to avoid any reaction between the crucible material and sample.

(3) To eliminate free-surface driven Maragoni convection in the sample-melt under µg in space, a spring controlled lid-movement mechanism is incorporated inside the cartridge.

3.3 Scientific Results

The optimum chemical composition range for the formation of icosahedral quasicrystalline phase (I-Phase) in Ga-Mg-Zn ternary system, is Zn44-40Mg40-38Ga16-

22. To achieve stoichiometry in the desired range, Ga-Mg-Zn samples were prepared in a graphite crucible using induction heating method. The stoichiometry of the resultant ternary alloy determined by ICP technique, confirmed the composition within the range.

Quasicrystalline phase forms and grows in Ga-Mg-Zn system as a result of the following two peritectic reactions:

i)   Liquid + MgZn2 = Quasicrystals

     (when Mg : Zn > 1)

ii) Liquid + Mg32Zn52Ga16 = Quasicrystals (when Mg : Zn < 1)

As-cast samples were subjected to XRD analysis using CuK_ (1.5418 A0) radiation and 0.020 step-width. Scanning electron microscopy (SEM) of the samples was carried out mainly to observe the morphology of the growth of quasicrystals. Characterization of GaMgZn samples processed under one-g conditions leads to following observations:

1) Peritectic reaction does not reach completion as it is evident from:

    a) presence of primary MgZn2 (primary solid phase) in XRD profiles

    b) presence of peritectic rim (solute depleted layer) around the quasi-crystalline phase as revealed by SEM micrographs (Fig. 5)

2) As further revealed by OM and SEM micrographs (Fig.5), morphological growth of quasi-   crystalline phase did not take place with sharp faceted features confirming the non-completion of peritectic reaction.

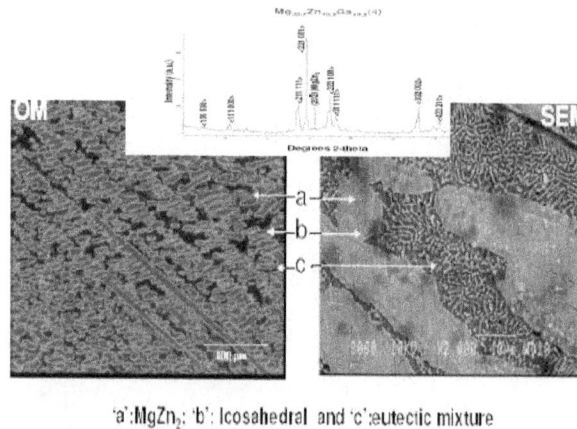

'a':MgZn$_2$; 'b': Icosahedral and 'c':eutectic mixture

Fig. 5 OM and SEM micrographs and XRD profile of 1-g reference sample

A two-soak temperature-time cycle was devised for the experiment. First soak at 790 K ensures the formation of primary homogenous liquid phase. Second soak at 690 K facilitates the growth of quasicrystalline phase. During micro-g experiment, temperature 790 K for the first soak was achieved. As after about 12 minutes, temperature started rising, the IHF furnace was put off in auto mode. Thus, temperature remained above 790 K for a total duration of about 32 minutes. Subsequently, during passive cooling, temperature passed very slowly through the second soak temperature of 690 K.

Results of the characterization of the micro-g processed samples reveal the following:

1. Completion of peritectic reaction which is evident from:

   a) absence of MgZn2 (primary solid phase) in the XRD profiles.

   b) absence of peritectic rim around the quasicrystalline phase as determined from the SEM micrograph (Fig. 6).

2. Features of the Massive growth morphology of the quasi crystalline phase as shown in SEM micrograph (Fig. 6).

3. Presence of one GaMg2 phase as revealed by selected area diffraction (SAD pattern in Fig.6 .

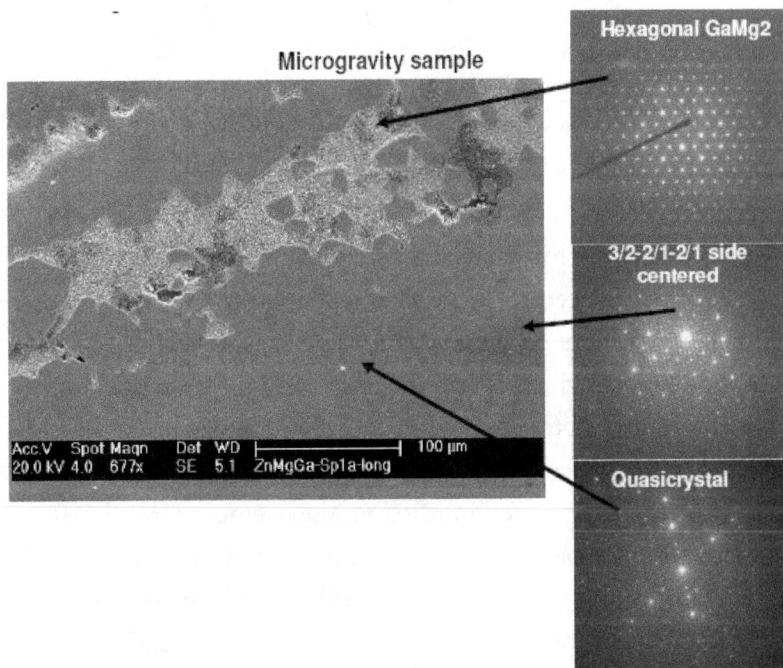

Fig. 6 SEM and SAD patterns of Micro-g processed sample

41

4. Bio-Mimetic Synthesis of Inorganic Nanoparticles (BMPR Payload)

Microgravity environment offers new opportunities to develop a deeper understanding of the relationship of secondary bond induced polymorphic conformation, which controls the morphological features of inorganic nanoparticles. The present micro-g experiment aimed at studying the effect of microgravity conditions on the morphological features of nanosized hydroxyapatite (HA) produced by bio-mimetic route.

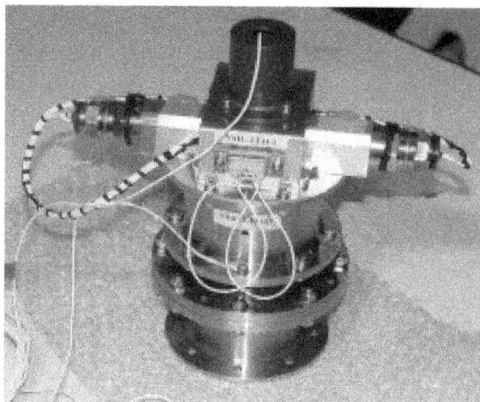

Fig. 7 Biomimetic Material Processing Reactor (BMPR)

To conduct this experiment, a Bio-mimetic Material Processing Payload (BMPR) was developed (Fig. 7). The reactor facilitates chemical mixing with bolt cutter actuation mechanism. HA-Liquid (placed in lower cylinder) the protein gel (placed in upper gel cylinder) are initially separated by glass separator and gasket sealing. The glass is broken by a spring-loaded plunger, which is released by the cutting of actual bolt by a Pyro-bolt cutter. The force and displacement provided by disc springs break the separation glass into pieces. The command for the actuation for bolt cutter mechanism is transmitted through tele-command incorporated in Mission Management Unit (MMU) and Electro-Explosive Device (EED) packages. One bit actuation telemetry (TM) scheme is implemented for confirming the actuation of mechanism in space. The TM provision is available in MMU. Throughout the reaction the reaction chamber temperature is monitored. Shri A. Sinha of National Metallurgical Laboratory, Jamshedpur, India was the principal investigator of this experiment project.

5. Conclusions

(i)   ISRO's Space capsule Recovery Experiment (SRE) validated an orbital platform or free flyer for effectively conducting micro-g experiments.

(ii)   Two micro-g experiments conducted aboard recoverable space capsule employing payloads viz. IHF and BMPR, developed for the first mission of SRE, yielded valuable scientific results.

(iii)   Quasicrystals grown under microgravity mark the absence of solute depleted layer or peritectic rim around the quasicrystalline phase.

(iv)   As a result of the above (iii), the quasicrystal grown under equilibrium conditions in space show highly faceted growth morphology.

(v)   Microgravity environment free from effects of sedimentation / buoyancy facilitated the predominated growth of quasicrystalline phase.

Acknowlegement

Permission granted by Dr. K. Radhakrishnan, Director, VSSC to publish this work is gratefully acknowledged.

References

1.   Correspondence with Prof. K. Urban, Institute of Solid State Physics, Juelich, Germany on 27 April 2001.

2.   Cyrot-Lackmann, F., *Quasicrystals as potential candidates for thermoelectric materials* Mat. Sc. Engg. Vol. 294-296, 2000, p. 611.

3.   Steurer, W, *Structure of Quasicrystals,* in Physical Metallurgy (ed). Cahn & Hasen, North Holland, 1996, p 394.

4.   Guyot, P. Nature 326, P. 640 (11987)

5.   Sharma, SC, Nair, KS, Mittal, MC and Chattopadhyay, K, *Growth of Quasicrystal in Space: Possible Fallouts and the Indian Experiment,* Proc. 8th Int. Conf. on Quasicrystals (ICQ-8), Bangalore, Sept. 8-13, 2002.

6.   Ohashi, W. and Spaepen, F., Nature 330, 1987, p. 550.

Nanomorphological Induced Electromagnetic Properties of Radar Absorbing Material in Ku-Band

Rahul Sharma[a], R.C Agarwala and Vijaya Agarwala

Surface Engineering Lab, Metallurgical and Materials Engineering Department,
Indian Institute of Technology Roorkee, IITR Roorkee (Uttarakhand), INDIA
[a]rahuldmt@gmail.com

Abstract *Nanomorphological M-type barium hexaferrite i.e., $BaFe_{12}O_{19}$ nano radar absorbing material (NRAM) were synthesized by modified flux method that combines the controlled chemical co-precipitation process for nucleation and complete growth during in-situ microwave annealing (MWA). Uniform morphological transformation of nano crystals from spherical to pyramidal shapes were observed under FESEM and TEM. The effect of such systematic nano morphological transformation of NRAM was observed on electromagnetic (EM) properties. Maximum reflection loss (RL) was improved to 38.25 dB at 16.00 GHz for MWA at 760 watt with continuous increasing absorption range under -10 dB for 2 mm thick coating layer in the Ku Band (12.4-18.0 GHz). Excellent microwave absorption properties are a consequence of accurate EM match in the nano morphological planes, a strong natural resonance, as well as multipolarization. This process of crystal growth, morphology evolution and RL enhancement with respect to the heat treatments were also explained in terms of Ostwald ripening and quantum size effect.*

Keywords:Nano radar absorbing material, NRAM; modified flux method; Ostwald ripening; microwave annealing, MWA; reflection loss, RL.

1.0 Introduction

Since II World War to update, radar absorbing materials (RAM) are the area of interest especially for defence applications. Currently, the world is confronting Global Warming, electromagnetic (EM) pollution is one the main contributors to it. EM pollution is arising from rapidly expanding use of communication devices. Hence a great interest is generating to develop microwave absorbing material with strong electromagnetic (EM) properties in a wide frequency range. To improve performance of various RAM, there has been increasing attention onto the morphology control of

44

materials as it plays very important role in determining chemical and physical properties of materials that is attributed to the novel application [1-20]. The emergence of Nanoscience and Technology has opened the door for new opportunities to further improving the EM absorbers by transforming RAM to nano NRAM by controlling the morphological properties [2, 5-8]. Various techniques have been developed to prepare micro/nanosized RAM for these purposes. These include the glass crystallization method [9], the wet method [10], the liquid mix technique [11], colloidal methods [12], and many others. The common feature of these methods is the intimate mixing of ions at the atomic level so that subsequent nucleation and crystallization can occur and induce the phase transition at relatively high temperatures. In recent works it has been shown that the aerosol synthesis technique [5, 13] and modified co-precipitation method [7, 14] also appear to be promising methods for the preparation of nanosized particles of unique composition and EM behavior [8, 15, 20].

This study deals with the development of pure single phase barium hexaferrite nano crystals under MWA to get the various nanomorphologies. Systematic uniform growth of nano crystals with respect to increasing MWA power 160-760 watts were investigated and characterized under XRD and FESEM. The effect of such systematic morphological transformation of nanocrystals was observed for EM properties in Ku-band.

2. Experimental Procedure

2.1. Synthesize Route

Nanocrystals of M-type barium hexaferrite i.e. $BaFe_{12}O_{18}$ were synthesized by using modified flux method. The systematic details of nucleation and growth nano crystals were explained elsewhere [2]. The in-situ MWA with increasing microwave power from 160, 360 to 760 watt were carried out for 5 minutes only to get the spherical, hexagonal-pyramidal, hexagonal and pyramidal nanocrystalline morphologies respectively.

2.2. Characterization Studies

The samples for transmission electron microscope (TEM) analysis were dispersed in ethanol with the aid of ultrasound, and then applied to a copper grid, where they were allowed to dry and later viewed under the TEM (TECHNAI 20G2-S-TWIN) at an operating voltage of 200 kV. Surface morphologies were seen under field emission scanning electron microscope, (FESEM, QUANTA FEG 200). X-ray diffraction, (XRD, Bruker D8 Advance) was carried out for phase analysis with Cu K_α radiation at diffraction angle ranging from 20 to 100 ° (2θ). MWA were carried out in Microwave Oven (LG Company, 2.45 GHz, 160-760 watt). Magnetization measurements were carried out in vibrating sample magnetometer, (VSM) (155, PAR). For RL measurements, all the powder samples were uniformly dispersed into the polymer (epoxy + resin) with a constant weight ratio of 1:10 respectively and then applied a thickness of ~ 2 mm onto the standard size aluminum sheet. Microwave properties for all the samples were studied by a HP 8720B vector network analyzer in anechoic chamber.

3. Results and Discussions

3.1. Nanomorphological Study

The FESEM and TEM micrographs with selected area electron diffraction (SAED) patterns of in-situ MWA at 160, 360, 600 and 760 watt for 5 minutes are shown in figure 1 (a-f) respectively. With increasing microwave power nano crystalline volume is increasing and finally transforming into single nano crystals of pyramidal shape. A systematic growth of nano crystals can be seen with increasing power giving sharp planes. The average particle size of NRAM increases with increasing MWA at 760 watt irradiated for 5 minutes is found to be 15 nm. A typical lattice image corresponding to the layered structure of hexaferrite obtained on one of the crystallites is shown in the insert Figure 1 (e).

Figure 1 (f) reveal the spacing of 5.9 Å between successive white fringes as of the magnetoplumbite unit cell along *a*-axis after MWA at 760 watt. The lattice image of a crystal with the beam parallel to [100] direction is shown in figure 1 (e-f) and corresponding SAED patterns are also inserted in these figures. There are no intergrowths seen as revealed by the HRTEM of selected area (doted circles) are shown in figure 1 (e and f). NRAM powders after both type of annealing, exhibit superlattice reflections. The electron diffraction (ED) pattern of NRAM powder MWA at 760 watt, taken with incident electron beam normal to (001) plane, is shown in figure 1 (f). The bright spots in the pattern are basal reflections of hexaferrite phase of NRAM particle.

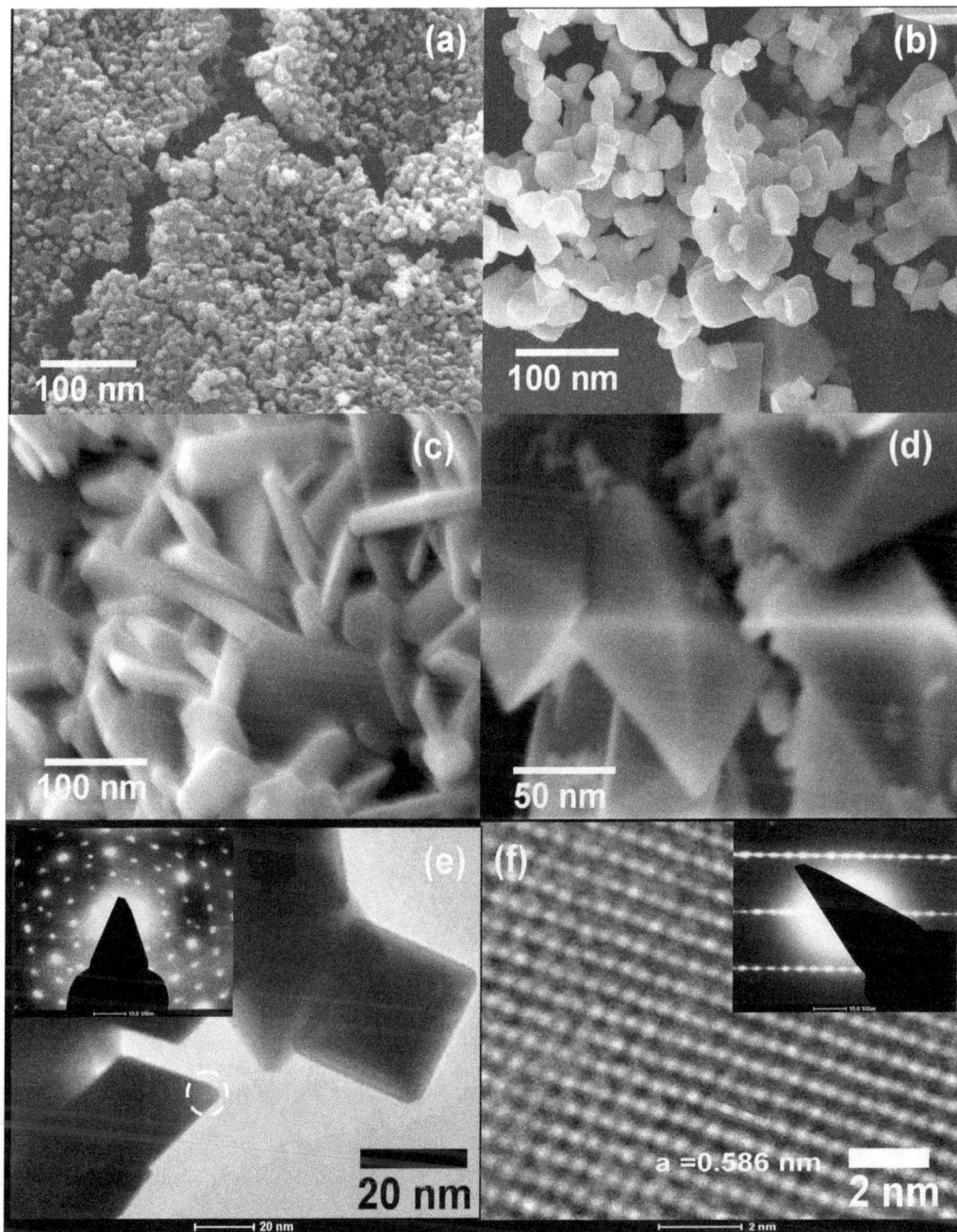

Figure 1 Development of pure M-type barium hexaferrite (NRAM) grown with increasing microwave annealing (MWA) from 160 -760 watt to get various nanomorphologies seen under FESEM (a) spherical (b) hexagonal-pyramidal (c) hexagonal and (d) pyramidal and (e) TEM images with its SAED showing the MWA at 760 watt with (f) HRTEM showing the inter planar spacing of a=5.86 nm of a-axis.

It is observed that all spherical nano crystals have transform into the rectangular prism morphology in the size rage of 20-40 nm and are shown in figure 1 (a-d). This process of crystal growth and morphological evolution can be described in terms of Ostwald ripening [16,17].

3.2. Phase Analysis

Figure 2 (a-d) is showing the direct development of pure NRAM single phase hexagonal $BaFe_{12}O_{19}$ (JCPDS Card No. 84-757) hexaferrite during in-situ MWA. The XRD patterns for the in-situ MWA powder at 160, 360, 600 and 760 watt are shown in figure 2 (a-d) with spherical, hexagonal-pyramidal, hexagonal and pyramidal respectively.

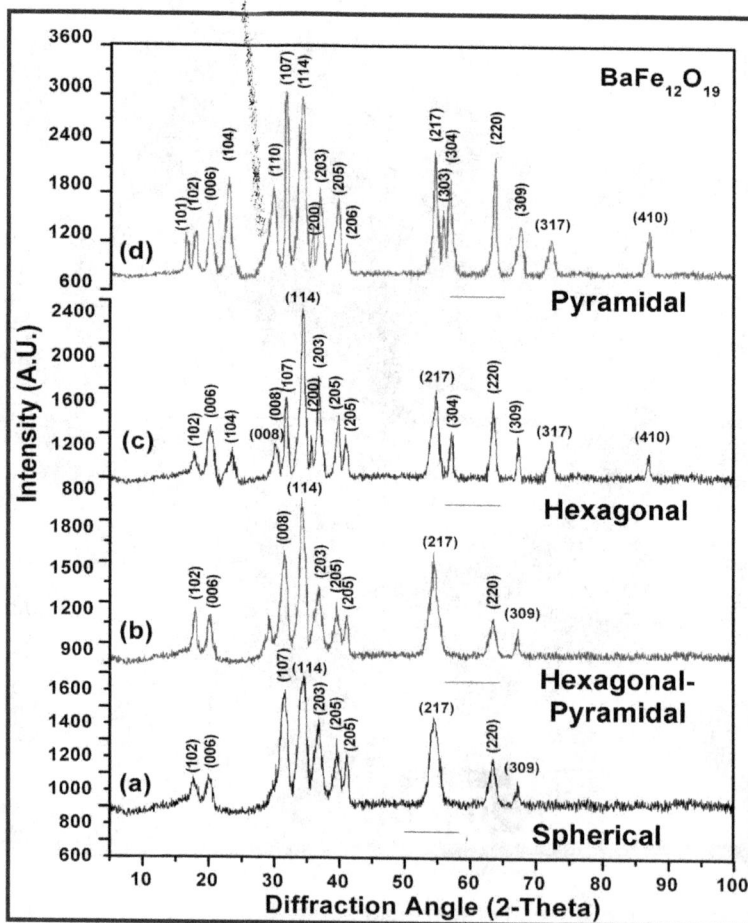

Figure 2 XRD patterns of pure single phase BaFe12O19 hexaferrite nanocrystals in-situ MWA at with nanomorphologies of (a) spherical (b) hexagonal-pyramidal (c) hexagonal and (d) pyramidal with increasing microwave power from 160-760 watt.

XRD patterns shows that the in-situ MWA powder is pure nano-crystalline and all the peaks are well matched with hexagonal $BaFe_{12}O_{18}$ that has suppressed the formation of α-$BaFe_2O_4$, γ-$Fe2O_3$, or any other intermediate ferrite phases that have been generally observed during annealing indicated the direct formation of pure NRAM crystals. With increasing MWA, the intensities of all the diffraction peaks increased but the broadening (full width at half maxima, FWHM) remain almost same during MWA. During MWA, the electric and magnetic field components of EM radiation interact with NRAM particles that are nucleate new diffraction planes without further growth of NRAM particles. In MWA mainly the electrons (less atoms) are moving inside the NRAM particles that attributed to the increase in nano crystallinity, although the particle sizes increase slightly from 7 to 13 nm (average particle size, calculated by Scherrer's formula for all the reflection peaks). More diffraction peaks from (101), (200), (303) and (410) planes were observed during MWA. Hence, it can be easily ascribed from the XRD analysis that the MWA is recommended while comparing with the other types of annealing as the particle size remains almost the same in the former case, with larger increase in the intensities of the diffraction peaks, giving rise to a reduced heat treating time.

3.3. Reflection Loss and Magnetic Study

The normalized input impedance Z_{in} of a metal-backed microwave absorption layer could be obtained from the following expression: $Z_{in} = \sqrt{(\varepsilon_r/\mu_r)} \tanh [-j (2\pi fd/c) \sqrt{(\varepsilon_r \mu_r)}]$, where ε_r and μ_r are the relative permittivity and permeability of the composite medium, c is the velocity of electromagnetic waves in free space, f is the frequency of microwaves, and d is the thickness of the absorber. Accordingly, the reflection loss is associated with Z_{in} as RL (dB) $=20 \log [(Z_{in}-1)/(Z_{in}+1)]$ [17, 18]. Figure 3 (a-b) reveals the magnetic and dynamic EM properties of in-situ annealing NRAM particles. It is observed that with increasing MWA the RL and magnetic properties are increasing systematically and shown in figure 3 (a-b) Maximum RL is improved to 38.25 dB at 16.00 GHz for MWA at 760 watt with continuous increasing absorption range under −10 dB for 2 mm thick layer in the Ku Band. All the RL characteristics are listed in Table 1.

Table 1 The effect of nanomorphologies attributed to enhance the reflection loss (RL) characteristics in Ku-band

Increasing Annealing MWA (in watt) With morphology (shape/size)	RL in K_u band range at (in dB)		The widest bandwidth with range under 10 dB RL	Strongest RL (dB) with resonance frequency
	12.4 GHz (Min.)	18.0 GHz (Max.)		
160 (spherical, 7 nm)	-04.8	-9.7	--	-7.80 (at 14.00 GHz)
160 (hexagonal-pyramidal, 8 nm)	-6.7	-11.1	0.75 (-13.25 to -14.00 GHz)	-21.45 (at 13.10 GHz)
360 (Multiple 9 nm)	-7.9	-13.7	2.25 (-13.50 to -15.75 GHz)	-28.65 (at 14.75 GHz)
760 (pyramidal, 12 nm)	-12.5	-28.2	3.00 (-14.25 to -17.25 GHz)	-37.78 (at 16.00 GHz)

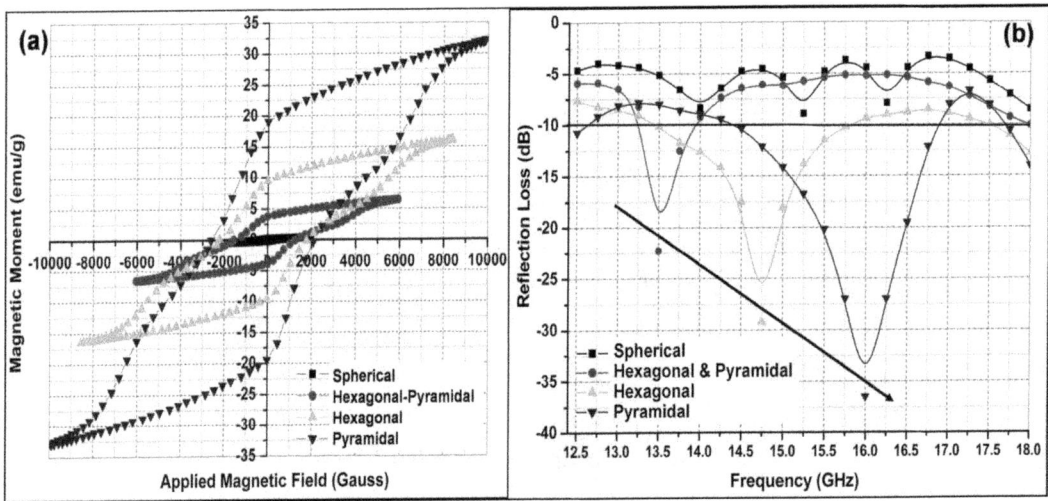

Figure 3 Magnetic and reflection loss (RL) of in-situ MWA powder with nanomorphologies of (a) spherical (b) hexagonal-pyramidal (c) hexagonal and (d) pyramidal with increasing microwave power from 160-760 watt.

It is reported that the microwave absorption would be improved when particle size is reduced from micron to nano range [2, 18-20]. It is observed that MWA is giving better RL characteristics because the larger number of nano planes are interacting with EM wave. As it is well known that the quantum size effect in nanocrystalline makes the electronic energy level split. The spacing between adjacent energy states increases inversely with the volume of the particle. At the same time, with the decrease of the particle size, the number of incomplete molecules and the defects of surface and interface increase rapidly, which will lead to the multiplication of discrete energy levels [20]. If the particle size of absorber (NRAM particles) is small enough and the discrete energy level spacing is in the energy range of microwave, the electron can absorb the energy as it leaps from one level to another, which may lead to the increment of RL properties. When the uniform morphology in nanoscale is subsisted, the structure of crystalline phase may form the multiple magnetic domains to single domain and the coercive force of the material increases largely as observed in figure 3 (a). This may lead to big hysteresis attenuation and the absorbing properties can be improved.

4. Summary

Pure nanocrystalline morphologies of spherical to pyramidal were under controlled in-situ MWA. Such nano morphological transformations has improved the maximum RL to 38.25 dB at 16.00 GHz for MWA at 760 watt with continuous increasing absorption range under −10 dB RL for 2 mm thick layer in the Ku Band.

Acknowledgements

The authors acknowledge Heads of Electronics & Computer Engineering Department and Institute Instrumentation Centre (IIC) for providing characterization facilities at IITR Roorkee, Uttarakhand India.

References

1. S. Sugimoto, T. Maeda, D. Book, T. Kagotani, K. Inomata, M. Homma, H. Ota, Y. Houjou, and R. Sato, J. Alloys Compd. ,Vol. 330, 2002, p. 301.

2. R. Sharma, R. C. Agarwala, and V. Agarwala, J., Nano Res., Vol. 2, 2008, p. 91.

3. R. C. Che, C. Y. Zhi, C. Y. Liang, and X. G. Zhou, Appl. Phys. Lett., Vol. 88, 2006, p. 033105.

4. A. Wadhawan, D. Garrett, and J. M. Perez, Appl. Phys. Lett., Vol. 83, 2003, p. 2683 .

5. L. J. Deng and M. G. Han, Appl. Phys. Lett., Vol. 91, 2007, p. 023119.

6. H. M. Kim, K. Kim, C. Y. Lee, J. Joo, S. J. Cho, H. S. Yoon, D. A. Pejakovic, J. W. Yoo, and A. J. Epstein, Appl. Phys. Lett., Vol. 84, 2004, p. 589.

7. X. G. Liu, D. Y. Geng, and Z. D. Zhang, 92, Appl. Phys. Lett., Vol. 10, 2008, p. 24-31.

8. R.Sharma, R.C.Agarwala, and V.Agarwala, J. Alloys Compd., Vol. 467, No. 1-2, 2009, p. 357.

9. B.T. Shirk and W.R. Bussem: J. Am. Ceram. Soc., Vol. 53, 1970, p. 192.

10. K. Haneda, C. Miyakawa and H. Kojima: J. Am. Ceram. Soc., Vol. 57, 1974, p. 354.

11. M. Vallet-Regi, P. Rodriguez, X. Obradors et al.: J. Phys. (Paris), Vol. 46 , 1985, p. 335.

12. E. Matijevic: J. Colloid. Interface Sci., Vol. I17, 1987, p. 593.

13. Z.X. Tang, S. Naris and C.M. Sorensen: J. Magn. Mater., Vol. 80, 1989, p. 285.

14. G.L. Messing, S.C. Zhang and G.V. Jayanthi: J. Am. Ceram.Soc., Vol. 76, 1993, p. 2707.

15. V.V. Pankov, M. Pernet, P. Gelani and P. Mollard, J. Magn. Mater., Vol. 120, 1993, p. 69.

16. W. Z. Ostwald: Phys. Chem., Vol. 34, 1900, p. 495.

17. P.A. Miles, W.B. Westphal and A.V. Hippel, Rev Mod Phys., Vol. 29, 1957, p. 279

18. K. Ishino, Y. Narumiya, Ceram Bull, Vol. 66, 1987, p. 1469.

19. S. Ruan, B. Xu, H. Suo, F. Wu, S. Xiang and M. Zhao, J. Magn. Mater., Vol. 212, 2000, p. 177.

20. R. Sharma, R. C. Agarwala, and V. Agarwala, Advanced Materials Research, In press.

Strain Mapping at the Nanoscale using Electron Back Scatter Diffraction

Angus J. Wilkinson

Department of Materials University of Oxford
Parks Road, Oxford, United Kingdom
Email: angus.wilkinson@materials.ox.ac.uk

Abstract_*Electron back scatter diffraction (EBSD) is a key quantitative microstructural analysis technique available on scanning electron microscopes. Hough transform based analysis is used for automated crystal orientation mapping at an angular resolution of ~10^{-2} rads (~0.5°). Recently we have dramatically improved the angular sensitivity to ~$\pm 10^{-4}$ rads by using a cross-correlation based analysis. This allows small lattice rotations and even elastic strains to be mapped at nanoscale resolutions. The EBSD methodology will be briefly described. Measurement of the elastic strain tensor coupled with knowledge of the elastic constants allows the local stress state to be fully described. Dislocation storage during plastic deformation can also be assessed as the local lattice curvatures present in the measured rotation fields provide links to the underlying geometrically necessary dislocation content.*

The technique has been applied to a growing range of materials systems, both functional and structural. Illustrative examples will be taken from work on strain relaxation in SiGe/Si mesas, assessment of tilt and twist mosaics in GaN films, strains and lattice curvatures near indents, thermal strains near carbides in a Ni-based superalloy, and deformation near crack tips after monotonic and fatigue loading.

1. Introduction

The need to know the local state of stress and strain in a region of material is of the utmost importance in a wide range of applications spanning from semiconductor structures and devices to structural engineering components. Over the years many techniques to measure and map stress and strain have been developed and implemented. The strong interaction of electrons with matter leads to small interaction volumes compared to x-ray or neutron probes. This coupled with the ability to make fine electron probes make electron diffraction techniques attractive for high spatial resolution strain mapping. Electron back scatter diffraction (EBSD) is a key quantitative microstructural analysis technique available on scanning electron microscopes. Hough transform based analysis is used for automated crystal orientation mapping at an angular resolution of ~10^{-2} rads (~0.5°), see for example reviews by Wilkinson and Hirsch (1997), and Humphreys (2001). Recently we have dramatically improved the angular sensitivity to ~$\pm 10^{-4}$ rads by using a cross-correlation based analysis. This allows small lattice rotations and even elastic strains to be mapped at nanoscale resolutions (Wilkinson et al 2006). Measurement of the elastic strain tensor coupled with knowledge of the elastic constants allows the local stress state to be evaluated fully. Dislocation storage during plastic deformation can also be assessed as the local lattice curvatures present in the measured rotation fields provide links to the underlying geometrically necessary dislocation content.

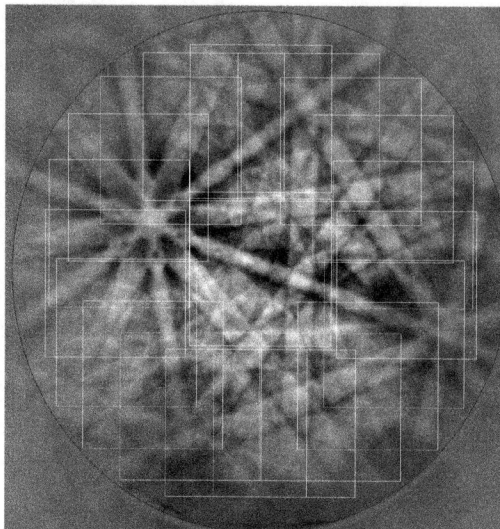

Figure 1 EBSD pattern from Fe sample with
sub-regions used for cross correlation analysis
marked in white.

2. Methodology

EBSD patterns are captured with the usual wide capture angle (~70°) geometry (see figure 1), but at the full 1300 x 1030 pixels resolution of the 12 bit deep peltier cooled CCD camera (Digiview 12) using TSL/EDAX OIM software. All patterns are record to hard disc for subsequent off-line batchwise cross correlation analysis using BLGproductions CrossCourt2[*] software. Typical operating conditions for the SEM are 20 keV beam energy and probe currents from 2 nA to 10 nA. Exposure times are of the order of 1 second which is considerably slower than for orientation imaging. This is due to the requirement for low noise and no binning so as to optimise the angular resolution.

As the electron beam is scanned over the sample lattice rotations and elastic strains cause small shifts in the positions of zone axes and other features in the EBSD patterns obtained. Two dimensional cross-correlation analysis is used to measure these small

---

[*] www.blgproductions.co.uk

pattern shifts in an automated image processing procedure. The shifts at four or more sub-regions widely dispersed across the EBSD patterns are sufficient to directly calculate 8 of the 9 degrees of freedom contained in an arbitrary strain and rotation. The remaining degree of freedom is the hydrostatic dilatation of the lattice, to which this method is insensitive. However, this too can be recovered if we, not unreasonably, assume that plane stress conditions are met within the diffraction region close to the sample surface. For the results presented here pattern shifts were measured at 20 sub-regions (see figure 1) and least squares methods used to calculate the best fit displacement gradient tensor. The quality of this solution was assessed by calculating the angular difference between the shift expected at each of the sub-regions on the basis of the best fit solution and the actual measured shift. These differences were averaged over all 20 sub-regions to give the mean angular error, which provides a useful means of assessing how good the best fit solution is.

2.1 An Example Application: Indent in Fe

The elastic strain (and hence stress) and plastic flow fields near to indents are complex and not understood in detail. Here we look at high sensitivity EBSD maps obtained near a 500 nm deep Berkovich indent in a large grained Fe-0.01%C polycrystal. Figure 2 shows and SEM image of the indent and EBSD generated maps of the elastic radial and hoop strain fields. We see that the radial strains are generally compressive and hoop strains generally tensile in line with expectations from an expanding cavity model. Anisotropy in the elastic constants and restriction of plastic deformation to slip on crystallographic slip systems complicates the situation compared to simple models treating the material as isotropic.

Although the elastic strains presented in figure 2 are considerable when compared to the elastic limit, but they are relatively small compared to the rotations that are shown in figure 3. The fact that the lattice rotations are large compared to the elastic strains is evidence for plastic deformation and accumulation of geometrically necessary dislocations (GNDs) around the indent.

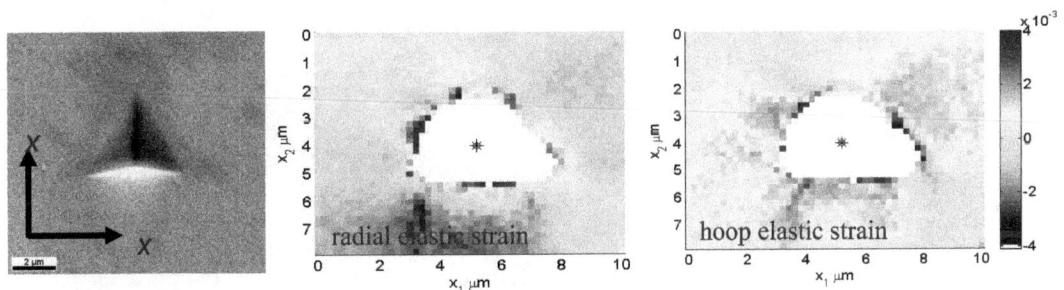

Figure 2 *Secondary electron image of 500 nm deep Berkovich indent in Fe and maps showing radial and hoop components of elastic strain field measured by EBSD.*

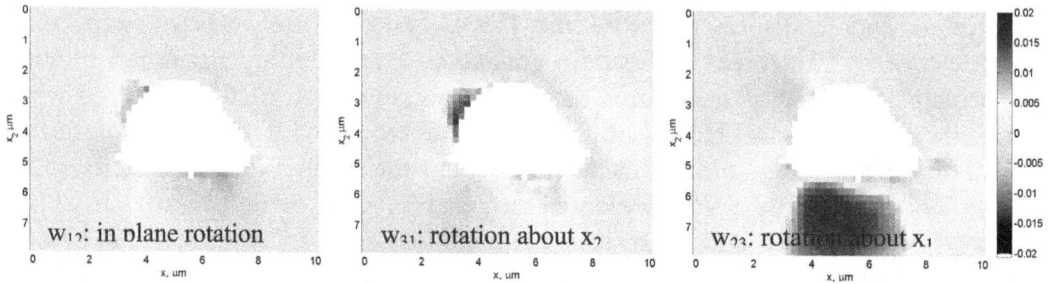

Figure 3 Lattice rotations (in rads) near 500 nm deep Berkovich indent in Fe measured by EBSD.

The lattice rotations shown in figure 3 give a means of estimating the density of GNDs present in the crystal. At a simplistic level one can think of a series of very low angle grain boundaries separating each of the points in the crystal at which the EBSD measurements are made (see figure 4). If one then thinks of these boundaries as being rather diffuse arrangements of dislocations a simple relationship between the average density of a single type of dislocation (defined by its Burgers vector and line direction) and the resulting rotation gradient can be written. Nye (1953) put such concepts into a formal mathematical framework that we can use to construct lower bound estimates of the GND density from the measured lattice rotation fields and knowledge of the crystal orientation and possible dislocation types. Figure 5 gives such a lower bound estimate of the GND content near this indent.

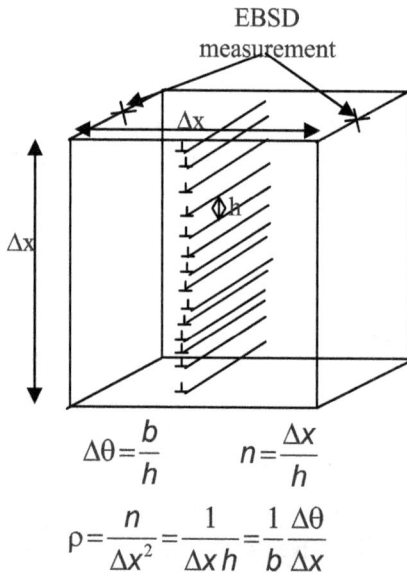

$$\Delta\theta = \frac{b}{h} \qquad n = \frac{\Delta x}{h}$$

$$\rho = \frac{n}{\Delta x^2} = \frac{1}{\Delta x\, h} = \frac{1}{b}\frac{\Delta\theta}{\Delta x}$$

Figure 4 schematic showing link between lattice rotations and dislocation content

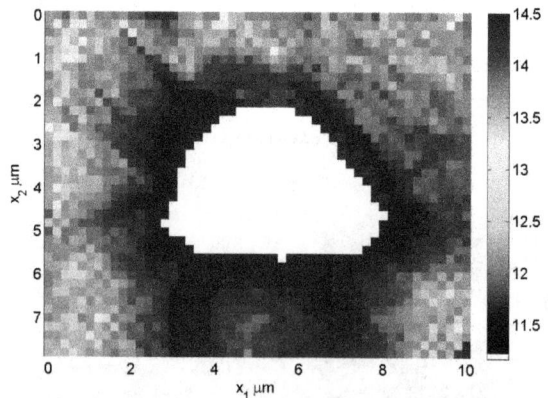

Figure 5 distribution of GND density near indent in Fe. Colour-scale shows $log(\rho)$ with ρ in lines/m^2.

3. Summary

Development of cross correlation based analysis of EBSD patterns has come to fruition over the last few year and software is now commercially available that can deliver strain measurements at a sensitivity of $\sim 10^{-4}$. The method gives access to variations in the strain and rotation tensor at the high spatial resolution offered by EBSD. This coupled with the relative ease of use and wide availability of EBSD systems on FEG SEMs is likely to lead to a considerable increase in research activity in this area over the next 10 years or so.

References

1. Humphreys F. J., Review, *Grain and subgrain characterisation by electron backscatter diffraction*, J Mater Sci., Vol. 36, **2001, p.** 3833.

2. Nye, J. F., *Some Geometrical Relations In Dislocated Crystals*, Acta Metall., **1**, 1953, p. 153.

3. Wilkinson A. J. & Hirsch P. B. *Electron diffraction based techniques in scanning electron microscopy of bulk materials*. Micron **28**, 1997, p. 279.

4. Wilkinson A. J., Meaden G., and Dingley D. J., *High-resolution elastic strain measurement from electron backscatter diffraction patterns: New levels of sensitivity*, Ultramicroscopy **106**, p. 307.

5. Wilkinson A. J., Meaden G., and Dingley D. J, *High Resolution Mapping of Strains and Rotations using Electron Back Scatter Diffraction*, Materials Science and Technology, **22,** 2006, p. 1271.

Scanning Electron Microscope – A Versatile Investigative Tool

GMK Sarma

Retired Professor in Metallurgy, Banaras Hindu University
 # 58, Om Nagar
Hyderabad, INDIA
Email: gmksarma@yahoo.co.in

Abstract: The Scanning Electron Microscope (SEM) is a truly versatile instrument in the investigation of various types of materials. It was conceived almost simultaneously as its more famous sibling - the Transmission Electron Microscope (TEM). Yet it took nearly a quarter century after the first TEM to commercially produce an SEM. However, since then it has withstood the test of time and continues to come to the aid of not only Materials Researchers but also Biological Scientists. It competes favourably even with more recent and exotic offsprings like the Atom Probe and the Scanning Tunneling Microscopes. It is proposed to discuss briefly the reasons for its popularity that makes the SEM an "A to Z Microscope"

1. Introduction

ABC s of SEM

1.1 A for Ardenne (1907-1997)

Manfred von Ardenne was a self-taught Engineer as he could not complete even high school education due to financial problems. He developed an experimental Scanning Electron Microscope (SEM) in 1938 in wartime German. After World War II, he worked in USSR for nine years in their atomic energy projects. During that period he developed a table-top SEM. Ardenne received Stalin Prize of USSR in 1947 for his researches in Isotope Separation

After his repatriation to East Germany in 1954, Ardenne was engaged in research in various fields including electron microscopy, medical technology, plasma physics, nuclear technology, Radio &TV technology, taking about 600 patents.

Binnig & Rohrer of ScanningTunneling Microscope(STM) fame received Nobel Prize in 1985. Ernest Ruska, the pioneer in Transmission electron microscope (TEM) which he built in 1934 also shared it the same year. But M v Ardenne whose SEM is based on radically new ideas of imaging, was sadly forgotten !!!

1.2 B for Broers (1938-)

Born in 1938 in Calcutta, Professor Sir Alec Nigel Broers graduated from University of Melbourne, Australia, in 1960 with a first degree in Physics followed by a further year of studying electronics. He decided to go to Cambridge, the then Mecca for electronics. Despite already having two first degrees, he was advised to enrol as an undergraduate before undertaking a PhD, which turned out to be "one of the best moves I ever made". He then started the serious business of his PhD, working on a prototype scanning electron microscope.

After his Ph.D., Broers moved to the IBM research laboratories in New York: "I went for two years and stayed for twenty". There he spent around 16 years in research in one of the best "playhouses for electronics" in the world, building scanning microscopes and equipment for the fabrication of miniature components.

Alec Broers was tempted back to Cambridge in 1984 to the Chair of Electrical Engineering. His return to Cambridge and academia meant a drastic reduction in salary (a cut of about 80%) and with two children in private schools he considers it now to have been "an irresponsible move".

In 1996 Professor Broers was appointed as Vice Chancellor of Cambridge University, the first time that such a post has been held by an engineer. "Those late nineteenth century opponents of the study of engineering at Cambridge must be turning in their graves"

Sir AN Broers was knighted for his services to education in 1998.
C for Charles Oatley (1904-96) & Cambrdige Instruments Company

Today many companies make and sell Scanning Electron Microscopes.Hitachi, JEOL, Philips are among the top suppliers of SEMs. But,after the war circumstances dictated that Ardenne played no further part in the development of the SEM, it was left to Professor Charles William Oatley & his research group (including Broers), "to take up Ardenne's baton once more". In collaboration with the Cambrdige Instruments Company, microscopes was manufactured in 1965. The first four production models were sold under the trade name "Stereoscan". For a long time thereafter, the words StereoScan and SEM were used synonymously.

In a sense, Charles Oatley & Cambridge Instruments, UK. are the pioneers in the commercialisation of SEM.

Back in 1960s,however, when the SEM was being considered as a commercial product, Cambridge Instruments sent out a Group of Marketing Experts to make an evaluation of the number of SEMs that could be sold. Hold your breath,...the Group came back with a figure ... probably between 6 (six), and 10 (ten) would saturate the market!

Today SEMs world-wide are in excess of 60,000 SEMs !

2. Main body : A to Z of SEM

2.1 Scanning Electron Microscope (SEM): an A to Z Microscope

SEM is useful in the study of a variety of experimental sciences Agriculture, Archeology, Ayurveda, Botony, Chemistry, Ceramics, Criminolgy, Dentistry, Environmental Studies, Metallurgy, Medicine, Physics, Polymers, Zoology.

In comparison to Optical Microscope and TEM, SEM can dispense with cumbersome and / or costly specimen preparation methods. Because of its greater depth of field, also called Vertical Resolution, SEM gives us amazing 3-D images which are very easy to interpret. The magnification range is in between the OM and TEM.

Unlike OM which can yield only surface microstructure, SEM can give, with suitable attachments, many other pieces of information, very much like the Analytical TEM, at much lower cost and lesser effort. These include compositional & crystallographic information.

In short, the special features of SEM are as follows:

Polishing and / or etching is unnecessary
Resolution can be as good as 30 microns
Depth of field can be as much as 300 microns at 100 X
Limit of Magnification may go as high as 100000 X

2.2 How does an SEM work?

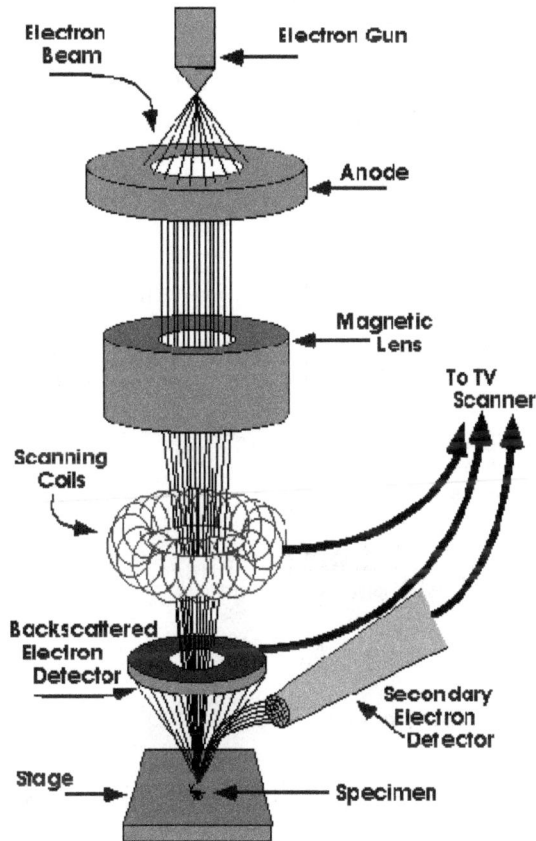

2.3 Types of SEM

Conventional SEM (CSEM) : This is the older type of SEM. Here the Electron source (Gun), electron lenses and specimen are under the same high vacuum. This is a disadvantage with biological sample which collapse under high vacuum. To overcome this problem, we resort to elaborate "specimen fixing" methods which are not always satisfactory.

Environmental SEM (ESEM) : This recent version of SEM is a boon for biologists. Here, the specimen is in a special gas-filled chamber. The secondary electrons emitted by the specimen are attracted to the positively charged detector electrode. As they travel through the gaseous environment, collisions between an electron and a gas particle result in emission of more electrons and ionization of the gas molecules. This increase in the amount of electrons effectively amplifies the original secondary electron signal. At the same time, the positively charged gas ions are attracted to the negatively biased specimen and offset charging effects.

2.4 Signals arising from Electron – Specimen interactions

SEM Setup
Electron/Specimen Interactions
When the electron beam strikes the sample, both **photon** and **electron** signals are emitted.

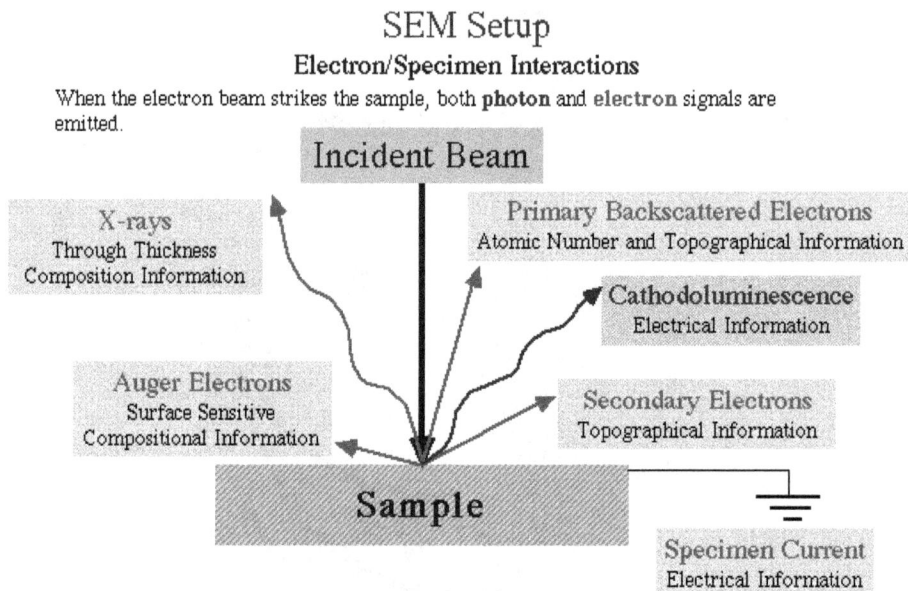

Incident Beam

X-rays
Through Thickness
Composition Information

Primary Backscattered Electrons
Atomic Number and Topographical Information

Cathodoluminescence
Electrical Information

Auger Electrons
Surface Sensitive
Compositional Information

Secondary Electrons
Topographical Information

Sample

Specimen Current
Electrical Information

2.5 Various Imaging Techniques in SEM

Signal	Technique	Information	Resolution,(nm)
1. SecondaryElectrons (SE)	Emissive (SEI)	Topographic	10
2. Back Scattered Electrons (BSE)	Refective (BEI) ECP / SACP	Compositional Crystallographic	100
3. X-rays Emitted	WDX / EDX	Compositional	1000
4. Visible Light from Specimen	CL	Mineralogical	100
5. Electron Beam Induced	Conductive (EBIC)	Semiconductivity	100
6. Currents absorbed	Absorptive	Topographic	1000
7. Auger Electrons	AES	Compositional	1000

With so many techniques available that can help in study of a variety of samples (metallic, non-metallic, semi-conducting or biological) pertaining to diverse branches of knowledge, SEM can claim to be truly an A-Z microscope !

3. Conclusions

Newer scanning microscopes such as Scanning Laser (Confocal) Microscopes (for sub-surface imaging), Scanning Acoustic Microscopes (Ultrasound instruments in non-invasive medical testing) and of course the sophisticated STMs , AFMs etc.,all have their origins in the humble SEM.

It is not just enough to have an expensive camera. To get good pictures, one has to have good idea about shutter speed, lens aperture, film speed and above all an aesthetic sense for frame composing.

Similarly, a lot of skill and experience is needed for proper utilization of SEM. And an uncanny ability to recognize the pitfalls in image interpretation so to get accurate and reliable results.

Above all, the microscopist, who lives in the world of images, has to remember the famous saying of Adi Shankara " Jagat Mithya, Brahma Satyam". (What we experience is Maya, the Truth is something else)

AEROSPACE MATERIALS

Development and Production of Super Alloys for Critical Publications

P Sarkar & K Ramesh

Mishra Dhatu Nigam Limited, PO: Kanchanbagh, Hyderabad.
E-mail: kramesh.midhani@ap.nic.in

Abstract_*Midhani has been engaged in the development and manufacture of several high performance alloys for different strategic sectors. A wide variety of alloys such as special low alloy steels, stainless steels, super alloys, titanium alloys, maraging steels, softmagnetic alloys have been developed and supplied in variety of mill forms. Alloys have been commercially manufactured using carefully selected raw materials and specialized melting techniques and subjected to controlled mechanical and thermal treatments in order to achieve a desirable combination of structure & properties. Amongst this range, superalloys occupy a pride of place in alloy development and manufacture.*

Superalloys, due to high temperature strength and stability, play an important role in several critical applications. Midhani manufactures all the three important variants of superalloys – namely, Iron, Nickel and Cobalt based superalloys such as, Superfer MDS, Superni 718 & Superco 605 etc. Based on its rich experience and diversified facilities available, Midhani has continuously been taking up the development of advanced variants of Superalloys for Nuclear, Aerospace and Aeronautical applications. Mechanical and metallurgical properties of the alloys are governed by the specific application based on which, in turn, the chemical composition and thermo-mechanical treatments are designed for the superalloys. Careful selection of raw material is also very important in order to reduce detrimental tramp elements. This presentation describes the development of superalloys, such as, Superni 690 M, Superco 605, Hastealloy C, ICNi 718, Superfer 286 and other cast alloys.

1. Introduction

Superalloys can be defined as a group of iron, nickel or cobalt base alloys which provide useful strength capabilities at temperatures exceeding 1000°C (540°C). They demonstrate, combined with mechanical strength, microstructure stability, higher corrosion and oxidation resistance at high temperatures. They are utilized at a higher proportion of their melting point than any other class of commercial metallurgical materials. Superalloys which have made very high temperature engineering technology possible are the materials leading the gas turbines that drive the jet aircraft and power sectors. In turn, these engines have been the prime driving force for the existence of these alloys.

Spectacular development on materials front took place during the war period (1939-45), the advent of the turbo super charger and gas turbine engine led to a rapid expansion in the use of superalloys. In the decade after the war, the development and improvement of many processing techniques for the alloys established the basis for the superalloy technology. Improved processing technique became available during the years 1950's and 1960's. Wrought thigh strength Nickel and Iron nickel base superalloys were invented and technologies such as Vacuum Induction Melting, vacuum arc remelting, investment casting, etc., were introduced. During the 1970's new alloy development continued at a modest pace.

Process development assumed important role.

2. Characteristics of Superalloys

The superalloys can be divided into three types. A) Iron base, b) Nickel base & c) Cobalt base. They are further sub divided into wrought and cast.

2.1 *Iron Base Superalloys*

The iron base superalloys are an extension of stainless steel technology and generally are wrought. These alloys are for application temperatures above 540°C (1000°F), have a FCC matrix and because of close packed lattice, are more resistant to time dependent deformation process. The strengthening phase is gamma prime in some typical alloys such as A286 (Superfer 286), AE696M etc. Typical applications of these alloys are jet engine and gas turbine components, cryogenic applications, non magnetic oil well equipments, springs, fasteners, and other cold headed products. Alloys such as Incoloy 903 (Superfer 903) are solid solution strengthened alloys and typical applications are gas turbine parts, rocket parts, constant, low coefficient of expansion components, etc.

2.2 *Nickel Base Superalloys*

Nickel base superalloys generally have greater resistance to high temperatures than low alloy steels and stainless steels. Nickel base heat resistant alloys contain nickel and up to 30% chromium. Iron contents range from relatively small amounts in most of these alloys to about 35% in alloys such as Inconel 706. Many nickel base alloys contain small amounts of aluminum, titanium, niobium, molybdenum, and tungsten to enhance either strength or corrosion resistance. The combination of nickel and chromium gives these alloys outstanding oxidation resistance. As a class, nickel base superalloys exceed stainless steels in mechanical strength, especially at temperatures above 650°C.

2.2.1 *Solid Solution Strengthened Nickel Base Superalloys*

Heat resistant nickel base superalloys are used in a wide range of applications that require high strength and/or oxidation resistance and high corrosion resistance.

Some typical alloys of solid solution strengthened type are Inconel 600 Superni 600). Inconel 625 (Superni 625). Hastealloy C276 (Superni C276) and typical applications are for manufacture of furnace parts and other heat treating equipment, and also as components in high temperature chemical processing equipments. In nuclear power plants, applications include steam generator tubing and structural components of reactor cores.

2.2.2 *Precipitation Strengthened Nickel Base Superalloys*

Precipitation hardening nickel base superalloys contain Al, Ti or Nb to cause precipitation of a second phase during appropriate heat treatment. The precipitated phase usually, gamma prime or gamma double prime, substantially increases the strength of the alloy. For example, Inconel X-750, a precipitation hardening version of Inconel 600, has yield strength about 3 times that of Inconel 600 at 540°C. Precipitation hardened Nimonic 80A shows a similar increase over its solid solution counterpart, Nimonic 75. Heat treatments for the precipitation hardening alloys generally consist of a solution treatment at 970°C-1175°C followed by one or more precipitation treatments at 600°C to 815°C. The typical alloys are Nimonic 80A (Superni 80A), Nimonic 90 (Superni 90), Nimonic C263 (Superni 263), Inconel 718 (Superni 718) etc. These superalloys are used in rocket engines and for aircraft gas turbine parts such as blades, discs, rings, shafts and various compressor and diffuser components. Other applications include bolts and springs in nuclear reactors.

The principal micro structural variables of superalloys are, a) the precipitate amount and its morphology, b) grain size and shape and, c) carbide distribution. Ni base superalloy properties are controlled by all these three variables. Structure control is achieved through composition, selection/ modification and by thermo mechanical processing.

Ni base superalloys typically consist of gamma prime dispersed in a gamma matrix, and the strength is a function of the volume fraction of gamma prime. The lowest volume fraction of gamma prime amounts are found in first generation Ni base superalloys. The gamma prime is commonly spheroidal in lower volume fraction gamma prime alloys, but often cubiodal in the higher volume fraction gamma prime nickel base superalloys.

Carbides exert a profound influence on properties by their precipitation on grain boundaries. If carbides precipitates as a continuous grain boundary film, properties can be severely degraded. At the other extreme when no grain boundary carbide precipitate is present premature failure will occur because grain boundary movement is essentially unrestricted. The carbon content of nickel base alloys ranges from 0.02% upward and metallic carbides can form in these materials both at the grain boundaries and within the grains. In most nickel base materials, the presence of gamma prime within the grains dominates the effect of any carbide in these regions and therefore only the inter-granular carbides are important. In Ni base alloys several types of carbides can form. Mono carbides of the general formula MC, where M is Ti, Ta, Nb or W are generally very stable and form during the melting of the material. They are difficult to dissolve in the solid phase and play an important role in restricting grain growth during solution treatment stage. More complex carbides having the general formula $M_{23}C_6$, M_7C_3 or M_6C where M denotes the metallic constituent are also observed in these alloys. In addition to Ni & Cr these superalloys contain Ti and Al and several other major alloying elements such as Co, Mo, W, Ta, Nb & V. Further, a number of important minor elements have to be controlled such as C, Zr, & B. The ranges specified for these elements are very limited for the attainment of optimum mechanical properties or tolerable hot workability.

2.3 *Cobalt Base Superalloys*

The cobalt base superalloys are strengthened by a combination of carbides and solid solution hardness. No inter- metallic compounds such as gamma prime is found. All alloys in the heat treated and softened condition have FCC crystal structure. Haynes 25 (Superco 605) alloy is widely used for hot sections of gas turbines, for nuclear reactor components, for
surgical implants, space applications, etc.

3. Melting

For most nickel base superalloys, vacuum induction melting (VIM) is required as the primary melting process. The use of VIM reduces interstitial gases (O_2, N_2) to low levels, enables higher controllable levels of Al & Ti to be achieved and results in less contamination from slag than air melting. Other advantages of VIM are good reaction kinetics, excellent decarburization, good degassing, excellent melt homogenization due to induction stirring, evaporation of harmful metallic impurities like Pb, Bi, Cu, Cd, Te, Se, etc., and effective temperature control.

VIM process generally is used as the initial melting process for superalloys and may be the only melting process used when material for investment casting is being produced. However, when the material is one of the higher strength superalloys which is to be hot worked to produce larger size gas turbine parts, a secondary remelting operation is required because VIM ingots generally have coarse and non uniform grain size, shrinkage and alloying element segregation. These factors restrict the hot workability of these alloys. These problems are resolved the use of vacuum arc remelting (VAR) which also acts to further refine the solidification structure of the ingot. At MIDHANI Nickel base superalloys are being melted through VIM + VAR route.

4. Deformation

The principal methods for deformation processes of super alloys are extrusion and/or forging, hot & cold rolling, etc. Hot workability is a direct function of alloy composition and solidification characteristics of the ingots.

Hot working operations are generally carried out on most of the nickel base alloys at temperature above which the hardening phases are taken into complete solid solution, since this results in a considerable reduction in strength and an increase in ductility both of these features are very essential for hot deformation. The gamma prime solvus temperatures increase with increasing content of precipitation hardening element. Thus, as alloy development has progressed, the useable hot working temperature range has narrowed considerably resulting for the more complex alloys in a narrow working range of only 10 to 20°C in terms of practical hot working limitations.

Improvements in hot workability have generally been sought in two ways.

1. The removal of deleterious impurities Pb & Bi by careful selections of raw materials and by using vacuum induction melting

2. By addition of beneficial elements in trace amounts to improve workability such as Mg, Ca.

Additionally, homogenizing treatment applied to ingots prior to hot working and to part forged billets eliminate segregation effects and is common practice for a no. of alloys particularly those used for gas turbine discs application example, Inconel 718.

The heating procedures adopted with nickel base superalloys must be strictly controlled in terms of rate of heating temperature and time held at temperature to ensure success in this important operation. LPG fired furnaces are used for hot working of all these alloys in order to ensure very low sulphur contamination.

5. Conclusion

Progress in such strategic applications as jet engines, turbine power generators, rockets and missiles is rate controlled by the development of structural materials with ever higher temperature capabilities and reliability. For the past forty years, Superalloys have been the core material system fulfilling such needs. Superalloys have gone through many process advances – from air melting to vacuum melting and refining, and on to double vacuum melting, directional structural manipulation and extra ultra clean alloys. Cast components are now enjoying not only higher yield, precision vacuum investment shaping and coring, but also the extra heat resistance benefits derived from directional heat extraction and the resulting directionally solidified grain structures and monocrystals.

Although the demand for superalloys, and in general the applications for superalloys have grown, serviceable high temperature limits for superalloys even with cooling schemes, are fast approaching. Accordingly, research and development in alternative high temperature systems is and has been in full swing for some time now. Such systems, like ODS and fibre reinforced superalloys (FRS), can be considered direct derivatives of superalloy technology.

Current alloys have been developed over fifty years to the point where further advances are becoming increasingly difficult and expensive to achieve at a time when resources are even more limited. Nickel base superalloys are now operating at temperatures of up to 85% of their melting point. It is clear that the melting point of Nickel imposes a natural ceiling to their potential and hence further improvement is limited.

There are two major classes of potential successor materials to the Nickel based superalloys, the intermetallics and the ceramic matrix composites. Two main categories of intermetallics are available viz., Titanium aluminides for use at lower temperatures and nickel aluminides for higher temperatures.

Multiphase Titanium Aluminides for Aero-Engine Applications

V V Kutumbarao[1], G. Sreenivasulu[1], T.K. Nandy[1], Vikas Kumar[1], D. Banerjee[1], Nidhi Singh[2] and J.-L. Strudel[2]

[1] DMRL, Kanchanbagh, Hyderabad, [2] Ecole de Mines, Paris, France

Abstract: *Phase equilibria, microstructure and mechanical properties of a Nb-rich O+B2 alloy (Ti-20at.%Al-25Nb-1Mo) were investigated in a range of heat treatment conditions. Phase identification and microstructural characterization involving optical, scanning / transmission electron microscopy and X-ray diffraction were carried out followed by mechanical property evaluation. Detailed structure-property correlation was attempted and the results were discussed in relation to the existing literature*

1. Introduction

A major thrust in the development of aero-engine materials is the replacement of Ni base alloys by the lighter Ti base alloys. Approaches towards substantially increasing the temperature capability of titanium primarily revolve around the development of engineering intermetallic alloys Ti_3Al and $TiAl$. However, poor room temperature tensile ductility prevented any commercial exploitation of these intermetallics. It has been found in the former case that ductility at room temperature could be achieved by Nb additions to this very brittle material. Several Ti_3Al-Nb alloys (Blackburn and Smith 1981, 1987, Gogia et al. 1990) were evaluated for the optimization of room as well as high temperature properties. While reasonable room temperature properties were realized, high temperature properties of the alloys fell short of target. The creep properties of the alloys, especially the creep life for 0.2 % strain, did not compare favorably with INCO718 alloy (Rowe, Gigliotti and Marquardt 1990, Materials Property Handbook 1994). Since then, the Nb content of the alloys has been increased substantially as it was found that Nb-rich alloys exhibit superior room as well as high temperature properties in comparison to Nb lean alloys (Rowe et al 1991, Row and Larsen 1996). It was discovered that Nb additions beyond a critical level result in a transformation of the hexagonal Ti_3Al structure to an orthorhombic phase (O) based on the Ti_2AlNb stoichiometry. The phase equilibrium of the alloys changes from α_2+B2 to α_2+B2+O to O+B2 (Rowe et al. 1993, Boehlert et al 1997, Gogia et al 1998). It was found that alloys containing the O phase exhibit superior combinations of strength and toughness but inadequate creep resistance. The superior properties of Nb-rich O+B2 alloys in comparison to Nb-lean α_2+B2 alloys have been attributed to higher solid solution strengthening of the O phase (Nandy 1999, Nandy and Banerjee 2000) - since it has higher amount of Nb in solid solution - and also the presence of additional '2c + a' slip systems (Banerjee 1995, Nandy 1999) that impart higher ductility and fracture toughness. However, even O+B2 alloys do not posses adequate high temperature properties, especially 0.2% creep life (Carisey et al. 1998, Materials Property Handbook 1994). This is attributed to enhanced primary creep that is also exhibited by other conventional titanium alloys even though the steady state creep rates are considerably lower. Therefore, an effort to develop new compositions to improve upon the existing achievable properties is an on-going process.

Significant efforts have been directed towards property optimization of different O+B2 alloys with the primary aim of maximizing creep properties. Quaternary additions of Zr, Nb and Si have been attempted with beneficial effects (Carisey et al 2000, Banerjee et al. 2001) and compositions have been developed that exhibit reasonable creep properties although still inferior to INCO 718.

2. Experimental

This article is a General Report on an exploratory study carried out as part of a joint Indo-French collaborative programme for development of new aeroengine materials. The study addressed the mechanical behaviour of one specific alloy of the Ti_3AlNb family, namely Ti-20Al-25Nb-1Mo (at%), hardened by the O phase and possibly by a small volume fraction of α_2. The Al content of the alloy has been kept relatively low to attain higher levels of ductility and also improved hot workability while Mo has been added for its beneficial effects on high temperature properties. Basic idea of the study is to derive a fundamental understanding of the mechanical behaviour of such alloys in relation with the various microstructures that can be generated in them by varying heat treatments. Optimization of the industrial processes and engineering properties of these alloys are expected to ensue from this approach. Mechanical behaviour of the alloy in three carefully chosen microstructural conditions has been investigated with the following main goals:

1. to identify three microstructures of fundamental interest namely, an O-lath microstructure called H_m, an equiaxed O microstructure called E_r and finally a partially recrystallized 3 phase microstructure of industrial interest called TNS 3.
2. to identify the tensile behavior of the 3 selected microstructures at room temperature, 550°C and 650°C and suggest an interpretation of the various hardening stages observed.
3. to characterize the low cycle fatigue (LCF) behavior of these microstructures at 650°C.
4. to establish the influence of strain amplitude, frequency and tensile hold periods on the LCF behavior.

5. to identify and characterize creep mechanisms in the temperature range 550°C-650°C and stress levels of 200 and 310 MPa.

3. Results and Discussion

The work, divided equally among the concerned laboratories in India and France, involved mechanical behavior study of the titanium aluminide alloy containing Ti-20Al-25Nb-1Mo designated as S8b in four major phases:

1. Melting, processing and preliminary micro-structural characterization.
2. Detailed study of microstructure, phase equilibria followed by tensile property evaluation in different micro-structural conditions to optimize the microstructure.
3. Fatigue behavior evaluation in optimized micro-structural conditions.
4. Creep and relaxation behavior evaluation of alloy S8b in optimized micro-structural conditions.

3.1 Microstructures

Three different heat treatments were explored as shown in Table 1

Table 1 Optimised heat treatments for a Ti-20Al-25Nb-1Mo alloy and the resultant microstructures

Treatment Name	Heat Treatment Details	Phases present	Phase Equilibria		
			Volume fraction %		
			O	α_2	B2
E_r	900°C/24hr/AC + aged at 650°C/24hr/AC	O+B	39	0	1
H_m	1080°/30min/FC@2°C/min to 900°C/24hr/AC + aged at 650°C/24hr/AC	O+B2	42	0	8
TNS3	1000°/2hr/FC@200°C/hr to 900°C FC@15°C/hr to 870°C/24hr/AC + aged at 650°C/24hr/AC	O+α_2+B2	47.3	11.3	1.4

The heat treatments E_r and H_m exemplify two opposite distributions and morphology of two-phase microstructure (O+B2). In case of E_r the O phase was found to be made of equiaxed spheroidal nodules, 2 μm in diameter uniformly distributed in the B2 matrix. On the other hand, in case of H_m, the O phase was found to be present in the form of laths (1 μm in width, 15 μm in length) rarely forming basketweave microstructure. To explore mechanical behaviour of a 3 phase material (O+α_2+B2), a third kind of microstructure was also developed with controlled grain size resulting from the presence of a small volume fraction of a third phase (α_2). The selected TNS$_3$ resulted in the most promising microstructure : colonies or bands of recrystallized fine B2 grains (5-10 μm) hardened by equiaxed O particles(as in E$_r$) alternating with areas made of larger recrystallized B2 grains (20-100 μm) populated with O lath colonies (akin to H$_m$). This scenery results from the non uniform distribution of equiaxed α_2 nodules: bands of closely spaced α_2 particles (Si rich) alternating with areas with loosely spaced α_2 particles (Nb and Mo rich); these particles are controlling the mobility of B2 grain boundaries during recrystallization and hence the final grain size. This point is of major engineering importance.

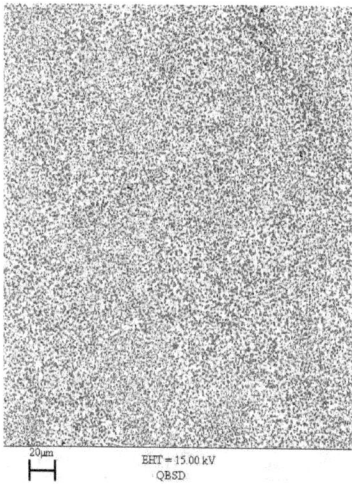

Fig. 1 SEM image Microstructure Er

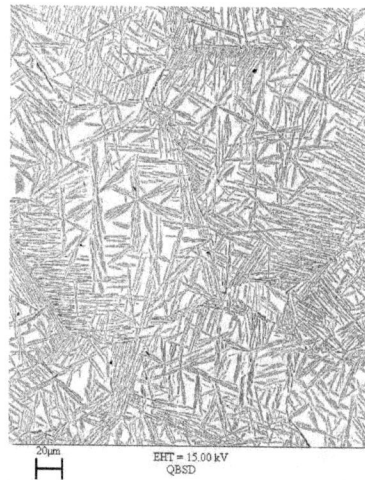

Fig. 2 SEM image Microstructure

Fig. 3 SEM image Microstructure TNS$_3$ microstructure

3.2 Tensile Behavior

Tensile tests were performed on the conditions E_r, H_m and TNS_3 at three temperatures viz. 20°C, 550°C and 650°C.

The tensile properties as a function of temperature are shown in Fig.4. The room temperature ductility of E_r and H_m was found lower than that of TNS_3. Further, at room temperature, no visible evidence of necking was seen on the fractured specimen.

Considerable increase in the ductility was observed at 550°C in all three conditions. While E_r still retained its ductility at 650°C there was a small reduction for TNS_3 and a large decrease for H_m. Therefore it can be deduced that there exists a ductility maximum at about 550°C in all three conditions. The UTS also showed appreciable differences in all the conditions and with the temperature. The best combination of tensile properties at 650°C was found with E_r which has the highest ductility as well as the highest strength. Microstructure TNS_3 exhibits a tensile behaviour intermediate between that of microstructures E_r and $_{Hm}$ that it may be thought to be composed of.

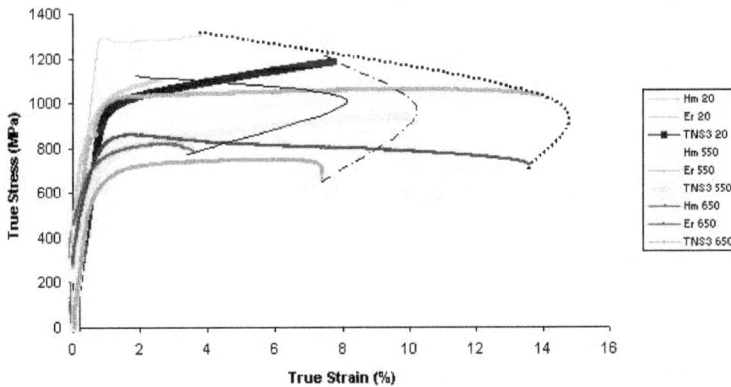

Fig. 4 Tensile curves of alloy S8-b for H_m, E_r and TNS_3 heat treated conditions at 20 °C, 550°C and 650°C

3.3 Low Cycle Fatigue (Lcf) Behaviour

LCF tests were performed on two microstructures E_r and H_m at a frequency of 0.2 Hz at 650°C. The results confirm that E_r microstructure offers higher LCF resistance as compared to H_m microstructures. Stress response curves (stress amplitude vs elapsed cycles) are shown in Figs.5-7

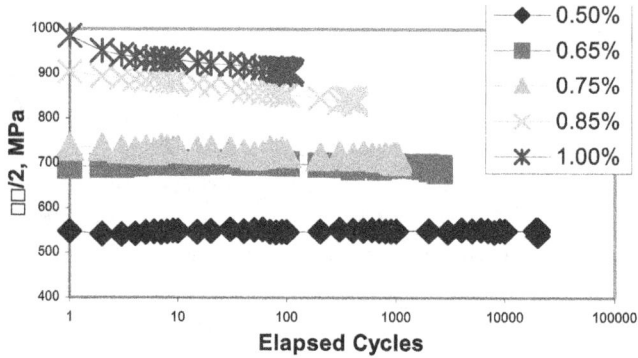

Fig. 5 Cyclic stress response curves of E_r microstructure at 650°C for different strain amplitudes.

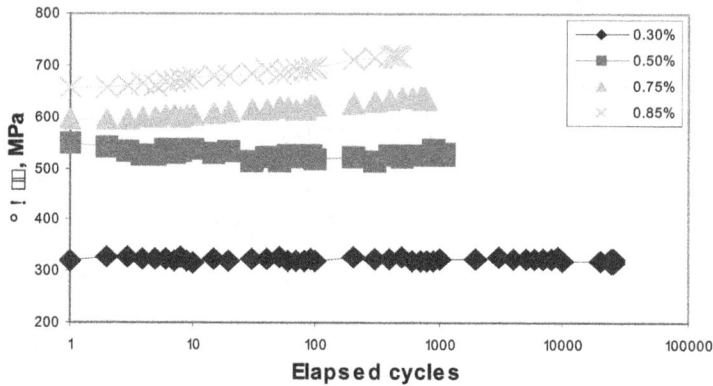

Fig. 6 Cyclic stress response curves of H_m microstructure at 650°C for different strain amplitudes.

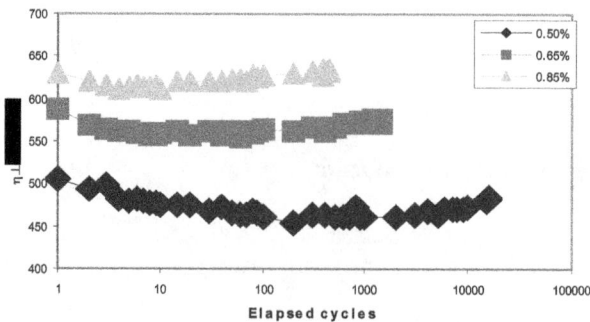

Fig. 7 Cyclic stress response curves of TNS_3 microstructure at 650°C for different strain amplitudes.

At low strain amplitudes, the microstructure E_r with equiaxed O phase has been observed to be much stronger under LCF conditions compared to the H_m microstructure with lath like O phase. However, at high strain amplitudes, the fatigue lives are comparable. The TNS3 microstructure shows an intermediate behaviour as expected. Cyclic hardening and softening behaviour is related to initial dislocation density in the material. Predominance of elastic strain controlled fatigue behaviour in these relatively low ductility alloys has been observed highlighting the role of microplastic deformation mechanisms in crack initiation and growth, aided by macroplastic behaviour at higher strain amplitudes. It is therefore expected that life enhancement methods employed under high cycle fatigue conditions could prove useful in this class of alloys under so called LCF conditions.

3.4. Creep Behaviour

The creep curves are shown in Fig.8 for the tests carried out at 550°C, 600°C and 650°C for the alloy in the three heat treated conditions. Out of all the microstructures E_r shows very large creep strain at all temperatures and large ductility(5 –6%). On the other hand, H_m shows very good creep resistance at all three temperatures with creep rates ranging from 10^{-10} s^{-1} to 10^{-9} s^{-1} under the applied test conditions, about 30 to 100 times lower than E_r. But steady state creep, when reached, is interrupted by a failure mechanism rather than terminated by a tertiary stage as observed with the E_r microstructure. As shown earlier by L. Germann (Thesis 2003) numerous cracks are initiated in the outer layer of the specimen which is hardened and embrittled by oxygen diffusion, thus inducing the growth of an α_2 layer.

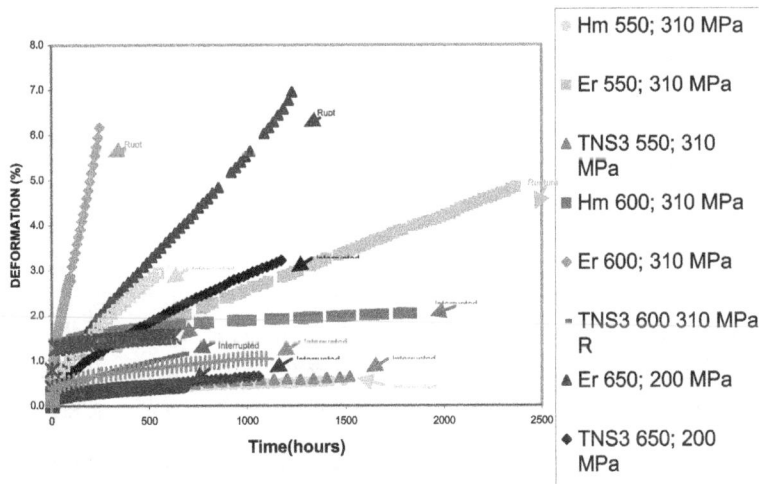

Fig. 8 Creep curves in terms of deformation vs. time of alloy S8-b for H_m, E_r, and TNS$_3$ conditions at 550°C, 600°C and 650°C.

Furthermore, crack propagation is stimulated by environmental interactions leading to a premature failure of the material. The alloy S 8, with a lower aluminum content (20% instead of the usual 24%), absence of Si additions and lack of zirconium, is especially prone to this kind of damage.

On the other hand, E_r shows very large creep strains ($4\% < \varepsilon_R < 7\%$) at all temperatures, thereby demonstrating intrinsic viscoplastic ability of the bulk material.

Microstructure TNS_3 appears more creep resistant than E_r but significantly weaker than H_m. At 650°C, its behaviour deteriorates. Again, the microstructure TNS_3 exhibits creep behaviours intermediate between that of E_r and H_m, yet closer to that of E_r because of the rather small " average size " of the O phase (fine equiaxed nodules in niobium-molybdenum rich zones, with microlaths in recrystallized areas) as opposed to the large lath microstructure present in H_m.

The primary creep behavior at 600°C of all three microstructures, with amplitude of 1% during the first 50 h of the creep life, appears as large and somewhat erratic. This phenomenon is likely to be resulting from the "transformation induced plasticity" (TRIP) associated with the growth of secondary O laths, on variants favourably oriented with respect to the local stress tensor (see thesis German 2003).

An activation energy of 150 kJ/mole measured for H_m with a stress exponent of 1.8 to 2, confirms that the (O- B2) interfacial viscous glide mechanisms are likely to be the creep controlling mechanisms. Hence, refining the microstructure of the O phase and fractionating it as in E_r, and that to a lesser extent as in TNS_3, results in multiplying the creep rates by a large factor (30 to 100). Bulk viscoplastic mechanisms are probably activated in E_r as well as TNS_3 microstructures since the apparent activation parameters increase respectively to about 292 kJ/mole and n= 4 in E_r with intermediate values for TNS_3.

4. Conclusions

1. Three different microstructures were developed: Er and Hm both represent 2-phase microstructures (B2 +O), with the O phase : equiaxed and nodular in Er, in the form of laths in Hm. In TNS3, the presence of O nodules resulted in a 3-phase microstructure (O, α2 and B2) composed of a mixture of the above microstructures. Controlled recrystallisation of B2 was obtained.

2. The best tensile properties amongst Er, Hm and TNS3 microstructures were associated with Er , which has the best combination of ductility and strength. At 650°C, the Er microstructure exhibits higher LCF resistance than Hm. Yet the creep resistance of Hm microstructure at 550°C-650°C is the best amongst

all the three materials investigated whereas for Er it is the worst. In stress relaxation tests at 650°C, two regimes, are evidenced: the high stress regime where plasticity is sensitive to strain and to morphological parameters of the phases and the low stress regime where viscoplastic flow is controlled by the density of interphase interfaces.

3. The most attractive compromise of Engineering properties for a specific application (blade, rod, casing etc.) will result from a careful adjustment of microstructures based on the newly established and reliable guidelines resulting from this study.

4. *Potential benefits: (a)* A specific processing path has been established, which leads to beta grain size control in the alloy studied. It can be easily extended to other Titanium Aluminides and possibly to a broad variety of Titanium alloys of industrial interest. *(b)* Drastically different microstructures were explored; as a consequence, relations between mechanical properties and microstructures could be established and understood for this alloy: a rare and novel type of achievement in Ti base alloys, with positive gains in engineering properties to be expected.

Acknowledgements

The authors express their gratefulness to the Indo-French Centre for the Promotion of Advanced Research (IFCPAR) for providing financial support to this work.

References

1. Andres C., Gysler A., Lutjering G., Z. *Metallkd., Vol. 88, 1997, p. 3.*

2. Banerjee D., Gogia A. K., Nandy T. K., Muraleedharan K. and Mishra R. S., *Physical Metallurgy of Ti₃Al alloys, Structural Intermetallics*, Darolia R., Lewandowski J. J., Liu C. T., Martin P. L., Miracle D. B., Nathal M. V. Edt., *The Minerals, Metals and Materials Society, 1993, p. 45*

3. Banerjee D., Philosophical Magazine A, *Vol. 72, 1995, p. 1559.*

4. Banerjee D., Gogia A.K., Nandi T.K., Joshi V.A., *Acta Metallurgica, Vol. 36, 1988, p. 871.*

5. Banerjee D., Baligidad R. G., Gogia A.K., Strudel J. L., Eds; Hemker K. J., Dimiduk D. M., Clemens H., Darolia R., Inui H., Larsen J. M., Sikka V. K., Thomas M., Whittenberger J. D., TMS *(The Minerals, Metals and Materials Society), 2001, p.43.*

6. Bania P.J. and Hall J.A., Titanium Science and Technology, *1985, p. 2371.*

7.	Bhattacharjee A., Joshi V.A., Hussein S.M., and Gogia A. K, Titanium'99 Science and Technology, Eds. Gorynin I. V. and Ushkov S. S., Central Research Institute of Structural Materials *(Crism), "Prometey", Vol. I, 2000, p. 164.*

8.	Blackburn M.J., Smith M.P., Titanium alloys of the Ti_3Al type, *Patent US 4.292.077, 1981.*

9.	Blackburn M.J., Smith M.P., *aluminum alloys containing niobium, vanadium and molybdenum,* US Patent 4.716.020, 1987.

10.	Boehlert C. J., Majumdar B. S., Seetharaman V., Miracle D. B., Wheeler R., Structural Intermetallics,. Nathal M.V., Darolia R., Liu C. T., Martin P. L., Miracle D. B., Wagner R., Yamguchi Y. Eds., *The Minerals, Metals and Materials Society, 1997, p. 795.*

11.	Carisey T., Banerjee D., Franchet J.M., Gogia A.K., Lasalmonie A., Nandy T.K., Strudel J.L., *Titanium Based Intermetallic alloys, US Patent No. 6.132.526, 2000.*

12.	Cho W., Jones J.W., Allison J.E., Donlon W.T., Sixth World Conference on Titanium, 1988, Eds. Lacombe P., *Tricot R. and Beranger G., vol. 1, p.187.*

13.	Cho W.S., Thompson A.W., Williams J.C., *Metallurgical Transactions A, 1990, 21A, p. 641.*

14.	Foster M.A., Smith P.R., Miracle D.B., Scripta Metallurgica et Materialia, 1995, 33, p. 975.

15.	Gogia A.K., Banerjee D., Nandy T.K., Metallurgical Transactions A, 1990, 21, p. 625.

16.	Gogia A.K., Nandy T.K., Banerjee D., Carisey T., Strudel J.L., and Franchet J.M., *Intermetallics, 1998, 6, p.741.*

17.	Gogia A.K., PhD thesis, Banaras Hindu University and Defence Metallurgical Research Laboratory, *India 1990.*

18.	Lipsitt H.A., High Temperature Ordered Intermetallic Alloys, *MRS, 1985, 39, p. 352.*

19.	Martin P.L., Lipsitt H.A, Nuhfer N.T and Williams J.C., Titanium'80, Science and Technology, Eds. Kimura H. and Izumi O., *Metallurgical Society-A.I.M.E., Pa, U.S.A. 1980, p.1245.*

20.	Materials Properties Handbook: Titanium Alloys, ASM International, Eds. Boyer R., Welsch G. and *Collings E.W., June 1994.*

21.	McAndrew J.B., and Simcoe C.R., 1960, *Investigation of the Ti-Al-Cb System as source of alloys for use at 1220°-1800°F,* Armour Research Foundation of Illinois, Institute of Technology, *WADD-TR-60-99, Wright Patterson Airforce Base, 1960.*

22.	Mishra R. S., Banerjee D., Material Science and EngineeringA, *Vol. 130, 1990, p. 151.*

Aircraft Materials: A DMRL's Perspective

G. Malakondaiah

Defence Metallurgical Research Laboratory, Hyderabad.

Materials used for aircraft applications can broadly be split into two categories; airframe materials and engine materials. Selection of materials for both applications is based on the design constraints. Typical design constraints include weight, stiffness, strength, fatigue performance (high/low cycle), corrosion resistance and cost.

Of major consideration of airframe materials is the specific stiffness (stiffness to density ratio E/ρ), and specific strength (strength to weight ratio σ/ρ). The broad aim for aircraft design is to maximise payload in relation to cost. By increasing specific stiffness and strength, the weight of a given component can be decreased and fuel consumption and running costs decreased. Consequently, aluminium alloys have been the most widely used materials for airframe constructions for decades. However, in recent years, composites have made considerable gains in utilization for airframe constructions in military aircrafts as well as in mid-size civilian aircrafts.

Varied operating temperatures encountered in the gas turbine engine require a range of high temperature materials. In the compressor section, where the temperature is low to medium, titanium parts are often used. In the high temperature turbine and combustor areas, nickel based superalloys are mostly used. Thermal barrier coatings are also employed to further increase the temperature capability of nickel based superalloys.

While giving the current status in each of these areas, the focus here will be on materials R&D activities at DMRL and perspectives of how these technologies are evolving.

1. Aluminium Alloys

A wide range of wrought Al alloys dominate both primary and secondary structures in all aerospace applications. The predominant commercial aircraft alloys have been the Al-Cu-Mg base 2xxx (such as 2324, 2524) when damage tolerance is the primary requirement, and Al-Zn-Mg-Cu base 7xxx (such as 7475, 7010, 7150, 7055, 7449) when strength is the primary requirement. The most recent development in high strength 7xxx series Al alloys has been the high strength, high damage tolerant 7085 thick alloy product with low quench sensitivity. 7085 plates and forgings [with Zn content comparable to that of alloy 7055, but reduced levels of both Cu and Mg] have attractive fatigue properties and 7050 equivalent stress corrosion and exfoliation properties. 7085 plates have found applications in wing spar and rib structures in Super-Jumbo Airbus A380. Substantial efforts were also made in the recent past to develop high strength Al alloys containing Li (the lightest metallic element) up to 2 wt%, offering typically up to 10% density reduction and 15% increase in stiffness, as gauge-for-gauge substitutes for the standard alloys (i.e. 2xxx and 7xxx). Technical problems associated with the first generation Al-Li alloys [e.g. 2090, 8090, 2091] have, however, restricted applications in the form of sheets and light to medium gauge plates or extrusions in military aircrafts. Further developments were, therefore, directed towards using lower Li concentrations. One such alloy 2195 containing about 1.0wt% Li and small additions of Mg and Ag (originally designated Weldalite 049) is being utilized for cryogenic tanks for space shuttles. Most recent work indicates that new Al-Li-Cu-Mg-Mn-Zn-Zr alloys 2099 (in the form of extrusions & plates), and 2199 (in the form of flat rolled products) have a superior combination of properties and could be explored further for aircraft applications. One major application of 2xxx and 7xxx alloys in the form of thin sheets lies in the successful development of Glass Fibre Reinforced Al Laminates (GLARE). GLARE consists of alternating layers of Al alloy sheets (typically 0.3 mm thick) bonded together with glass fibre reinforced resin (prepreg typically 0.25 mm thick). GLARE, a highly fatigue resistant material with good impact and damping properties, has found applications in the upper fuselage shell of the Airbus A380 and several applications in the fighter aircrafts.

DMRL has put in substantial efforts to indigenize selected commercial Al alloys in the form of sheets, plates and extrusions for aircraft applications. The wrought alloys indigenized at DMRL for aerospace applications include Al-Li alloy 8090, and Al-Zn-Mg-Cu alloys 7010, 7449 and 7055. DMRL has also developed technology for production of ≤ 0.30 mm thick 7xxx alloy sheets for GLARE. DMRL is currently engaged in the commercial scale production followed by type certification of a variety of wrought Al alloys for aerospace applications.

2. Titanium Alloys

Titanium alloys have firmly entrenched themselves as an important class of materials for aerospace applications. They are used for both airframe structural and aero engine applications due to their excellent balance of room and high temperature properties. For structural applications, as it competes and has an edge over aluminum alloys and steels, the primary reasons for use of titanium alloys are high strength to weight ratio, stiffness and excellent corrosion resistance. Titanium alloys have seen a steady increase in their use over last five decades and today the material share of titanium for airframe applications is about 7%. Ti-6Al-4V is the workhorse titanium alloy extensively used (80-90% of all titanium alloys) for airframe applications. It is used in all sections of aircraft that is fuselage, nacelle, landing gear, wing and empennage. A variant of Ti-6Al-4V, namely half-alloy (Ti-3Al-2.5V) has found key application for high pressure hydraulic lines replacing stainless steel with about 40% weight saving. Additionally, beta titanium alloys such as Ti-10V-2Fe-3Al, Ti-15V-3Cr-3-Al-3Sn find applications for airframes and landing gear.

The gas turbine engine constitutes the principal area for the application of titanium alloys. The alloys for aircraft engine are located in an operating temperature regime that is bound on the lower temperature end by polymeric composites and on the high temperature end by nickel base superalloys. Since the initial use of titanium alloys in the jet engines by Pratt and Whitney (USA) and Rolls-Royce (UK) in 1950, the titanium content in the engine has steadily witnessed an upward trend. Ti-6Al-4V is used primarily in the fan section of turbine engines, whereas $\alpha-\beta$ alloys such as Ti-17 and Ti-6Al-2Sn-4Zr-6Mo are used in the lower temperature envelope of the compressor. Near alpha titanium alloys such as Timetal 685, Ti-6Al-2Sn-4Zr-2Mo and Timetal 834 are employed at the higher temperature end of the compressor. For application beyond 600°C, where microstructural stability, oxidation and burn resistance of titanium alloys become a major issue, efforts to develop new alloys have gathered momentum in order to expand the temperature range of titanium applications. Titanium aluminide based alloys and Ti-matrix composites are some of the areas that continue to receive the focus of material scientists and technologists.

DMRL has successfully developed indigenous technologies for production and processing of high temperature titanium alloys for aerospace applications in collaboration with Midhani, Hyderabad and HAL, Bangalore. These include GTM 900, DMRT OT4-1 and Ti-3Al-2.5V. Currently, a near alpha titanium alloy, Titan 29A, is being developed for use in the compressor stages of gas turbine engines because of its excellent creep and fatigue resistance. Isothermal forging technology for manufacturing critical components (Discs and Shafts) using titanium alloys Titan 26A & 29A has been developed. Simultaneously, extensive studies have been undertaken on titanium aluminide based alloys. The highpoint of these studies has been the discovery of "orthorhombic phase" and its ternary ordering. This phase is key for providing enhanced creep and toughness in titanium aluminide based alloys. Following this discovery, the entire alloy development programmes on aluminide went through significant changes. Sensing the attractiveness and importance of the work, US and French collaboration also came up. The work has culminated in the development of a new orthorhombic alloy, which has now been patented. Compressor blades from titanium aluminide alloys were successfully fabricated using closed die forging at HAL. A collaborative work on boron containing titanium alloys with Air Force Research Lab., USA is currently underway.

3. Nickel-Based Superalloys

Nickel-based superalloys have served as the most competitive high temperature structural materials under highly stressed and aggressive operating conditions in a variety of applications for more than 60 years. The most demanding among all the applications has been the gas turbine aerofoil castings of modern aero-engines. These turbine parts operate in extremely aggressive environment of high velocity hot combustion gas–air mixture carrying highly corrosive ingredients at high pressure. Gas turbine aerofoil materials should possess adequate resistance to creep, fatigue and aggressive environment. Materials design for such application therefore has been extremely challenging, particularly since the engine designers always aim at higher turbine entry temperature (TET) in order to achieve greater engine thrust and better fuel efficiency. In spite of the enormous efforts made in the recent past towards developing ceramics and their composites, Ni-based superalloys continue to be most reliable blade and vane materials offering always the highest TET. This has been possible through better alloy design, directional solidification (DS) of either columnar grains or single crystals (SC), improved blade cooling schemes and protective coatings. During the last six decades, TET has gone up by about 500K.

Extensive research at DMRL has led to the development of new generation Ni-based superalloys, designated as DMD-4 and DMS-4 for DS and SC processing, respectively. Simultaneously, expertise has been developed to cast DS and SC components for aero-engines. Technology for ceramic cores, critical for making hollow blades and vanes with intricate cooling channels, has been established. Technology has also been established for pilot scale production of blades and vanes (Fig. 1).

Fig. 1 Directionally solidified hollow aerofoil castings (blades and vanes) for advanced aero-engines

4. Coatings

Most components of modern aero-engine gas turbines currently use a host of performance-enhancing and protective coatings. Among these coatings, thermal barrier coatings (TBCs) have particular relevance in high temperature sections of aero-engines. TBCs are multi-layer coatings which provide both oxidation resistance and thermal insulation to the superalloy components of gas turbine engines. While a zirconia-based ceramic coating provides thermal insulation, the underlying metallic bond coat enhances the oxidation resistance of the superalloy substrate. The importance of TBCs can be appreciated from the fact that their use has enabled dramatic increases in aerofoil temperature, far greater than that enabled by the switch from cast superalloy blades to single crystal blades over approximately 30 years.

Currently, the combination of Pt-modified aluminide bond coat and 7% yttria stabilized zirconia (7YSZ) ceramic coating (deposited by EB-PVD method) has emerged as the best candidate for thermal barrier applications for aero-engine gas turbine components such as blades and vanes. Over the years, there have been considerable developments in various aspects of TBCs including deposition methods, evaluation of physical properties such as thermal conductivity and coefficient of thermal expansion of the ceramic layer, and mechanical characterization of the 7YSZ coating. While improvements in their capabilities continue, there is a growing realization that new TBC systems will be required for the next generation turbines presently being designed. In this context, considerable research is being carried out worldwide on exploring newer coating materials, advanced deposition techniques, non-destructive methods of monitoring TBC degradation during its use and newer techniques of machining TBC coated components. Gadolinium-stabilized zirconia (GSZ) and gadolinium zirconate pyrochlores have shown promise as alternate ceramic coating materials since they have considerably lower thermal conductivity as compared to 7YSZ. The use of Pt-modified γ-γ' bond coats instead of conventional β-NiAl type Pt-aluminide bond coats is also being explored. Solution precursor plasma spray process (SPPS) and the process to achieve dense vertical cracked (DVC) coatings are among the emerging deposition techniques for the ceramic coating. Traditionally, for machining of cooling holes in TBC coated superalloy components, Nd-YAG type nanosecond (ns) lasers have been used. However, recently developed femto-second (fs) lasers have been shown to produce much superior cooling holes with virtually no collateral damage to the ceramic coating such as recast layer, micro-cracking, splashes and ceramic layer delamination. Therefore, various aspects of machining of TBC coated components using fs lasers are currently being researched worldwide. DMRL has undertaken extensive research in the area of Pt-Al bond coat. Currently, a major programme is underway, in association with ARCI, to develop thermal barrier coating technology for aero-engine components.

Statistical Analysis of Low Cycle Fatigue Properties of Aero Engine Disc Forgings

Hina Gokhale

Defence Metallurgical Research laboratory, Hyderabad, INDIA
hina@dmrl.drdo.in

Abstract: *Number of cycles to failure in low cycle fatigue of the cut-up samples of isothermally forged aero engine disc have been analysed statistically with an objective to arrive at engineering tolerance limits for the disc forgings. Two distribution models, namely Weibull and lognormal have been applied for analysis and both have been found statistically acceptable. 90, 95 and 99 percent reliability values have been arrived at based on these two models and their 95% confidence bounds have been also calculated. Based on the length of the confidence interval it was concluded that for the engineering tolerance limit of A-basis value Weibull is a preferable model, and for engineering tolerance limit of B-basis value lognormal is suitable. Accordingly, engineering tolerance limits for the property have been given.*

Keywords: Weibull distribution, lognormal distribution, Engineering Tolerance limits, Design Allowable, AD test based on Normalized spacing.

1. Introduction

Introduction of a new material or a new component into an aeronautical system raises the need to quantify the material or component quality for the system designers. In particular, the designers look for the statistical minimum property that the material or the component can offer. The present paper discusses the statistical analysis of low cycle fatigue property of isothermally forged aero engine disc component, with a primary objective to arrive at minimum fatigue life for the designers.

In statistics, number of cycles to failure in low cycle fatigue represents a typical example of "lifetime" data. Vast literature is available in the field of statistics to analyse such data. Excellent account of methodology applied can be found in Lawless [1], Meeker and Escobar [2], and Abernathy [3]. While the book by Lawless [1] provides complete account on theoretical treatment for lifetime data, the latter two books provide theory along with examples on actual application to analysis of mechanical properties of different alloys.

There are mainly three steps involved in such analysis. First, select appropriate parametric distribution model and test its adequacy. Second, estimate the parameters of the distribution and finally calculate statistical tolerance limits also called design allowable.

Choice of a model distribution depends on many factors [1]. Some of the factors considered when making a choice between different models could be the past experience with the data from similar experiments or specific nature of data or specific nature of the mechanisms that lead to such data. There are other statistical considerations that should be kept in mind: (i) statistical adequacy of model and (ii) awareness of consequences of the departure from the assumed model on any inference made.

There are several parametric distribution models available for the lifetime data. Among these Weibull distribution and lognormal distribution are the most widely used distributions to model the lifetime data arising from testing the durability of manufactured items [1]. Specifically for LCF properties, Metals Handbook [4] has indicated both these models. Abernathy [3] suggests lognormal distribution for materials property like LCF, however, he also clarifies that since Weibull distribution is more conservative in the lower tails, for sample size below 20, Weibull distribution is a better predictor. But, when the sample size is larger than 20, distribution analysis for selection between Weibull and lognormal is appropriate. Meeker and Escobar [2] have also given an example of modeling LCF data of an alloy using both Weibull and lognormal distributions and found the two fits to be close. Thus it appears that both lognormal and Weibull have been used or suggested to model LCF property. The present data was analysed using both lognormal and Weibull distributions.

2. The Data

There are 76 LCF tests conducted on disc components. LCF property is measured as number of cycles to failure in two different directions, namely tangential and radial, with 38 observations in each direction. Keeping time and cost factors in view, the low cycle fatigue tests were carried out up to 15000 cycles. That is, if the specimen fractured before 15000 cycles, than the cycles it withstood the test was taken as its LCF value, otherwise LCF value was taken as 15000. The life-time data obtained in this manner is called Type-I censored data. Table 1 provides distribution of 76 data points as censored and uncensored within tangential and radial directions. Figure 1 and Table 2, respectively, show histogram and descriptive statistics of the LCF data using statistical software SPSS 10.

Table 1 Sample Size

Data	#of observations not censored	# of observations censored	Total
Radial & Tangential	49	27	76
Radial	28	10	38
Tangential	21	17	38

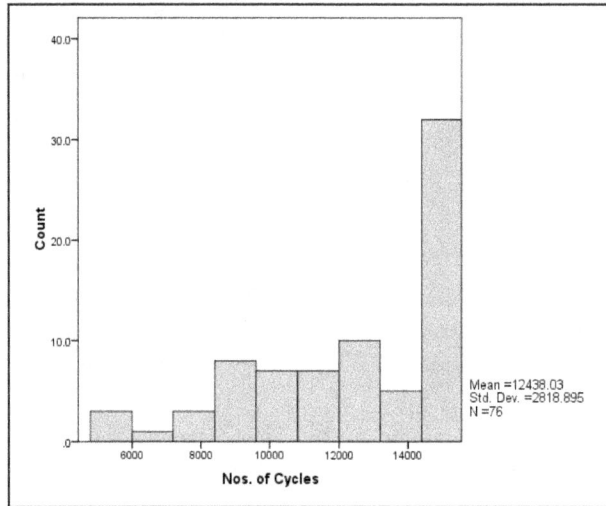

Figure 1 Histogram

Table 2 Descriptive Statistics

#of Obs.	Min	Max	Mean	Std.Dev.	Skewness	Kurtosis
76	4882	15000	12438	2818.90	-.878	-.080

The histogramme and the descriptive statistics clearly indicate the censored nature of the data. Also very high frequency at 15000 cycles indicates the Type-I censoring of data. Also, the histogram does not indicate any bimodality, implying that statistical behaviour of fatigue cycles in radial and tangential direction is not different. However, metallurgical considerations does not totally rule out the possibility of difference in behaviour in tangential and radial directions. Hence, the analysis was carried out for combined data on radial and tangential directions as well as for the LCF values in the two directions separately. As discussed in the previous section analysis was carried out using both lognormal and Weibull distributions.

3. Statistical Analysis

The analysis is carried out using both Weibull and lognormal distribution models. The analysis methodology is specifically chosen to suit the Type-I censored data. Accordingly, Anderson-Darling test based on normalized spacing is adopted for testing goodness of fit. Further, as it would be explained in section 4, the models are compared in the lower tail region using lower percentile values and their confidence bounds. The model description, parameter estimation, percentile estimation with its confidence bounds and goodness-of-fit testing methodology are explained below.

3.1 Weibull Distribution

A random variable X is said to follow Weibull distribution, if the probability that X≤ t is given by ([1], [2], [7])

$$P(X \leq t) = 1 - \exp\left[-\left(\frac{t-\tau}{\alpha}\right)^{c}\right] , \text{ for } t \geq \tau, \alpha > 0 \text{ and } c > 0 \qquad .....(1)$$

Anderson-Darling test of goodness-of-fit based on normalized spacing does not require estimation of shape (c) and scale (α) parameters of Weibull distribution. It requires estimation of only the threshold parameter (τ). However, in order to estimate the tolerance limits, estimates of all the three parameters are necessary. The threshold parameter τ is calculated based on normalized spacing, while scale parameter α and shape parameter c are estimated using maximum likelihood estimates (MLE) [7].

3.2 Lognormal Distribution

A random variable X is said to follow lognormal distribution if the transformed variable Y =ln(X) follows normal distribution. Lognormal distribution thus defined has two parameters corresponding to mean μ and standard deviation σ of normal distribution. Density function of the lognormal distribution with parameters μ and σ is given by

$$g_x(x) = \frac{1}{x\sigma\sqrt{2\pi}}\exp\left[-\frac{1}{2}\left(\frac{\ln x - \mu}{\sigma}\right)^{2}\right]dx, \text{ for } x > 0, -\infty < \mu < \infty, \sigma > 0 \qquad .....(4)$$

Anderson-Darling test procedure based on normalized spacing does not require estimates of the two parameters μ and σ. However, they need to be estimated for calculation of tolerance units and they are estimated using MLE.

3.3 Anderson-Darling Goodness-of-fit Test for Normalized Spacing

Let $x_1, x_2......x_n$ be an uncensored sample of size n. Let $x_{(1)} \leq x_{(2)} \leq...\leq x_{(n)}$ denote ordered sample and $x_{(1)} \leq x_{(2)} \leq...\leq x_{(r)}$ denote censored sample of size $r \leq n$. The normalized spacing y_i is defined as follows.

$$\text{Let } y_i = \frac{x_{(i+1)} - x_{(i)}}{E\left(x_{(i+1)} - x_{(i)}\right)}, \text{ for } i = 1, 2, ..., r\text{-}1; \text{ where } E(\text{ }) \text{ denotes expected values of}$$

the argument [6].

Further define $z_i = \dfrac{\sum\limits_{j=1}^{i} y_j}{\sum\limits_{j=1}^{r-1} y_j}$, for $i = 1, 2, \ldots, r\text{-}2$ .....(5)

Then Anderson-Darling test statistic based on normalized spacing is defined as [6]

$$A_s^2 = -(r-2) - \frac{1}{r-2} \sum_{i=1}^{r-2} \left[(2i-1)\ln z_i + (2r-3+2i)\ln(1-z_i) \right] \qquad \ldots\ldots(6)$$

Large values of A_s^2 indicate departure from the assumed distribution model. Asymptotic percentiles (cutoff values) at different significance levels for normal and extreme value distributions with different censorship proportions can be found in D'Agostino and Stephens [6]. Note that if a random variable X follows Weibull distribution then transformed random variable $(-\ln(X))$ follows extreme value distribution. Thus this table of percentiles can be used after natural log transformation for lognormal and Weibull distribution models.

3.4 *Statistical Tolerance Limits*

There are two types of statistical tolerance limits defined, A-basis value and B-basis value. 1% (or 99%) tolerance limit at 95% confidence is called A-basis value, and 10% (or 90%) tolerance limit at 95% confidence is called B-basis value. As the data is Type-I censored the tolerance limits are approximated by using the method in the MIL-HDBK-17-1F [8].

3.5 *Percentiles and its confidence bounds*

Let W_p and L_p denote $100*p^{th}$ percentiles of Weibull and lognormal distributions respectively. This implies that $P[X \le W_p] = p$ for Weibull distribution and $P[X \le L_p] = p$ for lognormal distribution. They are calculated as follows:

In the case of Weibull distribution

$$P[X \le W_p] = p = 1 - \exp\left[-\left(\frac{W_p - \hat{\tau}}{\hat{\alpha}} \right)^{\hat{c}} \right] \text{ from equation (1a).}$$

$\Rightarrow W_p = \hat{\tau} + \hat{\alpha} * \left[-\ln(1-p) \right]^{1/\hat{c}}$, where $\hat{\tau}$, $\hat{\alpha}$ and $\hat{c}$ denote the estimated values of

Weibull distribution.

In the case of lognormal distribution, L_p is calculated as

$$P\left(X \leq \frac{\ln(L_p) - \mu}{\sigma}\right) = p = \Phi\left(\frac{\ln(L_p) - \mu}{\sigma}\right), \text{ where } \Phi \text{ is defined as before,}$$

$$\Rightarrow \frac{\ln(L_p) - \mu}{\sigma} = \Phi^{-1}(p) \text{ or } L_p = \exp\left(\sigma * \Phi^{-1}(p) + \mu\right).$$

3.4.2 Calculation of Confidence Bounds of the Percentiles

In the present case there is no direct method of calculating the confidence interval for the pth percentile, say t_p. It is calculated through indirect method of testing hypothesis $H_0 : t_p = Q_0$ versus $H_1 : t_p \neq Q_0$, where $t_p = W_p$ or L_p depending on the distributions, and Q_0 is some pre-specified known value. The likelihood ratio test statistic for testing this hypothesis, Λ is approximately distributed as χ^2 distribution with 1 degree of freedom. Hence, 95% confidence intervals for the pth percentile is found by determining the set of values Q_0 for which $\Lambda \leq \chi^2_{(1,0.95)}$ [2, chapters 4 and 5]. The lower limits of these confidence intervals for t_p would provide the required engineering tolerance limits.

4. Results and Discussions

Statistical analysis was performed for three data sets:

- Combined data on radial and tangential measurements
- Radial data
- Tangential data

Further, analysis is performed using both Weibull and lognormal distribution models. Acceptance criteria for both the models are kept at 5% significance level. Graphical presentation of these results is given in figure 2 and parameter estimates and percentiles are given in Tables 3 and 4. Table 5 gives the 95% confidence intervals for these percentiles, while Table 6 shows statistical tolerance limits.

Table 3 Weibull Analysis

Data	Weibull accepted?	τ	α	c	1%	5%	10%
Radial & Tangential	Yes	3949	10919	2.30	5427	6951	8053
Radial	Yes	2146	11477	2.88	4469	6238	7400
Tangential	Yes	4490	11504	2.49	6317	7996	9166

Table 4 Lognormal Analysis

Data	Lognormal accepted?	μ	σ	1%	5%	10%
Radial & Tangential	Yes	9.49	0.38	5467	7083	8131
Radial	Yes	9.39	0.38	5014	6479	7427
Tangential	Yes	9.58	0.35	6369	8105	9216

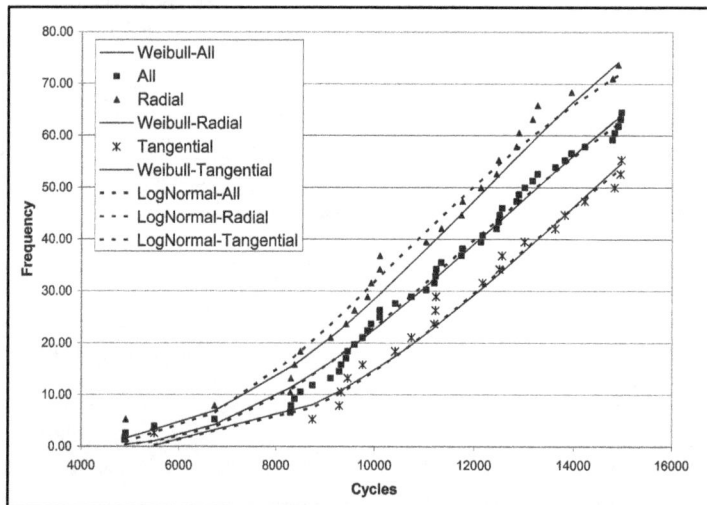

Figure 2 Distribution Functions for LCF

Table 5 95% Confidence Intervals for Percentiles

Data	Percentile	95% Confidence Intervals	
		Weibull	Lognormal
Radial .& Tangential	1%	[4775,6212]	[4536,6589]
	5%	[6005,7924]	[6152,8155]
	10%	[7011,9065]	[7214,9166]
Radial	1%	[3340,5720]	[3895,6456]
	5%	[4778,7616]	[5338,7865]
	10%	[5864,8769]	[6289,8772]
Tangential	1%	[5233,7605]	[4936,8219]
	5%	[6470,9446]	[6712,9788]
	10%	[7520,10615]	[7865,10,800]

Table 6 Design allowable (Statistical Tolerance limits)

Data	Design Allowable	Tolerance Limits	
		Weibull	Lognormal
Radial .& Tangential	A-basis	4937	4660
	B-basis	7228	7289
Radial	A-basis	3559	3884
	B-basis	6138	6246
Tangential	A-basis	5510	5134
	B-basis	7881	7951

The results show that both Weibull and lognormal distribution models are acceptable for LCF data. The percentile values for Weibull distribution are consistently smaller than those values for lognormal distribution. This is expected as Weibull distribution is known to be conservative in the lower tail. In the case of Radial direction, the one percentile value for Weibull distribution is noticeably smaller than that of lognormal distribution. Such a large difference can be explained as follows:

- Less proportion of the cases exceed 15000 cycles to failure in Radial direction as compared to only tangential direction and Tangential and Radial directions put together.

- Minimum number of cycles in Radial direction is 4882, while, in tangential direction it is 5484.

- The estimation of Weibull and lognormal distribution parameters, when the data is censored, depends on minimum number of cycles and proportion of data censored.

Thus for radial and tangential directions put together, minimum number of cycles is 4882 and proportion of censorship is 36% and the same values for tangential measurements are 5484 and 44%, respectively. Now, these values for radial measurement are 4882 and 26%, respectively. This results in difference in the probability density functions of Weibull distribution and lognormal distributions in the lower region before the densities peak. This can be seen from Figures 3, 4 and 5, generated using statistical software MINITAB 15.

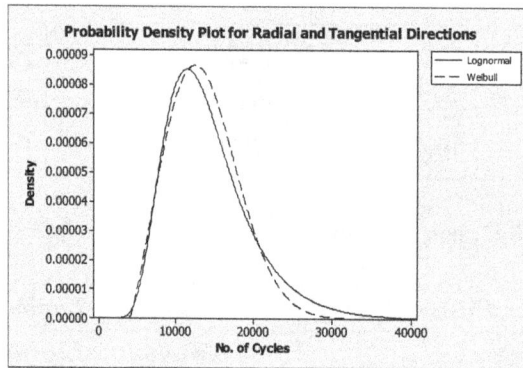

Figure 3 Probability Density Plot for Radial and Tangential Directions

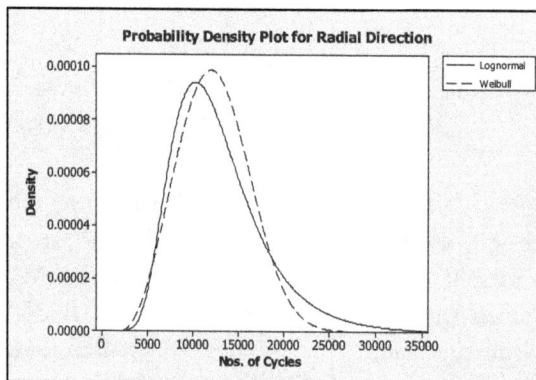

Figure 4 Probability Density Plot for Radial Direction

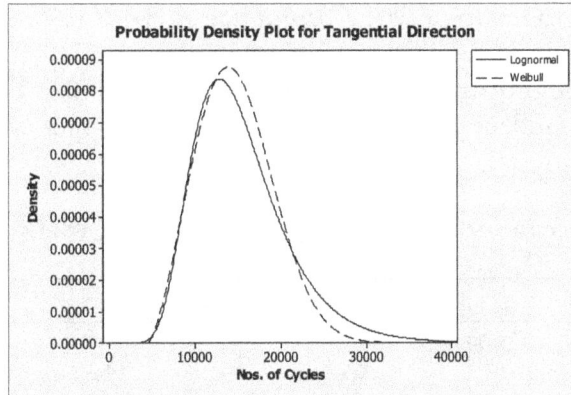

Figure 5 Probability Density Plot for Tangential Direction

This has showed up in large difference in one percentile values of the two distributions, and the difference becomes smaller for the fifth and tenth percentile values.

As mentioned earlier apart from statistical adequacy of the model another criteria for preferring one model over the other is awareness about the affect it may have on final conclusions made. Here the main objective of the analysis is to come up with lower tolerance limit. Thus, one need to compare the two models in the lower tail region. Therefore in the present case two more comparison criteria are used:

1. Statistical comparison of the percentile values

2. Statistical comparison of the 95% confidence bounds of the percentiles

4.1 *Statistical Comparison of Percentile Values:*

Here percentile values obtained from the two models should be compared statistically to see if the models are different. A way to statistically compare these values is through testing of hypothesis that they are same. However, this route in the present case is mathematically very cumbersome, hence another route, which is less formal but much less cumbersome is used, where, the confidence intervals of these estimates are compared. To carry out comparison at 5% level of significance, one need to estimate 95% confidence interval of these percentile points under both the models. If these intervals overlap it would imply that the two percentile points are not significantly different.

It can be seen from the Table 5 that the intervals are well overlapping, implying that as far as inference on the percentile values for LCF is considered, the two models are not different.

4.2 *Statistical Comparison of 95% Confidence Bounds of the Percentiles*

One of the desirable properties of the confidence interval is to be as short as possible. Table 7 gives the confidence intervals with its length d = upper bound – lower bound = ucb – lcb.

Table 7 Confidence Interval Length Comparison of Weibull and Lognormal Distributions

Direction	Percentiles	95% Confidence Bounds						Distn. with shorter conf. interval
		Weibull			Lognormal			
		lcb	ucb	d	lcb	ucb	d	
Radial & Tangential	1% (A-basis)	4775	6212	1437	4536	6589	2053	**Weibull**
	5%	6005	7924	1919	6152	8155	2003	Weibull
	10% (B-basis)	7011	9065	2054	7214	9166	1952	**Lognormal**
Radial	1% (A-basis)	3340	5720	2380	3895	6456	2561	**Weibull**
	5%	4778	7616	2838	5338	7865	2527	Lognormal
	10% (B-basis)	5864	8769	2905	6289	8772	2483	**Lognormal**
Tangential	1% (A-basis)	5233	7605	2372	4936	8219	3283	**Weibull**
	5%	6470	9446	2976	6712	9788	3076	Weibull
	10% (B-basis)	7520	10615	3095	7865	10800	2935	**Lognormal**

Note that the A-basis values will be in the neighbourhood of 1st percentile, and B-basis value will be in the neighbourhood of 10th percentile. Thus from Table 7 it can be seen that for A-basis values Weibull is preferable while for B-basis values Lognormal is preferable. Thus the A-basis values for combined data, radial data and tangential data are 4937 cycles, 3559 cycles and 5510 cycles respectively. While, B-basis values for combined data, radial data and tangential data are 7289 cycles, 6246 cycles and 7951 cycles respectively.

5. Summary and Conclusions

The data on LCF for isothermally forged Adour engine disc has been analysed primarily to obtain engineering tolerance limits or design allowable.

The data is a type – I censored lifetime data. Weibull and lognormal distribution models were chosen to model this data. Anderson-Darling test of goodness-of-fit using normalized spacing was applied.

The goodness-of-fit test found both the models adequate for the data. However, further comparisons of the models were carried out by testing equality of percentile values and by comparison of the length of 95% confidence intervals of these percentile values. The former did not show any difference between the two models, however, the latter comparison showed that for A-basis values Weibull distribution model gives shorter confidence interval than lognormal distribution and for B-basis value it is the Lognormal distribution that gives shorter confidence interval. This is consistent with the known fact that Weibull is conservative in tails. Thus at very tail end of 1% Weibull gives smaller confidence interval.

The analysis also provides a way to statistically compare the radial LCF values with tangential LCF values. Figure 2 clearly shows that radial measurements are smaller than tangential measurements, except in the lower tail end. However, at the lower tail end the percentile values and the tolerance limits support the above mentioned relationship between radial and tangential measurements.

In conclusion following points can be made:

(i) In general both Weibull distribution and lognormal distribution fit the LCF data adequately.

(ii) When the objective is to calculate statistical tolerance limits, Weibull model is preferable for A-basis values, and lognormal model is preferable for B-basis values.

(iii) Statistically it is shown that radial LCF values are smaller than the tangential LCF values. It is true for radial LCF data set as a whole and tangential LCF as whole. It is also true, when only percentile values or 95% tolerance limits are concerned.

(iv) The above conclusions (i) and (ii) remain true for the combined data of radial and tangential measurement as well.

(v) For the present data A-basis values for combined, radial and tangential data are 4937, 3559 and 5510 cycles respectively. And B-basis values for the same is 7289, 6246 and 7951 cycles respectively.

Acknowledgement

The author would like to thank Dr. G.G. Saha and Dr. T. Raghu for providing the data and giving necessary information on the genesis of the data. She would also like to acknowledge Dr. A. Venugopal Reddy, then Regional Director, RCMA(M) for providing guidance regarding the kind of statistical information was expected from the data.

References:

1. Lawless, J.F. statistical models and methods for lifetime data, John Wiley & Sons Inc, New York, 1982.

2. Meeker, W.Q. and Escobar, L.A., statistical methods for reliability data, John Wiley & Sons, Inc, New York 1998

3. Abernethy, R.B. The New Weibull Handbook, Gulf Publishing Company, Houston, 2001.

4. Metals Handbook, Ninth Edition, Vol. 8: Mechanical Testing; Section on Statistics and Data Analysis.

5. Goodness of Fit techniques, Eds. D'Agostino, R.B., and Stephen, M.A, Marcel Dekker Inc. New York 1986.

6. Ibid, Chapter 4.

7. Military Handbook- MIL-HDBK-5H, Chap.9 Dec 1998.

8. Military Handbook - MIL-HDBK-17-1F, Vol. 1, Chap. 8, 2002.

Effect of Quench Delay in Solutionising Treatment and Aging Heat Treatment on the Wear properties of Ti 6Al 4V

Sarala Upadhya* , B K Muralidhara

*Department of Mechanical Engineering University Visvesvaraya College of Engineering
Bangalore, INDIA*
e-mail: sarala.upadhya@gmail.com

The effect of different heat treatment parameters on the wear resistance of Ti 6Al 4V has been investigated in this study. For this purpose, the alloy is subjected to STA treatment of solutionising at 955 C followed by water quenching. The delay in quenching has significant effect on the microstructure and mechanical properties of the alloy. Quench delay of 3, 7, 10, 15, 25 and 35 seconds were adopted for the present studies. In the next step, aging was carried out at 540 C for 6 hours for all the specimens. When tested on dry abrasion wear testing machine, the alloys showed significant dependence of the wear rate on the quench delay time. Hardness, tensile strength and wear resistance of the alloy decreased with increase in quench delay. This effect is supported by the microstructural changes taking place in the alloy with quench delay.

Keywords: Ti 6Al 4V, STA, quench delay, wear resistance.

1. Introduction

Titanium and its alloys are widely used in many industrial fields from aerospace engineering to bio-medical applications. The vast range of application of Ti alloys is due to their high specific strength, low density, good mechanical properties, heat treatability and good corrosion resistance among many other properties. Ti 6Al 4V is one major alloy of the series, which has found extensive applications in the last two decades and it is likely to extend its usage in automotive field also due to large weight savings it promises with no loss of performance. The alloy is very sensitive to variations in heat treatment parameters because of its unique microstructural combinations of α and β phases developed in the alloy. This in turn affects the mechanical properties also. In industrial scenario, the control of heat treatment parameters is quite challenging to produce the precise properties required for a given component.

Solutionising Treatment and aging, generally called as STA treatment, is a common heat treatment procedure followed for Ti 6Al 4V for achieving a variety of mechanical properties accompanied by microstructures [1]. Significant work has been carried out in correlating structure and properties in this area. [1-3]. The general procedure consists of solutionising the alloy at a sub-β transus temperature followed by water quenching and aging at a lower temperature for a sufficiently long time to get a structure of α, β and α-martensite. The composition of the microstructure controls the mechanical properties of the alloy. However, the sensitivity of the alloy to any delay in quenching that may take place after solutionising is a parameter to be seriously considered in actual practice [1,4, 5] and it is a subject of interesting study also. As found by the earlier studies, any quench delay causes decrease in hardness and strength of the alloy with significant changes in microstructure.

The applications of the Ti alloy for machine components require the wear properties of the alloy to be taken into account. Early works in this area by Budinski [6] is note worthy, in which the author mentioned that the alloy showed a wear pattern "suggesting hard and soft regions" which is "not normal abrasion pattern". It was suggested that this could be due to "microstructural inhomogenieties" in Ti 6Al 4V. This alloy is a two-phase alloy with a stable α phase with HCP structure and a metastable β phase with BCC structure. The author concludes that, in comparison with CP Titanium, which has only α phase, Ti 6Al 4V has poor wear resistance. This early study is followed by many other studies [7-12] concerning the response of Ti 6Al 4V to a variety of wear tests. But the effect of microstructure on wear is yet to be understood.

It is the objective of the present work to make a comparative study of the abrasion wear response of Ti 6Al 4V to the microstructural changes effected by changes in STA treatment parameters [1] as it gives a good discernible combination of α and βphases. Different quench delays time periods are effected to get a variety of microstructure compositions.

2. Experimental Procedure

Ti 6Al 4V plates of 10mm thickness were obtained in annealed condition. Specimens were prepared by cutting and machining for tensile, hardness microstructure and abrasion wear tests. The abrasion wear specimens were of 75X25X10 mm size as per ASTM G65 standard.

The specimens were subjected to STA treatment in Argon atmosphere, in the following steps:

1.   Solutionising at 955°C for 1 hour
2.   Water quenching
3.   Aging at 540°C for 6 hours.

Quench delays of 3, 7,10,15,25 and 35 were affected between step 1 and step 2, to get a variety of microstructures. A set of specimens were air cooled after solutionising and aged. Microstructures of these specimens are examined in Nikon Type 108 Optical Microscope. Vickers Hardness tests were conducted on Leica VM HT MOT Micro Hardness Tester and tensile tests are conducted on Instron Machine. Abrasion wear tests were conducted on a Dry Sand Abrasion Rubber Wheel Tester (Ducom Make) following standards given in literature [13, 14]. These tests were conducted at two load levels viz. 28.6 and 77.84 N. The other parameters of the tests are as given in the Table1.

Table 1 Parameters of Abrasion wear test

Sl.no	Conditions	Quantity
1	Number of Revolutions of Wheel	2000
2	Speed of rotation	300rpm
3	Sand flow rate	250g/min

The weight loss due to abrasion in each case was determined and this was converted into volume loss by taking the density of the alloy as 4.3 g/cc. The abraded surfaces were observed under JEOL 6490LV Scanning Electron Microscope.

3. Results and Discussion

Effect of quench delay on the mechanical properties:

The variation of the mechanical properties of the alloy as a function of quench delay time in seconds is shown in Fig 1 to 3. It is found that Vickers Hardness and tensile strength decrease and percentage elongation increases with increase in quench delay time. The mechanical properties of alloy, which is air cooled is shown Table 2 for comparison.

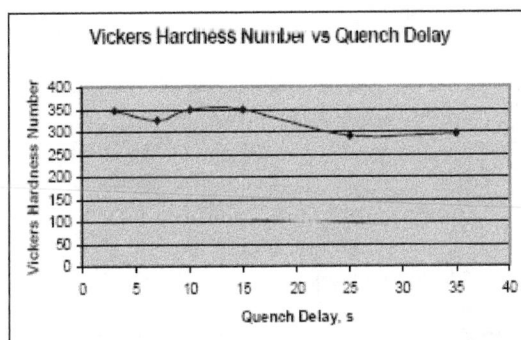

Fig. 1 Variation of Vickers Hardness Number with Quench Delay Time

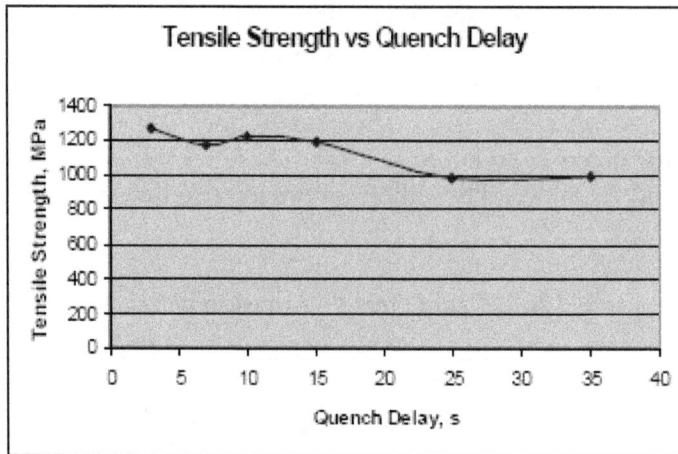

Fig. 2 Variation of Tensile Strength with Quench Delay time

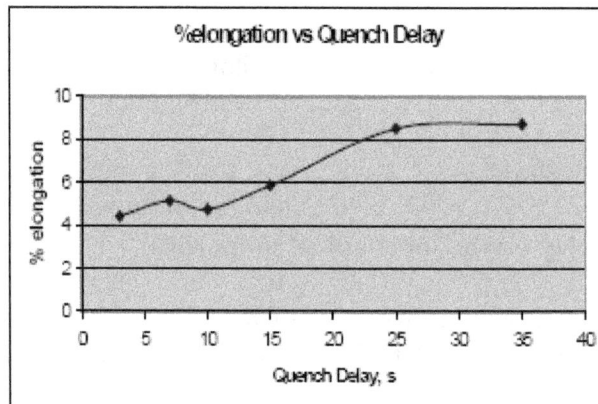

Fig. 3 Variation of % Elongation with Quench Delay Time

Table 2 Mechanical Properties of the alloy in as-received and air cooled conditions

Sl.No.	Mechanical Property	As received condition	Air cooled condition
1	Vickers Hardness Number	330	294
2	Tensile Strength, MPa	922.24	908
3	% Elongation	9.8	11

Effect of quench delay on the microstructure of the alloy:

The Microstructures of the alloy in as-received condition is shown in Fig 5 and in the various heat-treated conditions are shown in Fig 6 to 11. Fig 12 shows the same in the air cooled condition. All the microstructures are shown at a magnification of 500X.

Dramatic changes in microstructures were found to occur in the alloy with increasing quench delay time. Initially, with short delays of 3 to 7 seconds, quenching from the solutionising temperature produces primary α, α-martensite (α') and β structures. Subsequent aging gives rise to globular primary α in a matrix of hazy α'..

Sometimes the morphology shows elongated α' phase, if nucleated near grain boundaries. With increase in quench delay time, β phase tends to form acicular α. The structure formed is closer to Widmanstätten structure or basket weave structure. The Widmanstätten alpha can have several different morphologies. Slow cooling rates favour the formation of similarly aligned αplatelets in colonies, together with αformed at prior β phase. Faster cooling rates favour a more basket weave structure.

It can be generally said that when there is less quench delay, martensite α (α') is formed. This decomposes into α and fine β during aging. When there is more delay in quenching, formation of lamellar structure of a and b takes place and this stabilizes after aging.

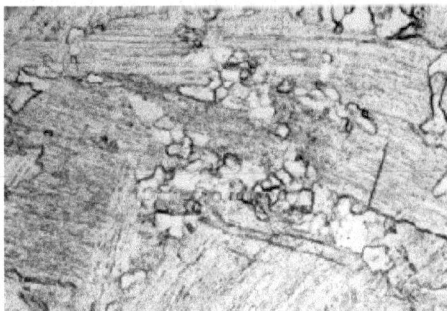

Fig 4 Microstructure in as-received Condition

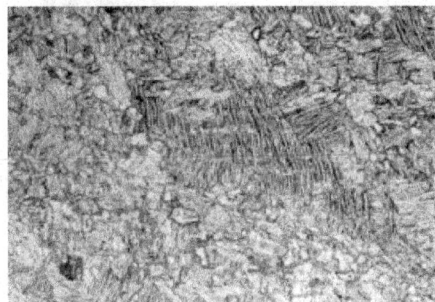

Fig. 5 Quench delay of 3 Seconds

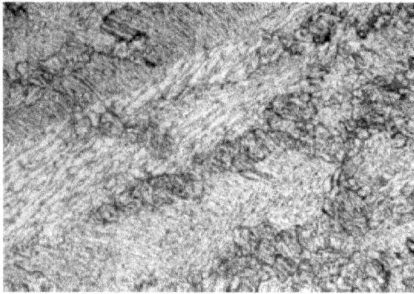

Fig. 6 Quench delay of 7 Seconds

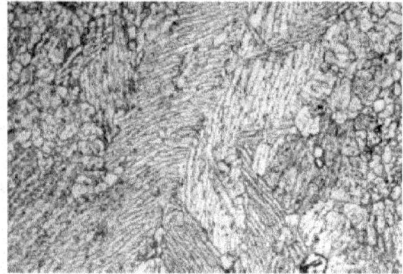

Fig. 7 Quench delay of 10 Seconds

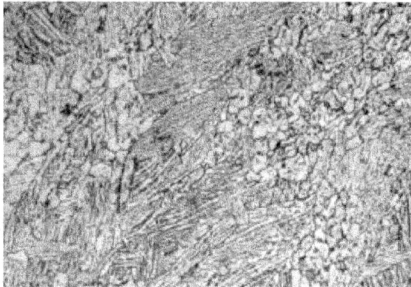

Fig. 8 Quench delay of 15 Seconds

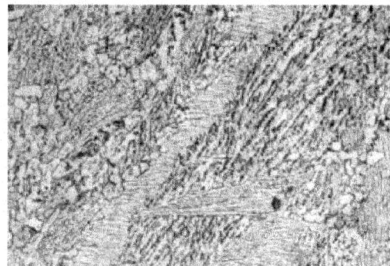

Fig. 9 Quench delay of 25 Seconds

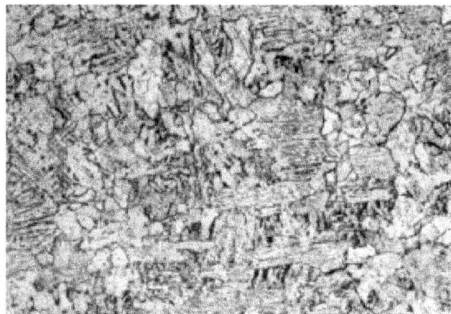

Fig. 10 Quench delay of 35 Seconds

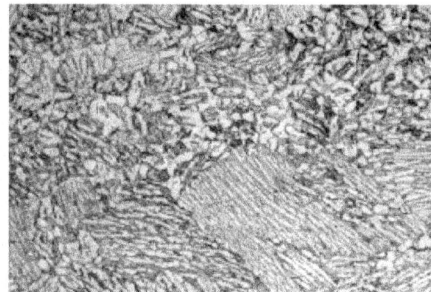

Fig. 11 Microstructure in air cooled condition

This effect is found to be similar to results in earlier work [15].

The microstructural changes described above have been to occur as shown in Fig 4 to 10. The quench delay structure for 35 seconds delay is similar to aircooled specimen shown in Fig 11. These changes have a direct effect on the Mechanical properties of the alloy. As shown in Fig 1 to 3, there is very little change in the mechanical properties up to a quench delay time of 10 seconds. After that, micro hardness and tensile strength decrease in value gradually, whereas percentage elongation increases gradually with increasing quench delay time.

Effect of Quench delay on the abrasion wear loss:

The abrasion wear rates in volume loss of material in mm^3 at different load levels, as a function of quench delay time are shown in Fig 12. It is found that the variations in mechanical properties have a bearing on the abrasion wear of the Ti 6Al 4V alloy. The wear volume increases with the quench delay time, for the both cases of loads of testing. This can be ascribed to the decrease in hardness of the alloy with quench delay with the nucleation and growth of alpha phase.

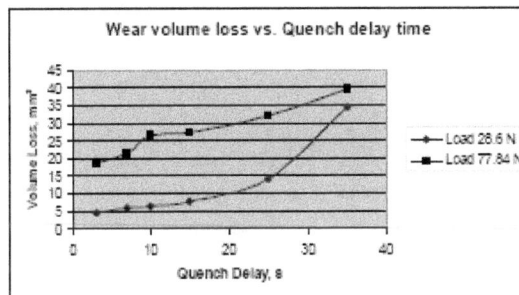

Fig. 12 Effect of Quench delay on abrasion wear loss at different loads.

A look at the SEM micrographs of the worn surfaces, taken for specimens tested at 28.6 N shows that at lower quench delay times of 3 seconds (Fig 13), deep furrows made by the abrading sand. Fig 14 and15 show the SEM micrographs of specimens with quench delay times of 10 and 35 seconds respectively. It may be observed that these surfaces to be more and more deeply gouged out. There is also a collection of wear debris sticking to the surface, in these cases. On the other hand, the air cooled specimen shows a ploughed surface due to the abrasion by the hard sand particles in Fig 16.

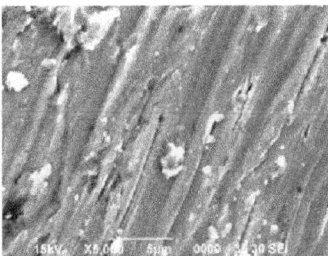

Fig 13: Abraded surface of specimen with quench delay of 3 seconds

Fig 14: Abraded surface of specimen with quench delay of 10 seconds

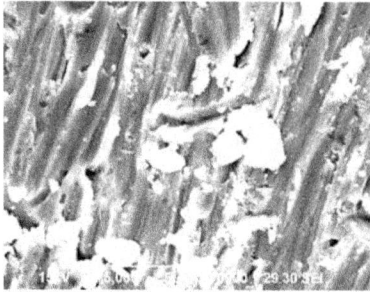

Fig 15: Abraded surface of specimen with quench delay of 35 seconds

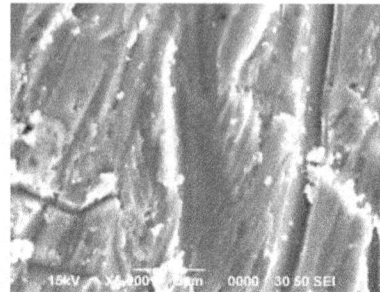

Fig 16: Abraded surface of air specimen

Conclusions:

Ti 6Al 4V alloy is found to be very sensitive to quench delays in the solutionising treatment and aging heat treatments. There are significant changes taking place in the transformation of βphase to α phase with the rate of cooling from solutionising temperature taken below the β-transus temperature. The associated effect on mechanical properties is to be noted carefully, particularly if heat treatments of components are carried out for crucial applications. The effect of quench delay on mechanical properties is also reflected on the abrasion wear resistance of the material. When there is less delay in quenching, the alloy shows more resistance to abrasion wear. But wear volume loss increases significantly as the quench delay time increases. The SEM micrographs show that more material is removed during abrasion from the softer specimens, which are the result of acicular structure of alpha due to slower cooling from solutionising temperature. Hence it can be concluded that the wear properties of Ti 6Al 4V alloy suffers further if there is a delay in quenching.

Acknowledgents:

The authors place on record their grateful thanks to ADA, Fig 16: Abraded surface of air cooled specimen

Bangalore -1 for sponsoring the Project on the Heat treatment parameters of Ti 6Al 4V. They are indebted to NAL, Bangalore for providing testing facilities. They acknowledge with thanks to World Bank for setting up SEM facilities at UVCE, Bangalore under TEQIP.

References

1 "Titanium-A Technical Giude" Second Edition,Ed. Mathew J Donachie. Jr. ASM International Ohio 2000, p55

2 "Light Alloys" Polmear. I J Arnold PublicationsThird Edition, London 1981

3 Robert Pederson, 'Microstructural and Phase transformation of Ti 6Al 4V', Licentiate Thesis, Lulea University of Technology 2002:30 ISSN: 1402-1757 ISRN:LTU-LIC 02/30 5E

4 H Ouchi Suenega, C Tetsu-to-Hagne, 'Decrease of strength due to delayed quenching' J. Iron Steel Inst. Japan 72)(1):131-137 1986: ISSN 00211575

5 S Hamai, Sugiura,Y, Netsu Shori, 'Mechanical properties of beta –STOA treated Ti   6Al 4V alloy-effect of Solution Treatment and delayed quenching' J. Japan Soc. Heat Treatment, 32(8): 157-163 1992 ISSN 02880490

6 K G Budinski, 'Tribological Properties of Titanium alloys' Wear,151 (1991) 203-217

7 Qui Ming, Zhan Yongzhen, Zhu Jun, Yang Jianheng 'Dry Friction Characteristics of Ti 6Al 4V alloy under High Sliding velocity' Journal of Wuhan University of Technology— Materials Science Edition, Vol 22 No.4 (2007) 582-585

8 Jun Qu, Peter J Bleu, Thomas R Watkins, Odis B Cavin, Nagaraj S Kulkarni: 'Friction and wear of Titanium alloys sliding against metal, polymer and ceramic counterfaces': Wear 258 (2005) 1348-1356

9 Peter J Bleu, Brian C Jolly, Jun Qu, William H Peter, Craig Blue, ' Tribological Investigation of Titanium Based Materials for Brakes' Wear, 263 (2007) 12021211

10 Md. Ohidul Alam, A S M A Haseeb, ' Response of Ti 6Al 4V and Ti 24Al 11Nb alloys to dry sliding Wear against Hardened Steel' Tribology International, 35 (2002) 357-362

11 A Molinari, T B Straffilini. , B Tesi, T Bacci 'Dry sliding Wear Mechanism of Ti 6Al 4V alloy' Wear 208 (1997) 105-112

12 Qui Ming, Zhan Yong-zhen, Zhu Jun, Yang Jian-heng, Zhu Jun ' Microstructure and tribological characteristics of Ti 6Al 4V alloy against G Cr 15 under high speed and dry sliding' Materials Science and Engneering A 434 (2006) 71-75

13 Stevenson A N J, I M Hutchings, ' Development of Dry Sand/ Rubber wheel Abrasion Tcst' Wear 195 (1996) 232-240

14 ASTM G 65-91, Standard Test Method for measuring Abrasion using the Dry Sand / Rubber wheel Apparatus in Annual Book of ASTM Standards. Vol. 103.02 ASTM, Philadelphia, PA 247259

15 J B Borradaile, R H Jeal, 'Critical Review- Mechanical Properties of Titanium alloys' Proceedings. Titanium'80 Vol 1 141-152

Experimental Study of Local Strains Near Carbides in A Superalloy

Phani S Karamched* and **Angus J Wilkinson**

Department of Materials, Oxford University, Parks Road, Oxford OX1 3PH, England, UK
phani.karamched@materials.ox.ac.uk

Abstract : *At the length scale of single grains, fatigue crack nucleation and growth depend crucially on microstructural features such as grain boundaries, triple points, crystallographic orientation, and inclusions. An insight into the micro-mechanics of crack initiation process and how it is influenced by microstructural features is necessary to make a better prediction of fatigue life in most engineering components.*

With recent progress that has been made towards elastic strain (and hence stress) mapping using electron back scatter diffraction, it is now possible to measure the plasticity exhibited around several microstructural features. The thermal strains (as a result of the regular heat treatment processes) have been measured around carbides in a directionally solidified material with near-prismatic grains. After deformation in bending, the strains have been measured around the same carbides and the differences recorded. The experimental results are described.

1. Introduction

The need to know the local state of stress and strain in a region of material is of the utmost importance, especially in structural engineering components. The local strain enhancements near microstructural features such as precipitates, inclusions, grain boundaries need to be studied if the deformation and failure processes are to be understood.

With recent progress made towards elastic strain (and hence stress) mapping using electron back scatter diffraction [1,2], it is now possible to measure the plasticity exhibited around several microstructural features. In this analysis, small shifts of features in test EBSD patterns relative to their positions in a reference pattern are measured. The pattern shifts across about 20 sub-regions, dispersed across the EBSD patterns are measured. With knowledge of the projection geometry, the pattern shifts are related to a general displacement gradient tensor **a** (i.e. the sum of strain **e** and rotation **w** tensors). The displacement gradient tensor is found using a least square error fit to the shifts measured at all of the sub-regions. This technique can measure only the differences between the three normal strains. However, EBSD measurements are made typically within 10-20 nm from the free surface of the specimen, where

110

equilibrium requires the stress normal to the surface be zero. Making this assumption that stress normal to the surface be zero, provides an additional linear equation relating the three normal strains, so that all three can be evaluated.

With the use of this cross-correlation based technique, we reveal the measured strains and rotations around microstructural features that are potential crack initiation sites. The sample is a directionally solidified Nickel based superalloy with near prismatic grains and carbide precipitates distributed randomly all over the matrix.

Thermal strains, as a result of regular heat treatment processes, have first been measured around the carbides. The heat treatment consisted of two steps, the first being one hour at 1100^0C in order to take the material over the gamma prime transus causing it to dissolve. This was followed by ageing for sixteen hours at 870^0C to precipitate a fine dispersion of gamma prime particles. The carbides are present throughout this heat treatment, and the final cooling from the aging temperature generates thermal contraction mismatch between the carbides and the metallic matrix. After determining the thermal strains in the material, it was then deformed monotonically in a three-point bend rig. Strains have been measured around the same carbide particles and the differences recorded.

In each case, the EBSD patterns were recorded using a ~1000 x 1000 pixel, peltier cooled CCD camera (with no binning). The scintillator screen was held in its usual position so as to subtend a large capture angle (~70°) at the sample. Typical SEM conditions used were 20 keV beam energy, and a beam current of about 10 nA, for which exposure time was about half a second. Patterns were recorded on hard disk using TSL/EDAX OIM DC 5.3 software for subsequent off-line batch-wise analysis using the strain determination software CrossCourt 2.

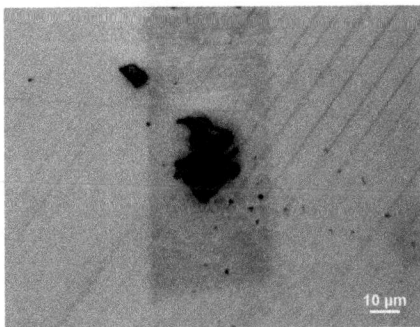

(a) Optical image

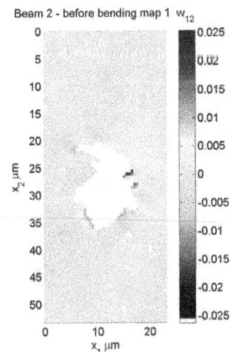

(d) rotation w_{12} about surface normal

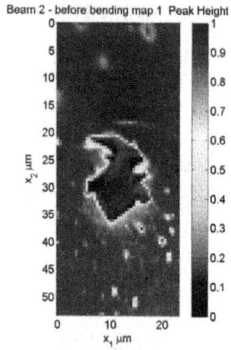

(b) Geometric mean of normalised cross axis

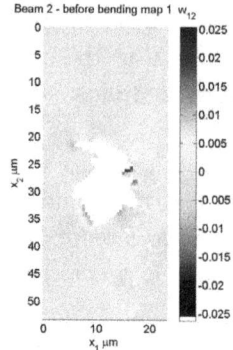

(e) Rotation w_{23} about horizontal correlation peak heights

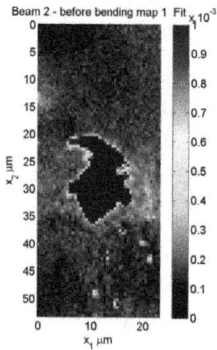

(c) Mean angular error in best fit solution

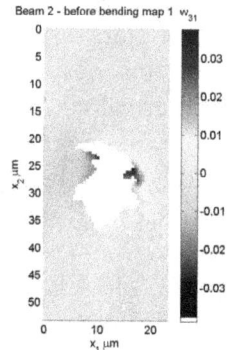

(f) Rotation w_{31} about vertical axis for strains and rotations.

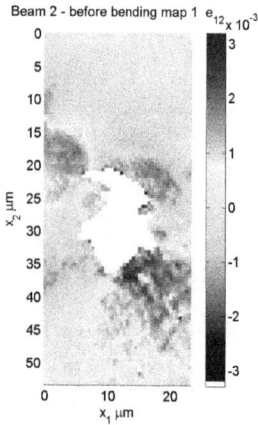

(g) Shear strain e_{12} within surface plane horizontal axis

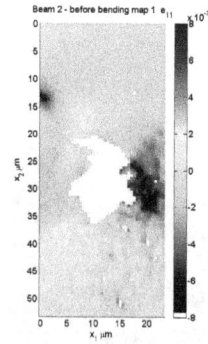

(j) Normal strain e_{11} along

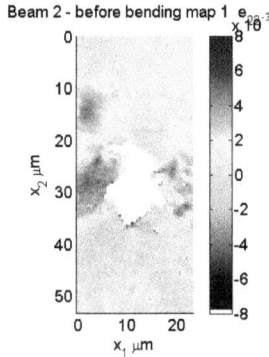

(h) **Shear strain e₂₃ out of surface**

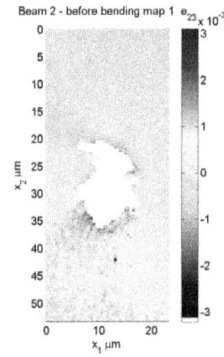

(k) **Normal strain e₂₂ along vertical axis**

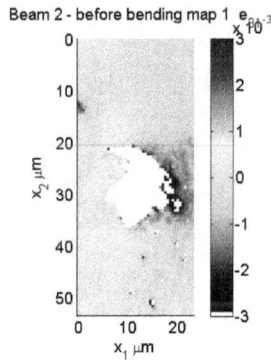

(i) **Shear strain e₃₁ out of surface**

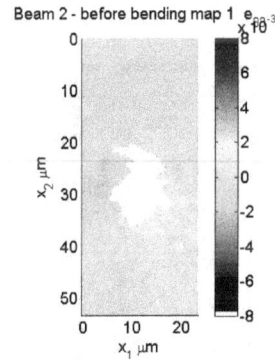

(l) **Normal strain e₃₃ along surface normal**

Figure 1 *Optical image of a carbide particle from the sample and ESBD generated maps of rotations and elastic strain fields (due to thermal treatment) near the carbide.*

The mean peak height map shows the shadowing of patterns obtained from within the carbide leads to poorer data quality. All other maps have been filtered so that results are only shown if the mean peak height is above 0.4. Such filtering successfully removes points with large mean angular error shows in figure 1(c). From the plots, it is very clear that e_{11} strain is mostly compressive, while the e_{22} strain is tensile, which is in good agreement with a rough prediction based on hoop and radial strains with the Ni matrix contracting around the carbide.

After deformation in bending, EBSD patterns were collected and analyzed for the same region around the same carbide particle. The following plots show the results thus obtained. The region containing the carbide of interest was subjected to tensile strain of about 5 - 7%, along the horizontal x_1 axis.

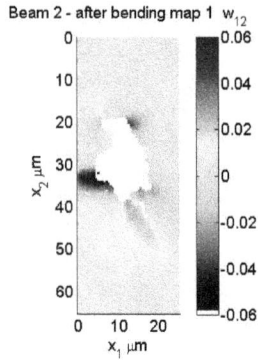

(a) rotation w_{12} about surface normal

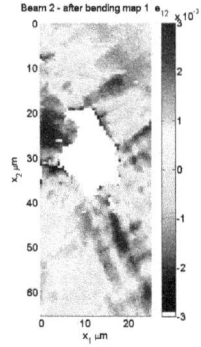

(d) shear strain e_{12} within surface plane

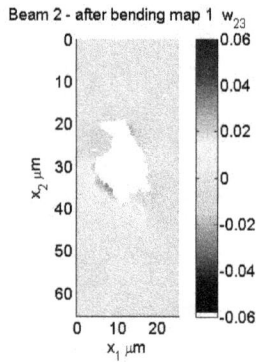

(b) rotation w_{23} about horizontal axis

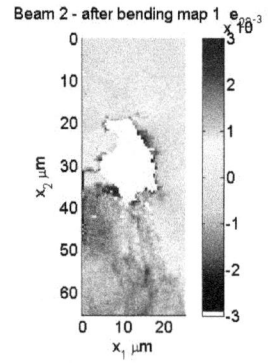

(e) Shear strain e_{23} out of surface

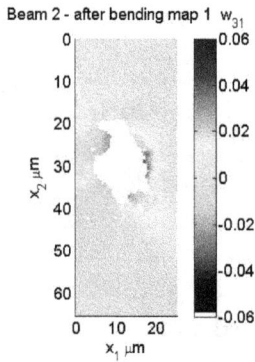

(c) rotation about w_{31} about vertical axis

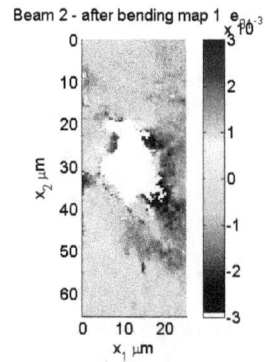

(f) Shear strain e_{31} out of surface

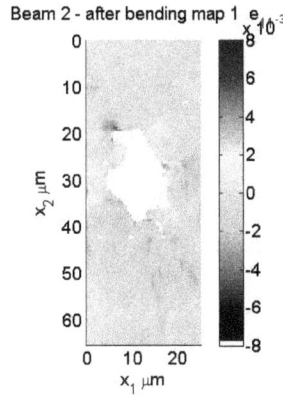

(g) Normal strain e_{11} along horizontal axis

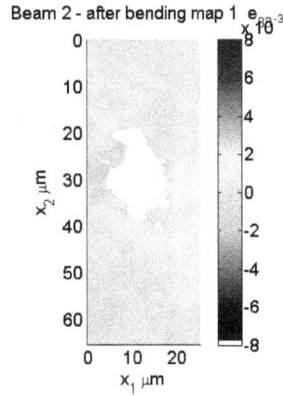

(i) Normal strain e_{33} along surface normal

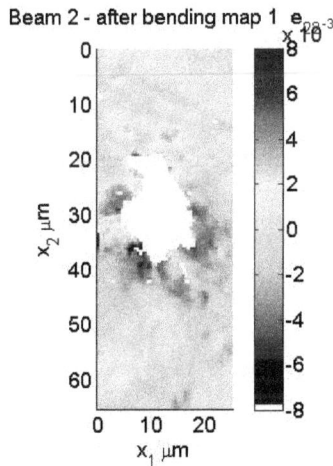

(h) Normal strain e_{22} along vertical axis

Figure 2 *ESBD generated maps of rotations and elastic strain fields (due to deformation in bending) near the carbide.*

The plots quite clearly show the increase in elastic strains and rotations caused by the imposed mechanical loading. From comparison of the maps in figure 2, it is clear that rotations dominate over elastic strains. This is a result of the significant plastic strain gradient around the carbide. It is also quite clear that in-plane shear strain distribution is more concentrated along the slip traces (which can clearly be seen in figure 3).

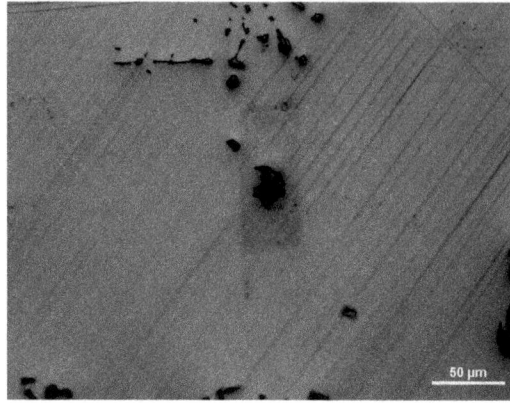

Figure 3 *Optical micrograph showing slip traces in the Ni-matrix around the carbide particle.*

In such circumstances, it is reasonable to use Nye's analysis [3] to estimate geometrically necessary dislocation (GND) content near the carbide. From the knowledge of three rotation components on the surface, it is possible to calculate six of the nine lattice curvatures that are required to form the Nye's dislocation tensor. Here we assume that pure edge and pure screw dislocations can be present on each of the <110>{111} slip systems, and find combination of densities of these dislocation types that support the measure lattice curvature whilst minimising the total dislocation line energy. No constraint is placed on the three missing curvatures. This gives a lower bound estimate of the total GND density. Figure 4 is a map (in logarithmic scale) which shows a lower bound estimation for the GND distribution near this carbide. GND densities of $\sim 3 \times 10^{14}$ m^{-2} are found in some regions close to the carbide.

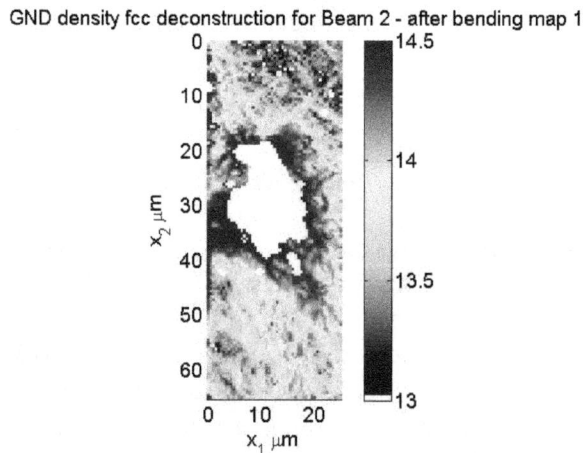

Figure 4 *lower bound estimate of geometrically dislocation distribution near a carbide in a Ni-superalloy. Colourscale shows log(ρ_{GND}) with ρ_{GND} in lines/m².*

116

References

1) Wilkinson, A. J., Meaden, G and Dingley, D. J., *High-resolution elastic strain measurement from electron backscatter diffraction patterns: New levels of sensitivity*, Ultramicroscopy, Vol. 106, 2006, p. 303.

2) Wilkinson, A. J., Meaden, G and Dingley, D. J., *High resolution mapping of strains and rotations using electron backscatter diffraction*. Materials Science and Technology, Vol. 22, 2006, p. 1271.

3) Nye, J. F., *Some Geometrical Relations In Dislocated Crystals*, Acta Metall., Vol. 1, 1953, p. 153.

COMPOSITE AND
CELLULAR MATERIALS

Coated Carbon Nanotubes Application for Metal Matrix Composites Design

G. P. Stovpchenko[1], I. I. Demchenko[2]

1 – E.O. Paton Electric Welding Institute of National Academy of Sciences of Ukraine, 11 Bozhenko str., Kyiv, Ukraine
2 - National Metallurgical Academy of Ukraine, 4, Gagarina av., Dnipropetrovsk, Ukraine

Abstract_Recently carbon nanotubes (CNT) are attracting scientific interest due to their strong physical and chemical properties. Known great properties of carbon nanomaterials promise wide range of application of these materials, among them carbon nanotubes (CNTs) are in the center of attention as a type of material which has really different physical properties in compare with similar type of materials. Promising application of CNTs in energy storage, molecular electronics and sensors, composite materials, and finally templates make it necessary to develop methods of synthesis and to know more about the structure and morphology of these materials to achieve desired properties.

It should be mentioned that factors like structural perfection of the CNTs, chemical inertness and low wettability by various liquids (in that number high temperature melts) make their input and dispersion very complicated. It complicates the use of the unique properties of CNTs at the creation and manufacture of practically any types of composite materials, where preparation of the CNTs mixtures with many different media is necessary.

CNTs coating by protective or functional films is one of prospective method to protect or intensify interaction between CNTs and contacting materials (solvents, high temperature melts, ceramic mixtures, glasses and so on). Nanometer-scale coatings of another material on the surface of CNTs make possible of formation of dispersive suspension in high temperature melts and this is the first factor of success of cast composite formation. Fine dispersion and exact positioning of CNTs is necessary for future integration with metal matrix.

Our investigations shown the use of the coated CNTs allow to overcome a few problems at cast metal matrix composite manufacture from high temperature melts: difference between metal melt and CNT densities as well as low wettability or possible dissolution with carbide formation. In many cases the carbides formation as the result of CNT carbon atoms displacement may be resulted in following: CNT could partially lose their mechanical properties up to the full destruction; composite can become brittle in whole.

As was shown in our theoretical and experimental researches, in most cases properties of the CNTs can be changed by coat them by metal or non-metal contenting compounds (oxides, nitrides, borides and so *on) to avoid the damage of CNT while contacting with high temperature metal and to improve their surfactant properties as well as to solve the problem of specific masses difference of melt and CNT.*

1. Introduction

Recently carbon nanotubes (CNTs) are attracting scientific interest due to their unique physical and chemical properties. There are many reports on the physical properties of CNTs such as Young's modulus (single-wall CNTs - from 1 to 5 TPa, multiwall ones - 1.8 TPa) under the rather high elongation (up to 16%). The application of CNTs as reinforcements can improve mechanical properties of metallic, polymer or ceramic matrix composite materials, as well as promise a wide range of application in energy storage, molecular electronics and sensors, composite materials. This makes necessary to know more about the structure and morphology of these materials to achieve desired properties of novel materials.

It should be mentioned that factors like structural perfection of the CNTs, their chemical inertness make their adding and dispersion into any matrix very complicated. It complicates the use of the unique properties of CNTs at the development and manufacture of practically any types of composite materials, where preparation of the CNTs mixtures with many different media is necessary. The penetration of carbon nanomaterials into high-temperature melts includes also problems such as difference between specific mass of CNTs and most metallic alloys, lack of CNTs wettability as well as possible destruction at CNTs interaction with liquid alloy components.

In our opinion the coating of CNTs surface (by protective or functional films) is one of prospective methods to protect or intensify interaction between CNTs and contacting materials (solvents, high-temperature melts, ceramic mixtures, glasses and so on). Nanometer-scale coatings of another material on the surface of CNTs make possible to form of dispersive suspension in high-temperature melts and this is the most important factor of success of cast composite formation.

1.1 Existing achievements

Reinforced by 2% CNTs magnesium was manufactured by a Swiss research group [1] that allow them to improve modulus of Mg by about 9%. Magnesium composites reinforced with 0.3, 1.3, 1.6 and 2 wt% of carbon nanotubes were fabricated in other research with the use of disintegrated melt deposition technique. The maximal value of Young's modulus was obtained under the 1.3 % CNT concentrations. The grain size decreasing for 1.25 % comparing to the initial magnesium was also noted [2].

Young's modulus improvement of more than 9 percents at magnesium reinforced by CNTs was obtained by authors [3]. The feasibility of manufacturing metal-matrix composites reinforced with carbon nanotubes has been assessed and the same processing steps, used for Mg-2wt%CN, were also used to produce Al 2wt%CNT. Aluminum-CNT compacts of density up to 96 percents of theoretical one have been fabricated and this research is in progress [3].

Some positive results on carbon nanotubes application in reinforced nickel matrix composite (high melting point metal) coatings were reported by authors [4].

The experience of CNT adding to pure plastic metals is presented in article [5], where authors using powder metallurgy methods have obtained the copper-matrix composite with 1 % of CNTs content as reinforcement material. It was found by mechanical testing of the consolidated 1 vol.% CNT-Cu that high mechanical strength could be achieved effectively as a result of the Cu matrix strengthening - 45 % of hardness increasing without significant plasticity decreasing.

Article [6] presents the experiment results concerning the adding of CNTs into aluminium. It must be noted, that the composite has been produced using the methods of powder metallurgy and CNTs were coated by zirconium fluoric salt. It is interesting enough the addition of K_2ZrF_6 as wetting agent results in an increase of 33.85% in the stiffness of the composite. The obtained material also possesses the higher Young's modulus comparing to the initial aluminium.

The authors of article [7] was discovered the formation of aluminium carbide under the heating at the appearing of liquid phase in aluminum – matrix composite which was reinforced by CNTs. The experimental results show that carbon nanotubes react with aluminium and form Al_4C_3 phases with needle shape. It means that CNTs should be protected from interaction with matrix components at high temperature.
For example, the Carbon Nanotechnologies, Inc (CNI) announced of U.S. Patent for coated single-wall carbon nanotubes and ropes of single-wall carbon nanotubes [8] with a nanometre-scale coating of another material that can include polymers and metals.

Thus, the surface treatment of CNT by coating them with a protective film of stable refractory components (like metal oxides or nitrides) may be an effective way of protection of CNT from dissolution in metal matrix or having interaction with liquid matrix components at high temperatures.

In this regard the investigation of a system composed of coated carbon nanomaterials and metal alloys for creation MMC materials with improved mechanical properties is an actual problem. This paper is a trial to clarify some problems of making such composite reinforced by CNT. Two main methods for producing of these materials seem to be - powder metallurgy and casting. In this research casting as a traditional method for manufacture of metallic parts is considered as a way of making Me-C-CNT materials.

2. Proposed principles of MMC-carbon nanocomposite design and manufacture

Application of carbon nanotubes (CNTs) as reinforcements for metal matrix composites gives the idea of preparing these materials according to the conditions, under which they will be used. We took into consideration that the highest melting temperature among known materials, one of the best heat resistance and chemical stability, high thermal conductivity, low thermal expansion and friction, very low wetting ability by most liquid metals are those properties that make usual carbon materials very perspective for various kinds of composites (MMC among them) for high-temperature application. It is also known that excellent thermal and service properties of high-speed steels and heat-resistant steels as well as of alloyed cast irons are provided by carbon and carbide microparticles located in the metal matrix. Usually, metals consist of carbon in three forms, such as solid solution, precipitated carbides or free carbon particles. The type and the amount of all carbonaceous phases are determined during metal solidification and due to phase transformation in both the liquid and solid states. Phase compound, amount and ratio are also limited by the thermodynamics and could be found from phase diagram of proper system.

The main idea of development of metal matrix - carbon nanocomposite arises from two known reasons: first one – alloys hardness and strength rises with increasing the carbon content (for example from steel to iron); second one – alloys plasticity changes vice versa.

There is a long-existent dream to design Fe-C materials with a plastic matrix and reinforcing particles that have a uniform distribution in a metal. The first factors of success in that composite manufacture is the formation of stable disperse suspension of metal with carbon nanosize particles. It should be mentioned that in many cases the carbides formation as the result of CNT carbon atoms displacement may lead to the following: CNTs could partially lose their mechanical properties up to the full destruction; composite can become brittle as a whole.

Therefore, the second idea is to use nanofilm coating on surface of CNTs that will help us to improve "wettability" of CNT by liquid metals and, from the other side, to protect carbon nanomaterials against interaction with carbide formation. Other important aspect is the fact that in practice some content of active oxygen is presented usually in metal melts. Free oxygen in liquid metal can react with carbon nanomaterilas that will result to their destruction. This requires quite full deoxidation of metal by vacuum treatment or adding of active deoxidation agents. The use of coatings of metal or non-metal contenting compounds (oxides, nitrides, borides and so on) allow avoiding the damage of CNTs while contacting with high-temperature metal, as well as solving the problem of specific masses difference of melt (2.7 g/cm^3 for Al and 7 g/cm^3 for Fe- alloys) and CNT (1.2-1.8 g/cm^3).

2.1 Experimental procedure

Carbon nanotubes were synthesized by $LowT_{CO}$ [7] method. By passing CO gas over the oxide catalytic particles, at the temperature range of 400-600°C carbon nanotubes were produced. Catalytic particles were Fe_2O_3, NiO and Co_2O_3. As the result the structure of CNTs has some amount of catalyst metal (Figure 1).

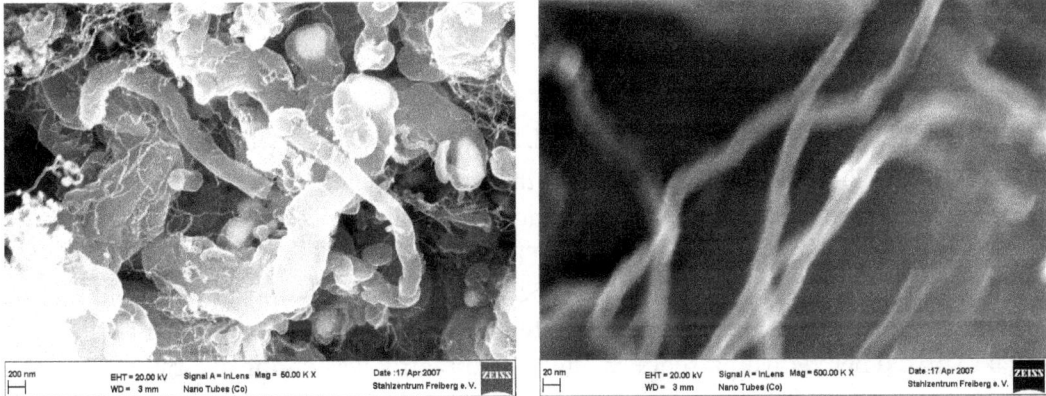

Figure 1 Images of MWCNTs are synthesized by $LowT_{CO}$ method. White spots in the CNTs caps are Ni catalysts' particles.

Synthesized carbon nanomaterials by CO disproportion are mainly in the form of multiwalled carbon nanotubes (MWCNT) and onion shape carbon nanoparticles. Content of impurities was minimum, so these products do not need to be purified before enter the metal matrix composites. Besides, we suppose that small amount of catalyst particles in inner cavity of CNT can improve the contact with metal as well as it has provided the additional increasing of CNT' apparent density. Removing purification step from production cycle of carbon nanotubes can reduce the costs and, as a consequence, the total price of these unique materials.

Synthesized CNTs were coated with metal bearing compounds (chromium oxide, chromium nitrides, and tungsten carbides) by chemical solution method and in some cases this process was completed by heat treatment. The use of chemical solution method as coating technique resulted in producing continuous nanoscale film on the surface of CNTs. The coating materials were chosen as results of thermodynamic calculations of probable reactions equilibrium between the components in systems "carbon-coating-metal matrix". The most stable high-temperature compounds, which are useful for composite matrix properties, were determined for both aluminium and ferrous matrix.

3. Results of CNT use in Al matrix

The thin tungsten and tungsten carbide layer was deposited on CNTs surface for their prevention from interaction with aggressive melt components as well as for total density increasing. Under the CNTs X-ray spectrum analysis, as it was expected, the peaks of tungsten and tungsten carbide (and nickel catalyst) were observed in phase composition of covered CNTs.

The technical aluminium (99.0%) was chosen as initial material. CNTs content (both the uncoated and coated) was 2 vol.%.

The diffraction pattern of sample Al+2 vol.% MWCNT (W coated) indicated that only aluminium α-solid solution is presented in the system. Peaks of carbides or CNTs were not revealed. At the same time, the allocation of small inter-metallic particles (up to 1 μm) is noted in obtained alloy structure. They are located in the grains body (not on grain boundaries). Their formation could be explained as only the result of aluminium interaction with CNTs coating, which indirectly testifies about CNTs' distribution into metal matrix. Their actual presence in aluminium matrix was confirmed by microscopic examination (Figure 2). Main result was achieved in these first experiments is that the hardness of Al-composite with CNTs reinforcement is on 37.5 % higher then for pure aluminium. This fact proves the reinforcing action of CNTs.

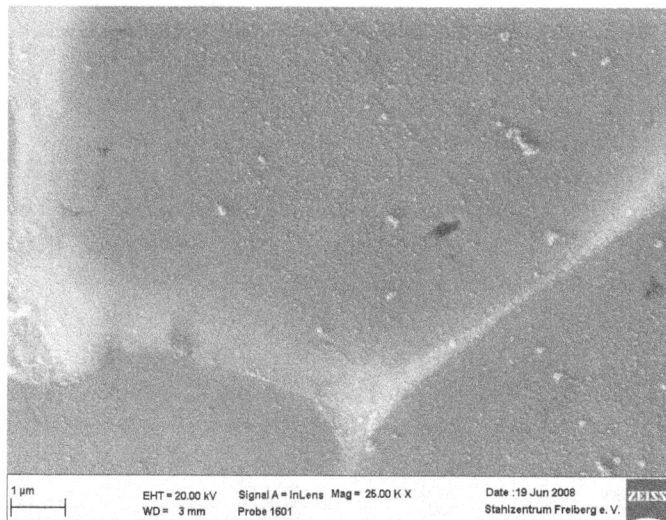

Figure 2 CNTs particles (dark spots) were discovered into Al matrix.

3.1 Results of CNTs addition into Fe-C melts

Carbon nanomaterials were added to ferrous melts gradually during the pouring into the mold. In first experiments the problem of nanotubes floating on the surface of liquid Fe-C alloys was observed because of difference between densities, which is more then 4 times. Besides, ferrous metals wetting behaviour on carbon substrate is very complex. Experimental results suggested that metals presaturated with carbon do not wet graphite while pure metals wet it well [8].

In order to organize enough good penetration of CNTs into metal matrix the high intensive mixing of metal due to high speed pouring into mould was used. Besides, rapid cooling process of melt was organized so that resulted in quick crystallization of melt. Both these techniques prevented the floating of CNT on the surface of melt in the case of high carbon and cast iron usage. In low carbon steel with the highest melting point and quite high temperature of pouring into mould, the main part of added nanotubes was floating on the surface of melt.

Three types of CNTs (uncoated CNTs, chromium nitride coated CNT and chromium oxide coated CNT) were added to melt. Each alloy was cast with and without CNT, so differences in structure and mechanical properties of both produced alloys may be determined and compared. Average content of added CNTs was 0.15 wt.%.

First results of structure investigations of nanotubes effect on high carbon steel as-cast structure have shown that grain size was reduced twice as minimum. This is the evidence of possible nucleation action of CNTs. It is interesting that grain size refining effect was found even in cases when CNTs floated up practically completely (case of low carbon steel IF type). Probably the cooling effect of CNTs displays as a suspension pouring. Detailed investigation of manufactured metal matrix composite microstructure shows that we succeed in adding of CNTs into ferrous matrix (Figure 3).

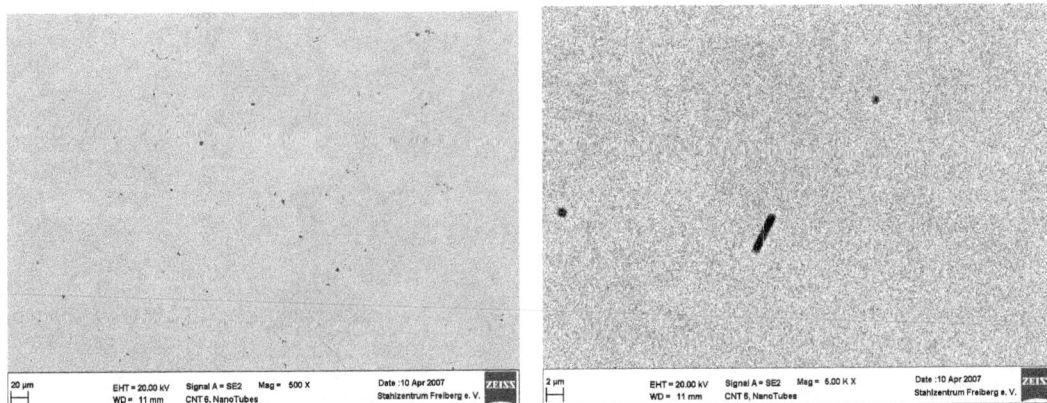

Figure 3 CNTs were found into carbon steel matrix (0.82% C).

4. Conclusions and Prospects

The possibility of adding CNTs into high-temperature melts to produce metal matrix composite was experimentally proved. Used heavy metal coatings on CNTs allow us: to improve "wettability" by high-temperature melts; to solve the problem of specific masses difference of CNTs and metal; to protect CNTs against aggressive components of melts and to avoid carbide formation.

A few positive effects of CNTs particles in aluminium and ferrous matrix as-cast: grain size refining; nucleation ability and reinforcing action.

Literature review shows that CNTs application in traditional commercial materials, like metal matrix composite, is quite limited. But rapid solidification of reduced thickness of metal products gives some interesting possibilities to modify the metal (steel/iron, aluminium, etc.) structure due to use of CNTs as reinforcement of strip or near net shape products. Undoubtedly that small amount of metal with added CNTs will crystallize in a very short temperature interval without phase transformation (with very restricted diffusive movements). The precise control of mechanical properties of MMC will be possible at good adhesion of CNT by metal as well as at the addition of exact amount of CNTs. Such "nano-amorphous" or "nano-crystalline" modified materials will have, most probably, challenging mechanical and/or service properties due to structure advances. Methods of CNTs addition at pouring (twin-rolls caster or rotating drum) in solidified melt are at the stage of development.

Acknowledgements

The authors would like to gratefully acknowledge Professor Piotr R. Scheller, Director of the Institute of Iron and Steel Technology, TU Bergakademie Freiberg for his beneficial co-operation and his department help provided for microstructure investigation.

References

1. Swiss research group, *Tiny tubes boost for metal matrix composites*, Metal Powder Report 2004, 59:40–3. www.metal-powder.net/features/archive/

2. Goh C.S., Wei J., Lee L.C., Gupta M. *Ductility improvement and fatigue studies in Mg-CNT nanocomposites*, Composites Science and Technology, No. 68, 2008, , p. 1432.

3. Shokuhfar T., Titus E., Cabral G., Sousa A.M.C. and Gracio J., *Fatigue and compression testing of carbon nanotube reinforced orthopaedic biomaterial coating*,

http://www.nsti.org/Nanotech2007/showabstract.html?absno=315Chen X. H.,Peng J. C, Li X. Q e.a. J. of Mat. Science Letter. Nos. 20-22, 2001, p. 2057.

4. Chen X. H.,Peng J. C, Li X. Q. et al., *Tribological behavior of carbon nanotubes— reinforced nickel matrix composite coatings*, Journal of material Science Letter, Nos. 20-22, 2001, p. 2057.

5. Quang P., Jeong Y.G., Yoon S.C. et al., *Consolidation of 1 vol.% carbon nanotube reinforced metal matrix nanocomposites via equal channel angular pressing*, Journal of Materials Processing Technology, Vol. 188, 2007, p. 318.

6. George R., Kashyap K.T., Rahul R., Yamdagni S., *Strengthening in carbon nanotube/aluminium (CNT/Al) composites*, Scripta Materialia, Vol. 53, 2005, p. 1159 .

7. Deng C.F., Zhang X.X., Wang D.Z., Ma Y.X., *Calorimetric study of carbon nanotubes and aluminium*, Materials Letters. No. 6, 2007, p. 3221.

8. Carbon Nanotechnologies Awarded US Patent for *Coating for Carbon Nanotubes and Ropes* http://www.azonano.com/details.asp?ArticleID=1049.

9. Prilutskiy O., Katz E. A., Shames A. I. e.a., *Fullerenes, nanotubes, and carbon nanostructures*. Powder metallurgy (Russian), Vol. 13, No. 1, 2005, p. 1.

10. Eustathopoulos N., Nicholas M.G., Drevet B., *Wettability at high temperatures*, Pergamon, An imprint of Elsevier Science, 1999, p. 419.

Modified Epoxy Coatings on Mild Steel: Tribology, Resistance to Scratch and Resistance to Wear

Witold Brostow a, Madhuri Dutta a and Reddy Venumbaka b

a Laboratory of Advanced Polymers & Optimized Materials (LAPOM), Department of Materials Science & Engineering, University of North Texas, 1155 Union Circle, Denton, TX76203-5017, USA; wbrostow@yahoo.com, madhuridutta@gmail.com
b Institute for Environmental and Industrial Science, Texas State University, Centennial Hall, San Marcos, TX 78666, USA; reddy@txstate.edu

Abstract: *We have studied a commercial epoxy modified by adding a fluorinated poly (aryl ether ketone) (12 F-PEK) and inturn metal micro powders (Ni, Al, Ti, and Ag) coated on mild steel. Static and dynamic friction, resistance to scratch, adhesion and morphology have been determined using techniques described in previous papers [1-4]. Two different types of curing agents have been used for the epoxy system; triethylenetetramine as a high temperature curing agent and hexamethylenediamine as a room temperature curing agent. The concentrations of fluoropolymer and metal powder in all the samples are 10 wt. % and 25 wt. % respectively. Variation in properties with varying metal powders and curing agents has been evaluated. The samples cured with the room temperature curing agent provide the lowest friction. While the presence of metal micro powder increases the friction by a small amount, the resistance to scratching also increases at the same time.*

1. Introduction

Modification of epoxies to decrease their friction, increase their scratch and wear resistance has been ongoing for some time. However, good values of all the three properties could not be achieved at one time [1-5]. In this work we have tried to achieve all these properties by using modifiers: fluorinated poly(aryl ether ketone) (12 F-PEK) and metal micro powders of size 1-5 μm (Ni, Al, Ag, Zn) [3, 5].

2. Experimental Work

We have studied a commercial epoxy - diglycidyl ether of bisphenol-A (DGEBA) modified by adding 12 F-PEK and in turn metal micro powders (Ni, Al, Ti, and Ag) coated on mild steel. Static friction, resistance to scratch, and resistance to wear have been determined using pin on disk tribometer, micro scratch tester, and scanning electron microscopy respectively. Two different types of curing agents have been used for this study; triethylenetetramine (TETA) as a low temperature curing agent and hexamethylenediamine as a high temperature curing agent. The samples of epoxy system cured using TETA have been cured at two different temperatures: 30oC for 48 h and 70oC for 3 h [2]. The samples cured using hexamethylenediamine were first cured at 30oC for 5 h followed by curing at 80oC for 24 h [3]. The concentrations of 12 F-PEK and metal powder in all the samples are 10 wt. % and 25 wt. % respectively. All the samples were stored at 30oC after their curing cycles. Variation in properties with varying metal powders and curing agents has been evaluated.

All tests have been performed at least 10 days after complete curing of the samples. Firstly, the samples have been tested for friction using a pin on disk tribometer at 5.0 N load for 500 and 5000 revolutions. We find that most of the samples cured at 30oC with TETA show less friction compared to the ones cured at higher temperatures. In particular, 12 F-PEK + Al powder sample shows the lowest friction of 0.034 in the entire set of samples (Figures 1, 2). TETA used with 12 F-PEK, 12 F-PEK + Ni powder, 12 F-PEK + Al powder samples cured at 30oC show good resistance to wear even after 5000 revolutions (Fig. 3). Also among these three, the sample containing Al powder shows better resistance to scratch for a progressive load scratch from 0 N to 30.0 N (Fig. 4). The scratch speed was 0.33 mm/min and the scan length was 5 mm. The volume loss of the same sample after the pin on disk friction test has been calculated as 0.303 mm3 by studying the wear area using SEM (Fig. 5) [6]. The samples cured using TETA at 30oC provide the lowest friction. While the presence of metal micro powder increases the friction by a small amount, the resistance to scratching and wear also increases at the same time.

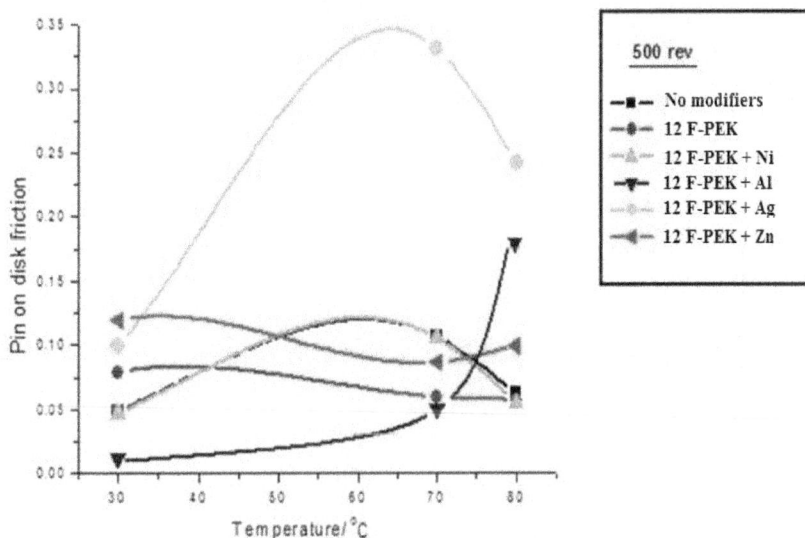

Fig. 1 Pin on disk friction for 500 rev as a function of curing temperature

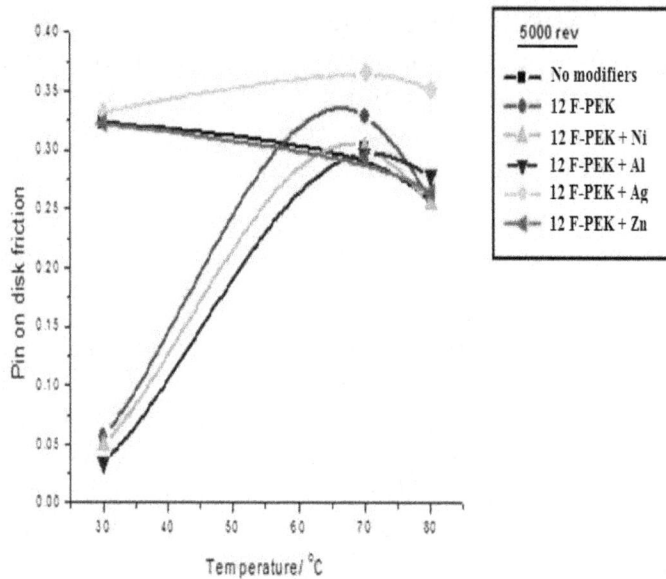

Fig. 2 Pin on disk friction for 5000 rev as a function of curing temperature.

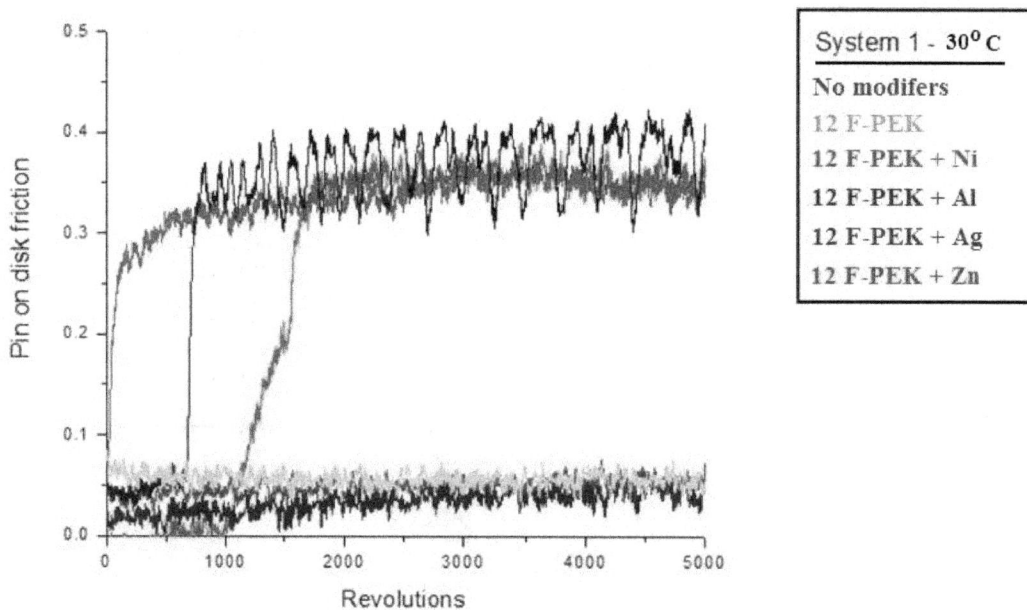

Fig. 3 Pin on disk friction as a function of revolutions for samples cured using TETA at 30oC

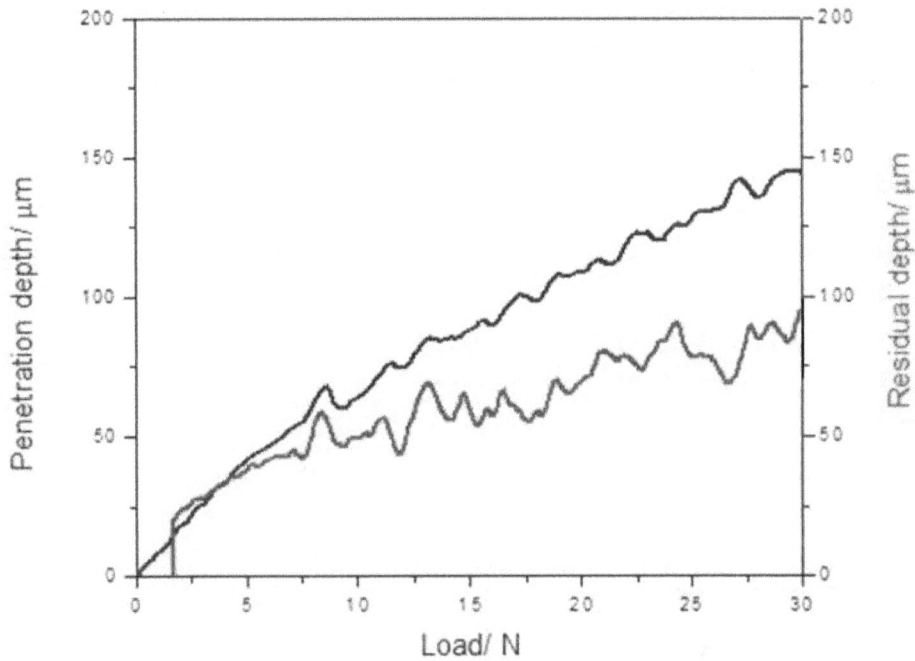

Fig. 4 Penetration depth and residual depth as a function of load during micro scratch test for 12 F-PEK + Al modified epoxy coating using TETA cured at 30oC.

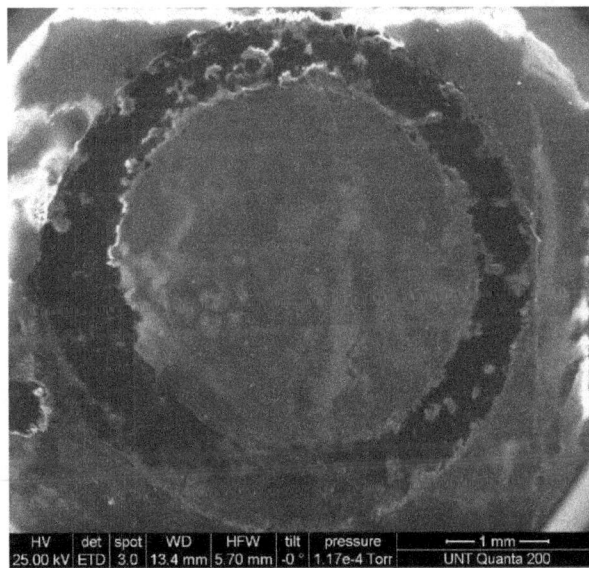

Fig. 5 SEM image of wear after pin on disk friction test at 5000 rev of 12 F-PEK + Al modified epoxy coating using TETA cured at 30oC.

Tribology of polymer-based materials (PBMs) involves a number of challenges [7]. We have defined hardness of materials and related it to recovery in scratch testing [8, 9]. This connection is also helpful in the development of PBMs with better tribological properties.

References

1. W. Brostow, P. E. Cassidy, H. E. Hagg, M. Jaklewicz & P. E. Montemartini, Polymer, Vol. 42 2001, p. 7971.

2. W. Brostow, B. Bujard, P. E. Cassidy, H. E. Hagg Lobland & P. E. Montemartini; Mater. Res. Innovat., Vol. 6, 2002, p. 7.

3. S. A. Kumar, M. Alagar & V. Mohan, JMEPEG Vol. 11, 2002, p. 123.

4. W. Brostow, P. E. Cassidy, J. Macossay, D. Pietkiewicz & S. Venumbaka, Polymer Int., 52, 2003, p. 1498.

5. W. Brostow, A. Buchman, E. Buchman & O. Olea-Mejia, Polymer Eng. & Sci. Vol. 48, 2008,p. 1977.

6. ASTM standard G 99 – 05.

7. W. Brostow, J.-L. Deborde, M. Jaklewicz & P. Olszynski, J. Mater. (Ed.), Vol. 25, 2003, p. 119.

8. W. Brostow, H.E. Hagg Lobland & M. Narkis, J. Mater. Res. Vol. 21, 2006, p. 2422.

9. W. Brostow & H.E. Hagg Lobland, Polymer Eng. & Sci. Vol. 48, 2008, p. 1982.

Aluminium Foam and Other Cellular Metals

Amol A. Gokhale

Defence Metallurgical Research Laboratory, Kachanbagh, Hyderabad

Abstract : *Among cellular metals, aluminium foams are the most commonly produced, providing a unique combination of properties such as: very low density, high energy absorption under static and dynamic compression, blast amelioration, sound absorption and flame resistance. Applications in automotive and defence sectors have been reported. Foams based on high melting point metals such as nickel and its alloys are also under active development throughout the world for applications requiring corrosion and oxidation resistance, relatively high thermal conductivity and high temperature strength. These cellular materials provide new options to designers of aerospace, transport and other defence systems.*

Development of good quality cellular metals requires effective synergy between concepts from thermodynamics (cell stability), kinetics (cell size evolution), solidification (shrinkage/ expansion effects), micromechanics (cell deformation and fracture) and shock theory (impact energy absorption) to name a few. Cellular metals provide unique opportunities to researchers specialising in various branches of science and engineering. Current status and future prospects of cellular metals will be described in the presentation.

1. Introduction

A cellular material or foam is a dispersion of gas in a solid matrix. The presence of a large proportion of gas phase in the foam helps in significant weight reduction, and imparts special mechanical properties vastly different from bulk metal. The use of cellular materials allows the simultaneous optimization of stiffness, strength, energy absorption, overall weight, thermal conductivity, surface area, and gas permeability (the last two applicable for open cell foams). As such, these materials are highly desirable for a wide range of engineering applications, and can be found commonly in the natural world (e.g. wood and bone). Until recently, however, only wood was used to any great extent. Only during the twentieth century have man-made polymer foams been available for insulation, cushioning, padding, and packaging, but the high structural efficiency of cellular materials has been little used. Techniques now exist for fabricating foams not only of polymers, but of metals, ceramics, and glasses as well. These newer foams are increasingly being used structurally, for insulation, and in systems for absorbing the kinetic energy of impacts [1].

Foams are commonly classified into two types: closed-pore foams and open-pore (or reticulated) foams. The closed-pore foams have structures resembling a network of soap bubbles and have higher compressive strength due to their structures. DMRL aluminium foam, e.g., belongs to this category. The open-pore foams are identical to the closed cell ones except the membranes have been removed which produces large channels of interconnected cells, the main advantage of which is its "flow-through" capability. Liquids and gases can flow through the structure with minimal resistance. If the nonsolid phases of the structure are non-random and in the form of closely spaced channels, it is usually called honeycomb or lotus structure. This honeycomb structures have one directional channel with anisotropic mechanical properties and permeability. Based on the average pore sizes (d_{av}), normally porous materials are of three kinds such as micro porous ($d_{av} < 1.6$ nm), meso-porous (1.6 nm $< d_{av} <$ 100-200 nm) and macro porous (($d_{av} >$ 100-200 nm), etc [2].

There is a special class of closed cell foams called syntactic foam. Syntactic foams are composite materials synthesized by filling a metal, polymer or ceramic matrix with hollow particles called microballoons. The presence of hollow particles results in lower density, higher strength, a lower thermal expansion coefficient, and, in some cases, radar or sonar transparency. Tailorability is one of the biggest advantages of these materials. The matrix material can be selected from almost any metal, polymer or ceramic. A wide variety of microballoons are available, including cenospheres, glass microspheres, carbon and polymer microballoons. The compressive properties of syntactic foams primarily depend on the properties of microballoons, whereas the tensile properties depend on the matrix material used in their structure. There are two ways of modulating properties of these materials. The first method is to change the volume fraction of microballoon in the syntactic foam structure. The second method is to use microballoons of difference walls [3].

These materials were developed in early 1960s as buoyancy aid materials for marine applications; the other characteristics led these materials to aerospace and ground transportation vehicle applications. Among the present applications, some of the common examples are buoyancy modules for marine drilling risers, boat hulls, and parts of helicopters and airplanes. New applications are coming up in sports industry, snow skis and Adidas soccer balls, to name a few [3].

2. Distinguishing properties, performances indices

In general, the properties such as plastic strength, elastic modulus, energy absorption, permeability and thermal conductivity depend on percent porosity and microstructure of cell walls. Mechanical properties such as crushing strength and Young's modulus are dependent on total porosity, average cell size and cell interconnectivity. Foams with different pore size distribution, pore orientation and pore morphology can be produced as per the requirements of different applications [4].

The reticulated foam process results in ceramic foams (not covered here) with porosity exceeding 90% and pore size ranging from 10's of micrometer to several millimetres. Foams prepared by this route have poor mechanical properties due to highly open pore structure and hollow struts left behind from burning of polyurethane. The microstructure of reticulated foams is most suited for applications requiring high permeability [5].

Closed cell foams have the lowest thermal conductivity of any conventionally non-vacuum insulation. Several factors limit heat flow such as low volume fraction of solid phase, small cell size which virtually suppresses convection and reduces radiation through repeated absorption and reflection at the cell walls and poor conductivity of the enclosed gas. Heat transfer has direct relation with cell sizes such that heat transfer increases with cell size due to less reflection of radiation as a result of less number of barriers. Cell size more than 10 mm has direct contribution to convection of heat through cell. Open cells affect heat transfer but to a less extent. Coefficient of thermal expansion is also another major factor towards selection of the materials [1,4].

3. *Aluminium and Other Metal Foams*

3.1 Manufacture

Though it is possible to foam almost all metals, manufacturing processes limit cell structure and properties. Among metals, aluminium is the most predominantly foamed metal commercially. DMRL has established an in-house process to make closed cell aluminium foams [6]. Fig. 1 shows typical foams and the cell structure in cross section.

Fig. 1 (left and mid) DMRL foam, (right) cross-section of foam revealing cell structure.

3.2 Mechanical Properties

Closed cell foams can absorb very high amount of compressive energy at a constant peak stress (see Fig.2). So, these are suitable for crash (impact) [7] and blast resistance application [8]. Good stiffness of these closed cell foams make them suitable for light weight structural applications. Cell size variation strongly influences variability of mechanical properties [9].

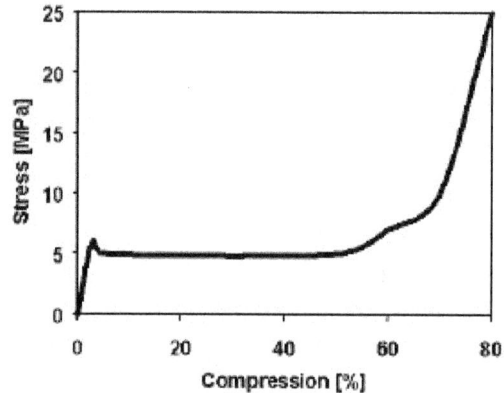

Fig. 2 Stress strain curve for a typical foam [4].

Stiffness of metal foam is a function of the modulus of the parent metal and density:

foam stiffness = constant (Young's Modulus of solid) * $(\rho_{foam}/\rho_{solid})^n$ $2.3 \geq n \geq 1.7$

[1,4] so different metals and alloys will have different stiffness and selection of foam may be based on the requirement.

3.3 Acoustic Properties

Metal Foams act as excellent dampers for vibration and also for sound absorption (see Fig. 3 a,b). The loss factor η for structure borne sound indicates what fraction of the vibratory (reversible) mechanical energy is lost (i.e., converted into heat) in one cycle of the vibration. In Fig. 2 (a) the loss factor of aluminium foam is displayed in dependency of the foam density. With decreasing foam density the loss factor increases. In comparison to the dense material aluminium foams show a significant higher loss factor which indicates the foams better applicability for structure borne sound sensitive structures.

In Fig. 3 (b) sound energy absorption is shown with frequency and the air gap between the sample and solid background. It is observed that foams have exceptionally higher sound damping capability in the mid-frequency range.

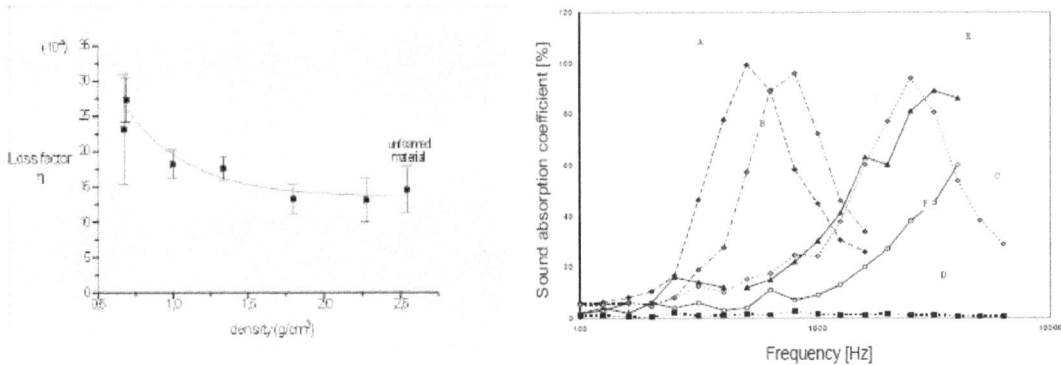

Fig. 3 (a) Loss factor for AlSi 12 foams with different densities, (b) Sound absorption coefficient of aluminium foam at various thickness of the air gap between the sample and the solid background
(A = 40 mm, B = 20 mm, C = 0 mm) compared with bulk aluminium (D), PU-foam (E) and glass fibre mat (F) [10].

3.4 *Thermal and electrical properties*

Metal foams exhibit reduced thermal and electrical conductivity in (somewhat non-linear) proportion to its density [1,11].

3.5 *Electromagnetic shielding*

Electromagnetic wave shielding is used to protect electronic devices and room interiors from the negative influence of electromagnetic waves. The ability to reflect the electromagnetic energy can be defined by the shield effectiveness. Recent experiments carried out on foams have revealed its excellent electromagnetic shielding properties (Fig. 4).

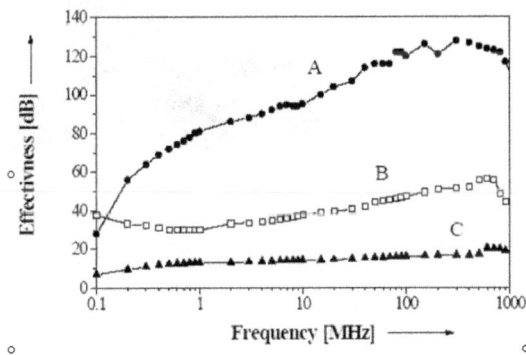

Fig. 4 Magnetic field shielding effectiveness as a function of frequency for aluminium foam (A), Si-steel (B), bulk aluminium (C) at the same weight of the samples [1,10].

139

3.6 Current Applications

Aluminium foams for their high stiffness and energy absorption properties are suitable for automotive industry and structural applications.

3.6.1 Light-weight construction

Aluminium foams have been used in utility transport vehicles to optimise the weight-specific bending stiffness of engineering components. Light-weight construction exploits the quasi-elastic and reversible part of the load-deformation curve (see Fig. 5 as an example).

3.6.2 Energy absorption

Aluminium foams can absorb a large quantity of mechanical energy when they are deformed, while stresses are limited to the compression strength of the material. Foams can therefore act as impact energy absorbers, which limit accelerations in crash situations. This mode exploits the horizontal regime of irreversible deformation in the load-deformation diagram (see Fig. 2). Fig. 6 shows that incorporation of foam in the tube enhances the number of folds during deformation, and the energy absorption substantially increases. For given impact loading conditions, an optimum density of foam may exist that absorbs energy to its fullest capacity, thus reducing impact force to the greatest extent [7].

Fig.5 Base of a lifting arm made from AFS sandwich panels [12].

Figure 6 DMRL aluminium foam filled pipes (below) develop more folds during impact than empty pipes (above), and absorb more energy and reduce impact force (impact tests carried out at BARC Facilities) [13].

3.6.3 Defence and Space

alm Germany, in collaboration with German Aerospace Society
(DLR) and French Aerospace Agency (CNES) built the transition cone of the Ariane Rocket V booster (see Fig. 7). The cone having a base diameter of 3936 mm was produced in 12 segments of aluminium foam sandwich (AFS), and welded. The incorporation of AFS has generated higher compressive strength per unit weight, isotropic behavior and steady crack growth. Another advantage with AFS in this application was its excellent electromagnetic shielding [14].

The radom shield of Airbus with a metal foam hybrid structure is now being tested for a high speed impact bird hit (Fig. 8) [15].

3.7 Future Application

In space sector, studies revealed that at component level, high efficiency heat exchanger system, cryogenic gas storage application and surface tension/ wick application might emerge as potential applications, while at system level, the applications as crushable mechanisms with thermal protection system capability and micrometeorite and orbital debris protection system for heat pipes may emerge soon [16]. Several applications in building, transport and acoustic industries have been described.

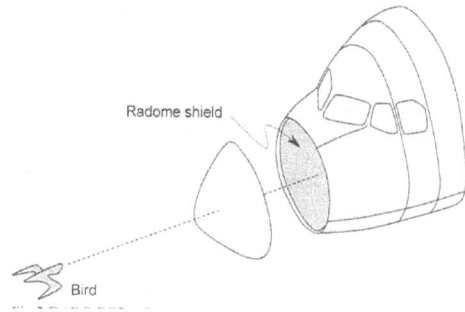

Fig.7 Ariane rocket cone on test facility. Arrow indicates one of the 12 AFS pieces that were joined along the circumference [14].

Fig. 8 Radom shield of Airbus to be replaced with foam [15].

3.8 Processing Methods for aluminium and other foams

Cellular metallic materials can be manufactured in many methods on the basis of their physico-chemical properties. An overview of possible production routes for different metals and alloys taken from literature is given in Table 1. Table 2 lists some important closed cell aluminium foam manufacturers and the methods of manufacturing followed [17].

In literature, three open cell foam production methods have been reported:

(a) impregnation of polymer foam by superalloy slurry and then burning off the polymer [2].

(b) electro-deposition of pure nickel foam on polymer foam and subsequent cell wall thickening by superalloy slurry impregnation method as above. A slight modification of the second process involves chemical vapour deposition of metal instead of electro-deposition, and is called the 'INCO' method [18]. The polymer foam is then burnt out and a sintering process follows to fill out the hollow ligaments left out by the deposition step. The CVD process has been extended to other refractory metals such as niobium, tantalum, tungsten and rhenium.

Table 1 Production processes for cellular metals [2, 17].

Category	Process	Achievable porosity (%)	Metals (examples)	Commercially available?
Liquid state processing	Direct foaming by gas injection	80-97.5	Al, Zn	Yes
	Direct foaming with blowing agents	91-93	Al, Zn	Yes
	Casting	5-75	Ni, Cu, Al, Mg	No
	Powder compact melting	60-90	Al, Zn, Pb	Yes
	Investment casting using polymer foams	80-97	Al, Zn	Yes
	Lattice block materials	-	Al	No
	Casting around space holders	≤65	Al, Zn, Pb, Cu	Yes
	Spray forming	≤60	Steel, Cu	No
Solid state processing	Sintering of powders and fibres	20-80	Bronze, steel	Yes
	Gas entrapment	≤45	Ti	No
	Foaming of slurries	≤93	Al	No
	Powder pressing around space holders	≤70	Ti	No
	Hollow sphere structures	≤80	Steel	No
	Powder/binder techniques		Fe, Cu	No
	Reaction sintering	≤50	TiAl, FeAl	No
Electro-deposition	Electro-deposition	92-95	Ni, Cu	Yes
Vapour deposition	Vapour deposition	91-97.5	Ni, Ni-Cr, Cu	Yes

Table 2 Methods and producers of aluminium foams.

Direct foaming	Melt alloy make alloy foamable create gas bubbles collect foam solidify foam	Indirect foaming	Prepare foamable precursor remelt precursor create foam solidify foam
Manufacturers (products)	Cymat, Canada (SAF) Foamtech, Korea (Lasom) Hütte Kleinreichenbach (HKB), Austria (Metcomb) Shinko-Wire, Japan (Alporas) (Distributor: Gleich, Germany)	Manufacturers (products)	alm, Germany (AFS) Alulight, Austria (alulight) Gleich-IWE, Germany Schunk, Germany

(c) mixing superalloy powder with sodium chloride, compacting and then dissolving sodium chloride (the 'space holder' method). The space holder method has limitations on porosity, due to the lower limit on the packing density of sodium chloride powder [19].

In addition, Shanghai ABC Material Science and Technology, China manufacturers foams of aluminum, copper, nickel, active carbon, Fe-Cr-Al/Ni-Cr/stainless steel which are widely used in sterilization, deodorization, heat exchange, air/water purification as filter, carrier, absorber and radiator. Ultramet USA makes open cell foams of niobium, tantalum, tungsten and rhenium by CVD/CVI technique. By pyrolyzing polymer foam, reticulated vitreous carbon (RVC) foam is created which has interesting mechanical and thermal properties. The company reported deposited material densities of up to 50% of theoretical values.

3.9 Bonding and joining

Adhesive bonding, laser welding, friction stir welding and ultrasonic welding have been reported, each having its own merits and demerits. Recent work showed that sheet metal and aluminium foam sandwich may be joined by ultrasonic welding technology without melting or other damage to the cell structure [20]. Parameters such as energy input, welding pressure and welding amplitude need to be optimized to produce sound joints. Adhesive bonding usually produces lower mechanical strength, but may be considered for its simplicity low capital costs.

4. Summary

Cellular metals (foams) offer a wide combination of properties, depending upon the material, pore volume and nature of cell structure. The range of properties includes high energy absorption, impact force reduction, blast amelioration, thermal insulation, electromagnetic shielding etc. A variety of processes have emerged to make foams. Metal foams are manufactured on commercial scale in various countries.

DMRL has been actively pursuing development of aluminium foams, and has established necessary laboratory scale facilities. The processes can be scaled up depending upon the requirements. Several possible defence applications have been mentioned in the paper, and a concerted effort by component designers and materials developers will result in early value assessment and possible utilization of foams.

Acknowledgements

Authors are grateful to Defence R&D Organisation of financial assistance. Colleagues at the Light Alloy Casting Group of DMRL are acknowledged with thanks.

References

1. L. J. Gibson and M. F. Ashby, Cellular Solids: Structure and Properties, Oxford: Pergamon Press, 1988, U.K., p. 52.

2. J. Banhart, Progress in Materials Science, Vol. 46, 2001, p. 559.

3. R. Erikson, *Syntactic Metals: Flight weight Materials*, Advanced Materials Processes, Vol. 160, No. 12, 2002, p. 44.

4. M. F. Ashby et al, Metal Foams: A Design Guide, Buttrworth- Heinemann, Boston, USA, 2000, p. 40.

5. L. A. Strom, T. B. Sweeting, D. A. Norris, and J. R. Morris, *Novel Application of Fully Sintered Reticulated Ceramics*, Mater. Res. Soc. Symp. Proc., No. 371, 1995, p. 321.

6. Amol A. Gokhale, S. N. Sahu, V.K.W.R. Kulkarni, B. Sudhakar, N. Ramachandra Rao, A Anand Rao and U Ramamurty, Aluminium *Alloy Foams Through Liquid Metallurgy: Processing and Properties*, Metals Materials and Processes, Vol. 19, No. 1-4, 2007, p. 181.

7. R. Rajendran, K. Prem Sai, B. Chandrasekar, A. Gokhale, S. Basu, *Preliminary investigation of aluminium foam as an energy absorber for nuclear transportation cask*, Materials and Design, No. 29, 2008, p. 1732

8. A.G. Hanssen, L. Enstock and M. Langseth, *Close-Range Blast Loading of Aluminium Panels*, Int. J. Impact Eng., Vol. 27, 2002, p.593

9. U. Ramamurty and A. Paul, Acta Materialia, Vol. 52, 2004, p.869

10. Report "Fominal® Properties and Design Guide", version 1.4, Fraunhofer IFAM report, Bremen, Germany, 2006.

11. J Geyda and S Padula II, NASA Glenn Research Centre, Cleveland, Ohio, Report No. NASA/TM- 2001- 211305, 2001

12. J. Banhart, Proceedings of the 9th International Conference on Aluminium Alloys, J.F. Nie, A.J. Morton and B.C. Muddle (Eds.), Published by Institute of Materials Engineering Australasia Ltd, 2004, p. 764.

13. R. Rajendran, K. Prem Sai, B. Chandrasekar, A. Gokhale, S. Basu, *Impact Energy Absorption of Slide Fitted Aluminium Foam in AISI 304 Stainless Steel tubes*, Materials and Design, Accepted for Publication, 2008.

14. H. W. Seeliger, Porous Metals and Metal Foaming Technology (METFOAM2005), H. Nakajima and N. Kanetake, (Eds.), Japan Institute of Metals, December 2005, p. 9-12.

15. G. Rausch and K. Stobener, *ibid*, 2005, p. 1.

16. J. Baumeister, D. Labuln and L. Pambaguian, *ibid*, 2005, p. 5.

17. J. Banhart, JOM, December 2000, p. 22.

18. V. Paserin, J. Shu and S. Marcuson, p. 317.

19. H Choe and DC Dunand, Mater Sci Engg A, Vol 384, 2004, p. 184.

20. C. Born, G. Wagner and D. Eifler, *ref. 14*, 2005, p. 631.

MANUFACTURING PROCESSES AND INDUSTRIAL PRACTICES

Biotechnology for Gold Production and Processing

K.A. NATARAJAN

Department of Materials Engineering
Indian Institute of Science, Bangalore
E-mail: kan@materials.iisc.ernet.in

Abstract *: Microbial processes are responsible for the solubilisation, migration and deposition of gold and thus result in gold mineralization. Gold particles encrusted with microorganisms have been isolated from mine sites. Microbially derived organic acids and amino acids form complexes with gold and microbial enzyme catalysis lead to gold deposition. Gold colloids have been immobilized by bacteria such as Bacillus subtilis. Biogeochemical cycle of gold in natural environments can be utilized for technology developments for gold metal production, processing and environmental protection. Indigenous microbiota present in soils can solubilise, precipitate and structurally modify gold particles. Such geomicrobial cycling of gold can be mimicked in laboratories to develop newer processing techniques for production of gold and processing for a variety of applications in modern materials, medicine and instrumentation. Biotechnology using Acidithiobacillus group of bacteria has now been commercialized to treat refractory gold-silver bearing concentrates in bioreactors to enhance the production of precious metals. Biosolubilisation of gold can ultimately replace currently used toxic cyanides and pave the way for environmentally-benign, cost-effective and energy-efficient gold-silver extraction. Similarly bioaccumulation of gold offers promise to treat gold-bearing effluents and wastes and synthesis of nano particles. Cyanide-free biocatalysed extraction of gold and silver using aerobic and anaerobic bacteria have become a distinct possibility. Biotechnology thus plays a definitive role in all facets of gold-silver production and processing. In this paper biotechnology for gold production and processing is analyzed. Utility of biotechnology is augmenting gold-silver production in India is also examined with typical examples.*

1. Introduction

Various types of microorganisms are indigenously associated with different types of gold ore deposits. Bacteria and archaea are involved in the biogeochemical cycles of gold which consists of not only gold mineralization, but also its dissolution, precipitation and mobilization. Microorganisms bring about the following beneficial consequences with respect to gold processing.

a) Biodissolution of gold-encapsulated pyrite and arsenopyrite leading to liberation of metallic gold (*Acidithiobacillus ferrooxidans, Acidithiobacillus thiooxidans,Leptospirillum ferrooxidans and thermophiles*)

b) Cyanide-free reductive dissolution of gold (Anaerobes such as sulfate reducing bacteria)

c) Chelation of gold (Amino acids produced by *Bacillus subtilis*)

d) Reductive precipitation of gold nanoparticles (*Bacillus subtilis, Shewanella, Rhodococcus and Chlorella vulgaris*)

e) Bioaccumulation of gold (Fungi, Yeasts and several bacterial biomass)

f) Detoxification and degradation of cyanides (*Pseudomonas spp.*)

Through a proper understanding of the biogeochemical gold cycles as occurring under natural environmental conditions, many processes relevant to gold processing can be 'mimicked' which would pave the way for the subsequent development of plant operations.

Various microbial processes relevant to gold extraction, processing and waste management are illustrated with examples in this paper.

2. Biogenesis of Gold

Microorganisms play a significant role in the geochemical cycling of metals such as gold. Through gold and its complexes are generally toxic to microbial life, some bacteria, archaea, fungi and yeasts are capable of solubilising and precipitating gold under natural conditions. 'Bacteriform' structures of secondary gold grains have been taken as evidence for microbial gold precipitation and biomineralization in the environment. Biogeochemical cycling of gold can be linked to biological processes of concentration in the environment. Both abiotic and biological mechanisms come into play in natural gold-bearing ore mineralization. Due to its poor solubility in aqueous solutions, various biologically-derived complexing agents are needed to solubilise gold. Weathering of base metal sulfides due to geochemical and biochemical forces lead to chemical mobility of gold which can complex with thiosulfate, bisulfides and chlorides. Biooxidation of sulfide minerals associated with gold occurrences generate the above solubilising reagents. Organic gold complexes are important in soils. Organic acids such as humic acids, amino and carboxylic acids are generated with many soil microorganisms, subsequently leading to gold solubilisation and precipitation under natural conditions [1-2].

Similarly, sorption of gold complexes and colloids to organic matter and clays need be considered. Bioaccumulation and biomineralization may lead to the formation of secondary gold particles. Bacteria and archea can be observed attached to ore particles under aerobic and anaerobic conditions under the earth's crust. Anaerobes such as sulfate reducing bacteria (SRB) and thermophiles can be associated with natural gold mineralization. Pyrite and arsenopyrite mineralization encapsulating gold particles can be biogeochemically formed.

Microbially mediated gold solubilisation is based on the ability of microorganisms to promote its oxidation and to generate complexing reagents. Chemolithotrophic iron-sulfur-oxidising bacteria such as *Acidithiobacillus ferrooxidans* and *Acidithiobacillus thiooxidans* can oxidize gold containing pyrite and arsenopyrite. Biooxidation of sulfide minerals leads to liberation of associated gold into the environment. Similarly, biologically generated thiosulfate can complex with gold leading to its solubilisation. Thiosulfate and bisulfides can also be produced by anaerobes such a sulfate reducing bacteria which participate in gold mobilization under natural conditions.

Heterotrophic bacateria such as *Bacillus subtilis*, *B.megaterium*, *Pseudomonas* and *B.mesentericus* are also implicated in gold solubilisation. A close correlation between gold solubilisation and microbial production of amino acids in soils under natural conditions can be drawn. Oxidation and complexation of gold with microbially-generated cyanides are also possible. Cyanogenic microbes are present in nature producing cyanide compounds.

3. Microbial Production of Gold Nanoparticles

Many kinds of bacteria such as anaerobes, aerobes, photosynthetic and cyano-bacteria are capable of producing hydrogen. Biohydrogen technologies can be used to produce gold nano particles under controlled conditions. Nano particles of gold formed by microbial weathering exhibits controlled growth with respect to shape and size. Microbial synthesis of gold nano particles can be achieved under neutral pH conditions using mesophilic bacteria or algae from gold chloride solutions. Suspension of dried cells of an alga, *chlorella vulgaris* in $HAuCl_4$ solution accumulates elemental gold in the cells. Different microorganisms useful in the synthesis of gold nano particles include *Bacillus subtilis*, *Shewanella algae*, *Rhodococcus*, *Chlorella vulgaris* and fungi such as *Verticillium* [3-4]. It is possible to harvest the metal nano particles formed with in the fungal or bacterial biomass.

4. Microorganisms in Gold Processing

Some important microorganisms relevant to gold processing are illustrated in Figure 1. Among them *Acidithiobacillus ferrooxidans* and *Leptospirillum ferrooxidans* are used in enhanced gold recovery from refractory pyrite-arsenopyrite concentrates containing gold. Sulphate Reducing Bacteria (SRB) can be usesd to solublise gold through bisulfide generation. *Bacillus subtilis* is known to directly solubilise gold. Fungi and yeasts are used as good biosorbents for gold from effluents and leach liquors. *Pseudomonas spp.* can biodegrade cyanides while other groups of bacteria are used in the production of gold nano particles and colloids.

Thermophiles

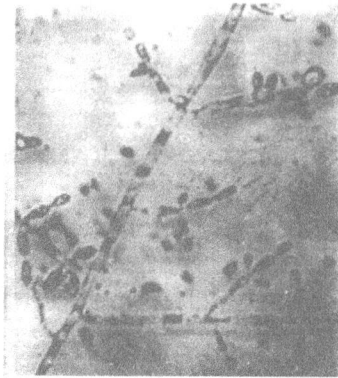

Fungi

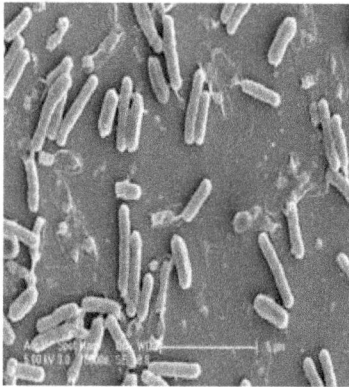

Bacillussubtilis

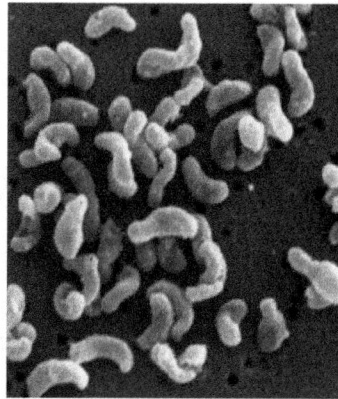

Leptospirillum ferrooxidans

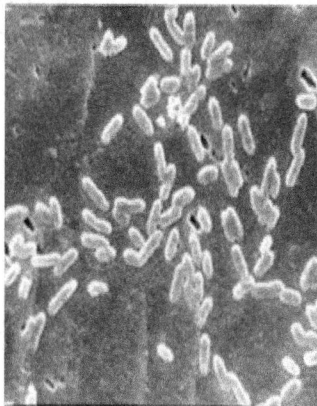

Acidithiobacillus ferrooxidans

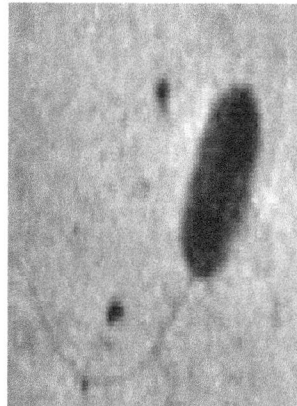

Sulphate Reducing Bacteria

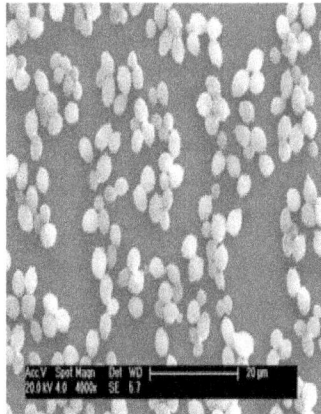
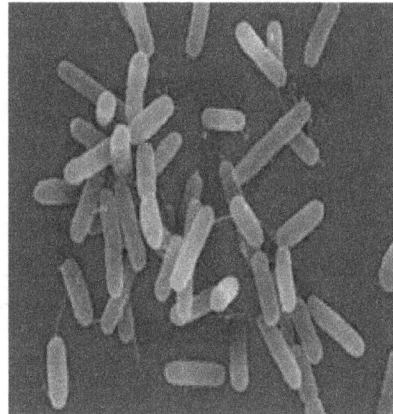

Yeast   *Pseudomonas*

Figure 1 Microrganisms relevant to gold processing.

5. Biooxidation of Refractory Sulphidic Gold Concentrates

Biotechnology could be effectively applied to recover gold from the following unconventional and waste resources [5-7]:

a) Tailing dumps accumulated at the mine sites over a period of several years amounting to several millions of tonnes, which in few locations contain as high as 1 gram / tonne of gold

b) Lean grade free milling ores containing less than 1-2 gram/tonne of gold, which cannot be economically processed through the current practice.

c) Refractory ores-sulphidic or carbonaceous, where finely disseminated gold particles are locked up, making them unsuitable for direct cyanidation

Gold occurs in nature essentially in its native state and can be considered to exist in three different types of ores, namely, free milling, base metal and refractory. Free-milling ores are those from which the precious metal can be efficiently liberated from the host rock through appropriate size reduction. Most of the current gold-processing plants around the world utilize such deposits to recover the metal through the century-old cyanidation process. Base metal ores containing copper, lead, zinc and iron as their sulphides often contain small quantities of gold which is recovered as a by-product metal. The third type, namely, refractory ores, is becoming increasingly important as a potential primary source of gold. Among the refractory gold ore occurrences, two types are relevant, namely, sulphidic and carbonaceous. In the former case, the precious metal is finely disseminated within sulphide minerals such as pyrite and arsenopyrite and as such the encapsulated gold particles cannot be satisfactorily liberated even after fine grinding making direct cyanidation very inefficient for metal recovery. In the case of refractory gold bearing carbonaceous ores also, gold particles remain locked-up, making recovery by gravity or cyanidation methods unacceptably low. Another problem with carbonaceous materials is 'preg-robbing' which means even if liberated gold particles are cyanided, they get absorbed onto carbon matter making recovery extremely difficult.

The use of mixed strains of bacteria containing *Acidithiobacillus ferrooxidans*, *Acidithiobacillus thiooxidans* and *Leptospirillum ferrooxidans* could prove to be more effective in sulphide mineral oxidation due to synergistic mechanisms in operation. The use of thermophilies such as *Sulfolobus* which have optimum activity at about 60-80°C would ensure enhanced leaching kinetics.

It should be understood that bioprocessing of sulphidic gold-bearing minerals is essentially a pretreatment process aimed at liberation of locked-up gold particles. The acid bioleached residues need be alkali treated before subsequent cyanidation. In the absence of prior biotreatment, even finer grinding of such refractory ores would result in very poor gold recoveries, even as little as 6-10% and at any rate never exceeds 30-40% even under ideal processing conditions. The improvement in precious metal recovery conferred by prior biooxidation would be significant, often exceeding 95%.

There exists a direct relationship between the degree of sulphide mineral oxidation and percent gold recovery. Complete oxidation of sulphides is not necessary to achieve significant enhancement of gold recovery. On the basis of the sulphide entity, high gold recoveries can be obtained with as low as 50% oxidation of the total sulphides. The entire biotreatment process is agitation leaching, carried out in stirred bioreactors or Pachuca type reactors. The extent of mineral oxidation depends on the type of sulphide involved, whether pyrite or arsenopyrite.

Though gold bearing sulphides can be directly bioleached, it is often preferable to treat a flotation concentrate of the ore since it enables easy handling of a reduced tonnage of enriched material. The following variables need be closely controlled to achieve optimum efficiency.

Choice of appropriate preadapted bacterial culture (adapted to sulfide concentrates and arsenic)

Particle size – fine size favours faster oxidation

Pulp density – 20-30% optimum

Oxygen and carbon dioxide availability

Agitation and homogeneous mixing of pulp

Temperature $30\,^{\circ}C$ to $45\,^{\circ}C$ (use of moderately thermophilic strains)

pH 1 to 2

Residence period 2 – 7 days

Eh 700-900 mV

Availability of nutrients

Removal of toxic and interfering oxidation products

Large-scale applications of biotechnology for refractory gold ores and concentrates have been coming up during the last 20 years.

In Table 1, a list of major commercial gold bioreactor plants in the world are given

Table 1 Some Commercial Gold Bio-oxidation Plants

Place and locations	Size (tonnes concentrate/day)
Fair view, South Africa	40
Sao Becco, Brazil	150
Harbour Lights, Australia	40
Wiluna, Australia	158
Ashanti, Ghana	960
Younami, Australia	120
Tamboraque, Peru	60
Beaconfield, Australia	70
Laizhou, China	100

6. Bioreactor Engineering for Refractory Sulphidic Gold Concentrates: An Indian Experience

India is the largest consumer of gold in the world while in terms of production of the precious metal, India produces only about 3.5 tons of the metal per year. The Hutti Gold Mines Company Limited (HGML) is the only primary gold producer in the country today. Most of the ore is mined by underground mining and at present gold is extracted from free milling ores by cyanidation using carbon-in-pulp method. The typical grade of the ore is about 4-5 gm/ton.

In recent years, HGML has undertaken major expansion plans and has identified several deposits for extracting gold in the coming years. G.R.Halli and Anjanahalli deposits are sulphidic ore deposits containing about 3-4 gm of gold per ton of ore. As expected, extraction of gold from such ore or its flotation concentrate by conventional cyanidation has not been encouraging. The concentrate needs to be oxidised before it is cyanided to enhance gold recovery. It is in this regard that HGML and Indian Institute of Science undertook a joint project to develop a bioreactor technology for biooxidation of the refractory G.R.Halli concentrate with financial support from the Department of Biotechnology, Government of India during 1997–2002 [8-9].

Biooxidation tests were run in a continuous mode, the reactor was first run in batch mode to bring the pulp density upto 10%. So initially the experiment was started with 1% pulp density and the iron leached was measured. When the iron leached out approximately reached the theoretical amount in the concentrate, then some more concentrate was added and this step leaching was continued until the pulp density reached 10%. Further increase in pulp density did not increase the leaching rate. The high iron and sulphur content in the concentrate induces a lag period thus slowing the leaching rate. This method of step leaching is more effective than starting with 10% pulp density at the beginning itself. Most of the iron leached was in the form of ferric while very little ferrous was present. The leaching rate of iron was consistent with that reported in literature for pyrite and arsenopyrite bearing gold ores. Then the reactor was operated in continuous mode with a feed rate of 1.5L /day and 4 day residence time. The slurry was pumped using a peristaltic pump, which was operated for 10 minutes every hour. On an average about 85 % of gold and 95 % of silver could be extracted from biooxidised concentrate. The bioreactor was operated for about 15 days in this mode and it was observed that consistently good recoveries could be achieved over the period of study. From these studies it can be definitely concluded that biooxidation as a pretreatment can be used effectively for treating G. R. Halli sulphidic concentrate at the HGML.

After intensive laboratory testing for nearly a year, it was decided to test the technology on-site at HGML. For this purpose, a demonstration plant was designed with a$_3$

capacity to process 100Kg of concentrate per day. Three bioreactors with a total capacity of 6 m was used. The concentrate was fed from the feed tank to the first two bioreactors, which were in parallel through peristaltic pumps. Then the outputs from reactors 1 and 2 was fed to reactor 3 from where the treated slurry was sent to a settling tank. The solid was drawn from the settling tank for cyanidation. The reactors were provided with water jacket for controlling the temperature. The start-up strategy was to get all the three bioreactors running in batch mode before the continuous operation started. The bacteria were cultured in the bioreactors itself in the presence of the concentrate by the step leaching technique with all the necessary nutrients for bacterial growth without ferrous sulphate. The feed was acid stabilized in the feed tank before it was fed to the bioreactors. The important process variables such as Eh, pH, temperature, pulp density, iron concentration, bacterial concentration were measured at regular intervals.

Fig. 2 A view of Demonstration Bioreactor at Hutti Gold Mines.

Two types of precious metal concentrates (pyrite-arsenopyrite) obtained from G.R. Halli and Ajjanahalli mines were tested. After biooxidation in the presence of *Acidithiobacillus ferrooxidans* more than 40% enhancement in gold and silver recovery could be obtained. A view of the demonstration Bioreactor assembly set up at Hutti gold mines is illustrated in Figure 2.

7. Biosorption of Gold and Biodegradation of Cyanides

Mining and mineral processing operation generate effluents containing varying concentrations of toxic metal contaminants. One of the most recent developments in environmental biotechnology is the use of microbe-based sorbents for the removal and recovery of precious metals from industrial waste effluents. Such biotreatment processes are of interest for the detoxification of industrial wastes. Moreover, the subsequent recovery of accumulated metals is possible after suitable treatment of the metal-loaded biomass [10].

Leaching in a cyanide solution followed by cementation of the dissolved gold with zinc or adsorption in carbon is being used by a number of gold-processing plants around the world. The cyanide effluents discharged from the plants contain small amounts of unrecovered gold and varying concentrations of base metals such as zinc, copper, iron and lead. These base metals enter the process stream through many unit operations. The recovery of gold from such effluents is of great economic significance, and the removal of heavy metals is environmentally mandated.

The removal of heavy metals would also enable efficient recycling of the used cyanide solutions within the process circuit. The use of microorganisms for the recovery of gold and for the removal of heavy-metal ions from gold-plant cyanide effluents is of great commercial and environmental significance.

Microorganisms are known to be active participants in the formation, conversion and accumulation of gold in mineral deposits. They are known to be capable of dissolving gold, reducing it to an elemental state and accumulating it within their biomass. The activity of *Aspergillus sp.* biomass often exceeds that of activated carbon by a factor of 10 to 12 in the recovery of gold from colloidal solutions. Microscopic fungi are capable of absorbing gold from aqueous solutions with the formation complex compounds. It has been earlier shown that the gold content of bacterial cells associated with gold ore deposits are 15 to 600 times higher than that of the mine waters. An almost complete removal of gold from synthetic solutions by algae could be achieved. Fungal biomass as well waste industrial biomass are very good gold biosorbants from cyanide solutions.

Destruction of cyanide by microorganisms through biodegradation from tailings solutions and other process effluents is a proven alternative to conventional chemical and physical processes. Degradation of cyanides by microorganisms in mill effluents is a natural process that can be readily exploited and engineered to commercial scales of operation [11].

Cyanide destruction by microorganisms was studied in the early twentieth century and was first commercially demonstrated in the gold mining industry at the Homestake

Gold Mine, USA in the middle 1980's. Since then, several other investigations of the microbial destruction of cyanide have been reported including laboratory, pilot and full-scale facilities.

Various industrial applications have been developed using both aerobic and anaerobic processes in full-scale active, passive, and *in-situ* treatment facilities. These aerobic and anaerobic biotechnological applications include active treatment of cyanide containing solutions from tailings ponds and barren solutions to passive treatment of gold heap leach pad drain using *in-situ* anaerobic systems. At the Homestake Gold Mine in the USA full-scale facility has been in continuous operation treating high volumes of tailings pond solution for nearly three decades. The aerobic attached growth fix film biological facility consists of five stages of forty-eight rotating biological contactors (RBCs) for the removal of thiocyanate, cyanide, ammonia, and metals.

8. Cyanide-Free Biodissolution of Gold

The development of a novel bioprocess for gold solubilization is based on the following anaerobic reaction, namely,

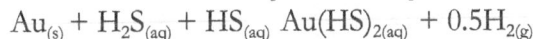

$$Au_{(s)} + H_2S_{(aq)} + HS^-_{(aq)} \; Au(HS)^-_{2(aq)} + 0.5H_{2(g)}$$

The use of naturally-occurring sulfate-reducing bacteria (SRB) to biocatalyze the following hydrogen gas-consuming/bisulfide-producing reaction is of great commercial significance in this regard.

$$SO_4^{-2} + 4H_2 + H^+ \; HS^- + 4H_2O$$

In the above reaction, the microorganisms can also neutralize the solution by consuming hydrogen ions. It would then become possible to develop a bioprocess for gold extraction which is both cost-effective and environment-friendly. The process will be novel and lead to technological innovations in precious-metals extraction leading to exploitation of ore deposits that are not amenable to conventional leaching processes. It would also eliminate the need to preserve waste tailings containing residual cyanide [12].

In the new bioprocess sulfidic ore is first aerobically oxidized to mobilize gold prior to bisulfide leaching through the use of mesophilic or thermophilic chemolithotrophs. In bio-oxidation, aerobic, acidophilic bacteria, such as *Acidithiobacillus ferrooxidans, Leptospirilllum ferrooxidans* and *Sulfolobus*, can be used to oxidize iron and sulfur minerals in which precious-metals are encapsulated.

The second step involves anaerobic bacteria. Bisulfide ions can be generated biologically at very low cost using an acidic sulfate waste product.

Microbial sulfate reduction is achieved by indigenous sulfate-reducing bacteria (SRB). In dissimilatory sulfate reduction, sulfate is used as the electron acceptor for the oxidation of an electron donor, such as an organic compound or hydrogen and the end product is hydrogen sulfide, which is excreted by the microorganisms. *Desulfovibrio*, *Desulfomonas* and *Desulfotomaculum* genera of SRB can be used.

A soil bacterium, *Bacillus subtilis* is capable of dissolving gold through the generation of amino acids that can complex with the precious metal.

Summary

Microbiology and biotechnology offer exciting opportunities in gold processing. Native microorganisms are involved in the biogenesis of gold, its dissolution, precipitation and mobilization. Many natural biogeochemical processes can be efficiently mimicked in a laboratory for the development of environmentally benign biotechnological processes. Recently developed bioprocesses relevant to gold processing include, a) biooxidation of refractory gold-bearing sulfidic concentrates in bioreactors, b) bioheap leaching for lean gold ores, c) biological degradation of gold-process cyanides, d) recovery of gold from effluents through bioaccumulation and e) microbial production of gold nanaparticles.

References

1. F. Reith, M.F. Lengke, D. Falconer, D. Craw and G. Southam, *The geomicrobiology of gold*, The ISME Journal, Vol. 1, No.18, 2007.

2. F. Reith, S.L. Rogers, K.C. McPhail and J. Brugger, *Potential for the utilization of microorganisms in gold processing*, World Gold Conference, October 2007.

3. K. Kashefi, J.M. Tor, K.P. Nevin and D.R. Loveley, *Reductive precipitation of gold by dissimilatory Fe(III) – reducing bacteria and archaea*, App.Environ.Microbiology, Vol. 67, 2001, p.3275-3279.

4. M. Gericke and A. Pinches, *Microbial production of gold nanoparticles*, Gold Bulletin, Vol. 39, No.1 2006.

5. K.A. Natarajan, *Bioprocessing for enhanced gold recovery*, Mineral Processing and Extractive Metallurgy Review, Vol. 8, 1992, p 143.

6. K.A. Natarajan, *Biotechnology in gold processing*, Bull. Mat. Sci., Vol. 16, 1993, p. 501.

7.   Natarajan K.A. Microbes, *Minerals and Environment*, Geological Survey of India (Bangalore) 1998.

8.   K.A.Natarajan, J.M.Modak and A.M.Raichur, *Bioreactor engineering for treating refractory gold-bearing concentrates: An Indian Experience*, V.S.T.Ciminelli and O.Garcia Jr(Eds), BIOHYDROMETALLURGY: Fundamentals, Technology and Sustainable Development, Elsevier, Amsterdam, 2001, p. 183.

9.   M. N. Chandraprabha, Jayant M Modak, K. A. Natarajan, Ashok M Raichur, *Strategies For Efficient Start-Up of Continuous Biooxidation Process For Refractory Gold Ores*, Minerals Engineering, Vol. 15, 2002, p. 751.

10. M. Balakrishnan, J.M. Modak, K.A.Natarajan and J.S. Gururaj Naik, *Biological uptake of precious and base metals from gold-process cyanide effluents*, Minerals & Met. Processing, Vol. 11, No. 4, 1994, p. 197.

11. Ata Akcil and T. Mudder, *Microbial destruction of cyanide wastes in gold mining*, Biotechnology letters, Vol. 25, 2003, p. 445.

12. R.M. Hunter, F.M. Stewart, T. Darsow, M.L. Fogelsong, D.W. Mogk, E.H. Abbott and C.A. Young, *New alternative to cyanidation, Biocatalysed bisulfide leaching*, Min. Process. Ext. Met. Rev. 19, 1998, p. 183.

ESR: Yesterday. Today. Tomorrow

L.B. MEDOVAR, A.K. TSYKULENKO, V.Ya. SAENKO, B.B. FEDOROVSKII

E.O. Paton Electric Welding Institute, 11 Bozhenko, Kyiv, Ukraine.

Abstract :_Abbreviation ESR as ElectroSlag Remelting or ElectroSlag Refining known for metallurgist world wide already 50 years: year 1958 – is the year of modern ESR commercialization. Since that time ESR became to be world wide industrialized as a method of superior ingots of alloyed steel and superalloy manufacturing. It is an attempt to analyze not just step by step development of ESR technology and equipment, but to highlight main ideas on the basis of which various ESR technologies where developed. From that point of view main targets of further ESR development to be discussed on the focus of following topics: ESR of Sophisticated Alloys including Titanium; Electroslag Processes in Current Conductive Mould, ESS LM – electroslag surfacing by liquid metal and ESR with two power circuit; ESR of huge forging ingots, hollow and solid, Arc Slag remelting.To be discussed new areas of possible ESR application as well as trends for standard technology and equipment perfection.*

1. State -Of-The-Art and Problems of Electroslag Technologies

In the up-to-date industrial production the technologies, based on an electroslag process, have found the widest application. The electroslag technologies, united by a single common name, are well-known for a long time including such technological processes as electroslag welding and surfacing, remelting and casting, hot topping and refining of metals.

The process of electroslag remelting (ESR) has been recognized by the metallurgists for more than 50 years as one on the basic methods of production of the high-quality ingots from steels and alloys. At the beginning of the fifties of the last century a new metallurgical method was suggested at the E.O. Paton Electric Welding Institute of NAS of Ukraine. The principle of this method was based on remelting of consumable electrodes by heat, generated due to electrical resistance in a slag pool during passing a.c. current through it [1-5]. One cannot but mention the Kellog-process which was patented by R.K. Hopkins as long ago as the 1930s and preceded the ESR [6]. In private talk with B.I. Medovar Prof. R.K. Hopkins witnessed himself that he was two steps short from an existing ESR, i.e. from understanding the difference of the slag process from and arc burning under the slag layer and application of alternating current.

The industrial application of the existing now ESR commenced since 1958 when at "Dneprospetsstal" Works in Zaporozhye and at "Novokramotorsk Machine-Building Works" in Kramatorsk the first industrial AC electroslag remelting furnaces were put into service. The unique flexibility and versatility of the technological ESR process made it possible to develop the technology of production of ingots of the widest assortment. Technological processes of production of ingots of a square section, slab and hollow ingots were created. The ESR found the widest application in industry for the production of the high-quality ingots from structural and tool steels, medium-alloyed and high-alloyed stainless steels, alloys on the base of iron and nickel, copper and its alloys, etc. [7-12]. The electroslag technologies were progressing continuously, accompanied by their constant improvement and appearance of new technological diagrams. Thus, an electroslag casting (ESC) appeared on the basis of experience of using electroslag remelting of production of ingots, for producing shaped castings being maximum close in sizes and shape to the ready components [13]. Electroslag technologies of melting in shielding atmosphere and under the pressure (ESUP), arc-slag remelting (ASR) and a number of other technologies were developed and found their place at the market [14-18]. At the junction with foundry the technologies based on an electroslag crucible melting (ESCM), i.e. centrifugal (CESC) and permanent mould (EPM) electroslag casting were developed [19-20].

At present hundreds and hundreds of thousands of tons of the high-quality electroslag metal have been produced in the world. Hundreds of industrial units for realizing different diagrams of the electroslag process are functioning and more upgraded ESR furnaces are developed and put into service [15, 16, 21, 22].

At the same time it should be noted that in spite of existing of a wide range of techniques and methods of melting and solidification of metal, united in a classical concept "electroslag technology", all they are characterized by several common features*, namely:

♦ presence of a consumable melting electrode;
♦ strict relation between the temperature parameters of the electroslag process and its efficiency;
♦ central type of heat supply to the billet solidified and periphery heat removal from it.

These peculiarities of a classical electroslag process are manifested in technologies based on it and they are not negative until the ingots have a comparatively small size, and the materials are sufficiently simple and low-sensitive to the processes of liquation, inheritance of structure, etc.

163

Under the present conditions, when new more complicated problems are put forward before the quality metallurgy, stipulated by changing in both the chemical composition of the materials used and also in the assortment and geometric

*Here, the electroslag technologies with non-consumable electrodes are not discussed sizes of the billets used, the above-mentioned peculiarities of the electroslag technologies begin, to a certain extent, to limit their capabilities and cause a number of problems which cannot be solved using the classic diagrams of the ESR.

Moreover, since the moment of appearance of this process in industry the problem of a high cost of consumable electrodes was very acute (not depending on the method of their production: rolling, forging, extrusion, continuous casting or mould casting). Sometimes, the cost of the consumable electrodes was 60 % of the ESR metal cost. In addition, the remelting of a solid consumable electrode is characterized by a high energy consumption that influence negatively the economic characteristics of the ESR.

From the very beginning of development of the electroslag remelting, many attempts were made to changing the above peculiarities of the classical diagram of the ESR and widening its capabilities.

Numerous works are known on introduction of different dispersed macrocoolers in the form of shot, cuts, wire, etc. to a slag pool during the ESR to decrease the strict dependence of the ESR temperature parameters.

The attempts directed to refusal of consumable electrodes were the works on ElectroSlag Remelting of metallized Pellets (ESRP) [25-29], Portion ElectroSlag Casting (PESC) of large ingots [29, 30]. One more method in the range of the ESR technological processes directed to the refusal of melting of a consumable electrode was a method of electroslag remelting of lumpy filler materials (shot, chip, etc.) in a current-carrying mould , suggested at the Paton Institute in the 1970s [31-33].

However, for any of several reasons, both of an objective and subjective nature it was not managed to reach an integrated solution of the problem of widening the capabilities of a classical ESR diagram.

The present paper describes the results of works which realized the ideas of academician B.I. Medovar and were fulfilled at his direct scientific supervision of company "Elmet-Roll - Medovar Group" in collaboration with the E.O. Paton Institute and many other Ukrainian and foreign enterprises and companies. These works were directed, first of all, to the widening of potentiality of technologies based on the electroslag process.

New technological diagrams of the electroslag process in a current-carrying mould using a liquid metal instead of a solid consumable electrode were suggested and named **ElectroSlag Technologies with a Liquid Metal (EST LM)** (Figure 1). The new technologies [34, 35] eliminated to a some extent some economical problems of a classic ESR connected with a production and preparation of the consumable electrodes and also, that is more important, could break the strict relation between the ESR process efficiency and its temperature parameters.

Taking into account the fact that the realization of the EST LM diagrams requires not always the simple means for the maintenance of the molten metal in a liquid state and its dosing during a long time, and also the moment that diagrams of EST LM cannot be always used due to absence of production capacities of the enterprise (for example, due to absence of means of preparation of a liquid metal in the majority of the ESR shops) one more method was suggested and named **ElectroSlag Remelting of the consumable electrode in a current-carrying mould according to Two-Circuit diagram (ESR TC)** (Figure 1). During ESR TC the supply is performed by two circuits, "consumable electrode - bottom plate" and "mould - bottom plate". In this diagram, as in EST LM the current-carrying mould plays a role of a non-consumable electrode and, here, the feasibility is provided to break the strict dependence between the temperature parameters and the ESR process efficiency and to widen significantly the capabilities of control of heat processes proceeding in slag and metal pools that is hardly achievable at the classic ESR.

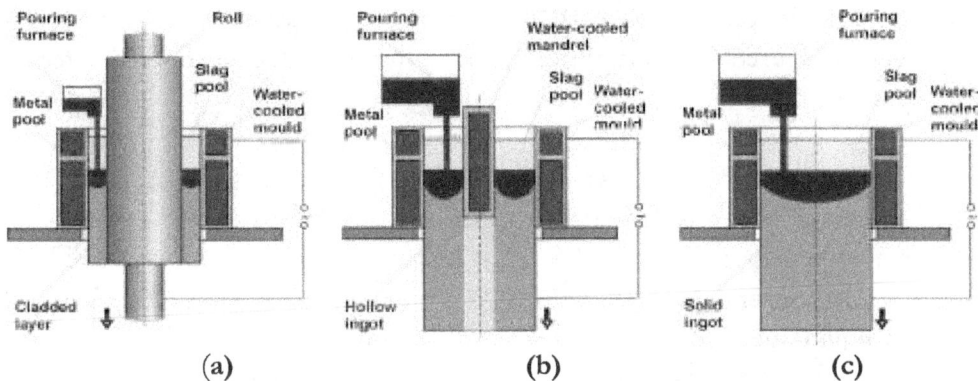

Fig. 1 Diagrams of electroslag technologies with a liquid metal
(a) - electroslag surfacing; (b) – melting of hollow ingots; (c) - melting of solid ingots.

165

2. Electroslag Technologies with a Liquid Metal

At present three basic diagrams of the electroslag processes without the consumable electrodes have been developed. These are: ElectroSlag Surfacing and ElectroSlag Remelting of solid and hollow ingots using the Liquid Metal — ESS LM and ESR LM.

2.1 Description of ESR LM processes

The diagrams of the new technological processes are given in Figure 1. The principle of these processes is as follows.

In realizing the technological diagram ESS LM (Figure 1, *a)* the billet to be clad is mounted inside the water-cooled copper current-carrying mould. The slag, melted in a separate vessel, is poured into a gap between the billet surface to be clad and the mould wall and then the voltage from the power source is supplied. As in this case the mould is not only a device, forming a deposited layer, but also the non-consumable electrode, which maintains the slag process, the heat generated in the slag pool melts a surface of the billet being clad. Then, the liquid metal of the required chemical composition is poured from a separately-located steelmaking unit into the above-mentioned gap in portions according to the preset program or continuously. The liquid metal forces out the slag upward and, taking its place, contacts the fused surface of the billet, thus forming the deposited layer. During cladding the billet is continuously withdrawn from the mould and next portions of the liquid metal are poured (or continuous pouring) until producing the deposited layer of the preset length.

The sequence of technological operations in ESR LM is almost the same. Only in case of realizing the diagram of production of hollow ingots a water-cooled mandrel is mounted inside the mould (Figure 1, b), while in production of solid ingots (Figure 1, c) no elements are required inside the mould. As in case of cladding, after pouring of liquid metal of the preset composition into mould in the necessary amount the ingot (solid or hollow) of the required sizes is formed.

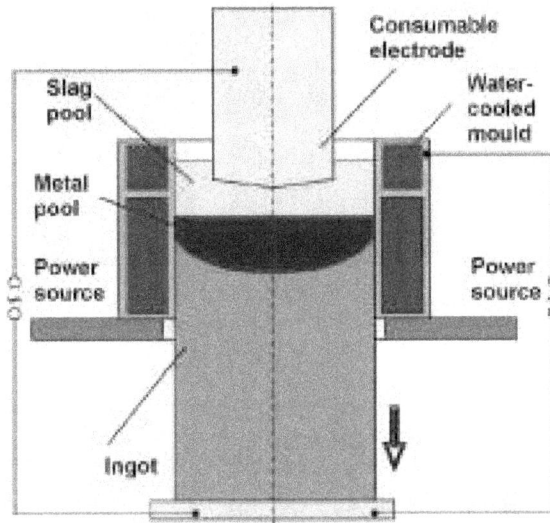

Fig. 2 Diagram of ESR TC process.

2.2 Metallurgical peculiarities of EST LM

As is known, the presence of pure metal without impurities does not guarantee the producing of the high-quality solidified metal. It can be solidified in a proper way.

The electroslag process is characterized by a high degree of overheating (150 - 200 °C) of the metal remelted above the liquidus temperature. And the metal is refined from gases and harmful impurities. Both these factors influence favourably the decrease of microheterogeneity of the metal melt. At the same time the EST with consumable electrodes are characterized by a comparatively low time of the molten metal staying in the overheated state. Only the drops of the molten electrode metal are strongly overheated, while the metal pool is cooled quickly due to an intensive heat dissipation to the water-cooled mould walls. The time of existence of the metal pool in a molten state can be approximately estimated as a quotient from division of the metal pool depth by a linear speed of melting. It is not difficult to calculate that this time is calculated in minutes that can be not sufficient for a complete elimination of the melt inhomogeneity. The another peculiarity of the ESR of consumable electrodes is a "hot pouring". i.e. the metal enters the front of crystallization directly from the overheated metal pool. This leads to the formation of a coarse-grained structure of the metal remelted.

167

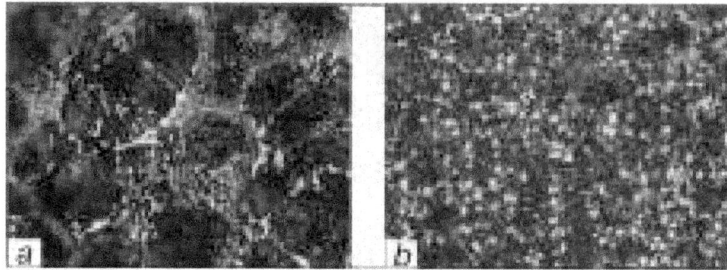

Fig. 3 Microstructure of deposited layer produced by a traditional casting (a) and using ESS LM (b), x100.

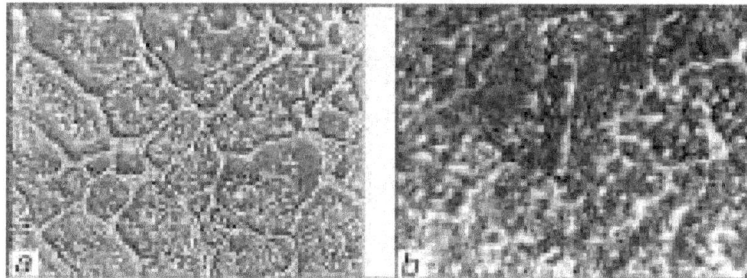

Fig. 4 Microstructure of steel M1 of conventional melting (a) and ingot produced by ESR LM, x 100.

The methods of EST using the liquid metal are a different matter. Here, the liquid metal prepared in a separate steel melting unit can be overheated to any preset temperature and soaked at this temperature for the required time. Pouring of this metal can be performed after its preliminary cooling by any known method to the required temperatures, even up to temperatures close to liquidus. In other words, during these methods of EST the conditions are created for suppressing the microstructural inhomogeneity and formation of the fine-grained structure. Microstructure (Figure 3) of 32 mm thick layer of high-speed steel of composition, %: 1.8 C; 0.3 Si; 0.7 Mn; 4.5 Cr; 1.5 Ni; 3.5 Mo; 6.1 V; 4.4 W deposited by using the method of ESC LM on 280 mm diameter billet illustrates the mentioned consideration. In this case the liquid metal was soaked in melting-pouring unit at temperature 1570 - 1600 °C for 1 h and fed in portions into the electroslag cladding furnace after preliminary cooling of the molten metal of each portion to the liquidus temperature. The comparison of this microstructure with the microstructure of the initial metal shows a favourable effect of the mentioned technology on the quality of the deposited metal. Similar results were obtained in electroslag remelting of the tool steel M1 in the mould of 110 and 215 mm diameter (Figure 4. In this case the carbide inhomogeneity of steel melted using the new

168

technology is much lower than that in steel of a traditional method of melting. The conclusion can be made from the data obtained that the electroslag technologies with a liquid metal provide most favourable conditions for the suppression of a microstructural heterogeneity and formation of a fine-grained structure of the metal.

2.3 Quality of metal of the billets and technological capabilities of EST LM

The integrated investigations of technological capabilities if different diagrams of EST LM and the metal quality of billets produced by their use, were carried out. The investigation of technological potentiality of ESR LM confirmed the data of preliminary investigations of the quite new possibilities which are opened up for the electroslag processes. As was above shown, the application of ESR LM makes it possible to create new optimum conditions for suppressing the microstructural heterogeneity and formation of the fine-graned structure of the ingot (Figure 4). In this case the use of the current-carrying mould and liquid metal enabled the metallurgist to separate the stages of melting and solidification of metal, to provide the periphery heating and averaged cooling of the solidifying ingot and also to decrease significantly the volume of the metal pool due to the fact there is no consumable electrode in the slag pool center and the heat does not enter the central zone of the metal pool with overheated drops of the electrode metal.

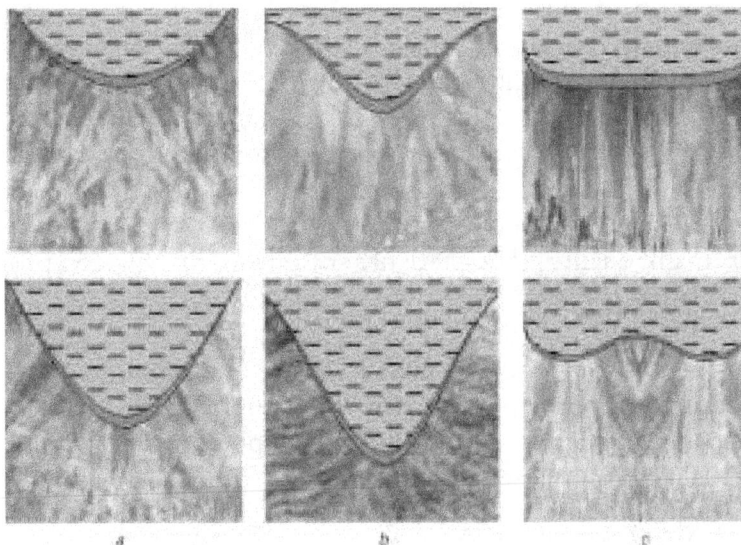

Fig. 5 Structure and shape of metal pool in VAR (a), classical ESR (b) and ESR in a current-carrying mold (c) using liquid metal.

169

With a control of the process of feeding the liquid metal and power at the mould it is possible to influence shape of the metal pool which can be obtained not only shallow but also to have a smaller depth in the center (shape of a "two-humped camel") (Figure 5). This shape will result in a crossing columnar structure which can provide the more dispersed primary structure of the metal. The comparison of macrostructure of hollow ingots produced by the traditional scheme of ESR and by using the ESR LM is also in favour of the latter. The examination of properties of metal of the ESR LM ingots showed that by their level they correspond to the level of properties of standard electroslag ingots and in separate cases they are superior to it.

One more advantage of the new system is the independence of the rate of liquid metal enter the front of crystallization of the power at the slag pool. It gives a possibility to melt the ingot at any preset rate not depending on the ingot diameter. If additional cooling of the central part of the metal pool is necessary there is no obstacles for introducing coolers. The macrocooler here can be not dispersed (wire, for example). In this case an aiming introducing of macrocoolers (for example, with the help of tribodevices) is provided. Moreover, with a change in a wire diameter and speed of its introducing it is possible to control the amount and temperature of the macrocoolers, supplied to the front of crystallization within the wide ranges. It is also important that the cost of these macrocoolers is much lower than that of the dispersed macrocoolers. With a low speed of feeding the liquid metal to the mould or combination of this feeding with introducing macrocoolers into the mould it is possible to provide any small volume of the metal pool in a current-carrying mould, not disturbing temperature conditions of formation of the ingot being withdrawn from this mould. This makes it possible to obtain such parameters of the metal pool crystallization which prevent the formation of crystalline cracks in the ingot metal, caused by liquation phenomena.

Let us consider main quality characteristics of metal of the composite billets produced by the ESS LM method.

Figure 6 presents a macrostructure of ESS LM billets of different diameters and different thickness of the deposited layer from tool or high speed steels. The metal of the deposited layer is sound, without mircopores, cracks, cavities or other defects; the fusion line is dense without lack of fusion defects in parent and deposited metal. The penetration in height and diameter of the billet is uniform.

One of the important conditions of producing the quality composite billet is a minimum changing in chemical composition of the deposited layer metal because of its dilution with metal of the billet being deposited. The examination of the isotropy of the chemical composition in height and section of ESS LM billets (the samples were cut from bottom, middle and upper parts of the billet) showed almost uniform distribution of main alloying elements both in height and section of the deposited layer. Similar results were obtained from measurement of hardness at the surface of the deposited billets, thus testifying the isotropy of properties of composite ESS LM billets.

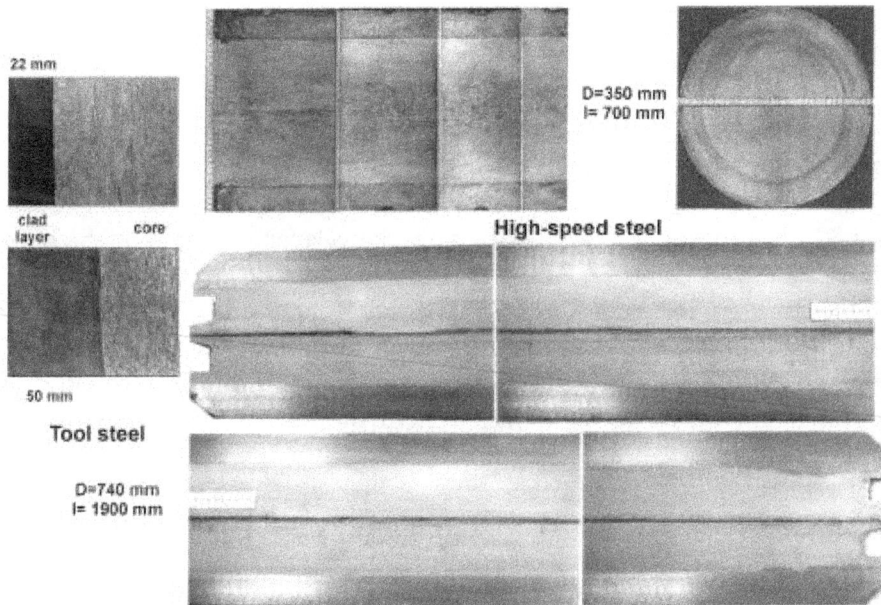

Fig. 6 Macrostructure of ESS LM billets with a different thickness of deposited layer from tool and high-speed steel.

The examination of the fusion zone of the ESS LM billets showed that distribution of the microhardness along the line "parent metal-fusion line-deposited layer" is uniform with a smooth increase from the beginning of the fusion zone into the depth of the deposited layer. The change in content of main alloying elements along the line of scanning "parent metal-fusion zone-deposited layer" has a rather smooth and uniform nature. The evaluation of technological capabilities of the electroslag cladding with a liquid metal showed that this method possesses a number of advantages as compared with traditional methods of cladding. The application of the ESS LM technological diagram makes feasible to perform cladding of external surfaces of cylindrical billets of almost any diameter, the thickness of the deposited layer is determined exclusively by the requirements of the customer and can amount to 20 - 100 mm and more. The ESS LM is characterized by a high process efficiency, which is ten times superior to the efficiency of the traditional methods of cladding and, depending on the sizes of components and cladding materials, is 200 - 800 kg/h and more. The use of materials of different chemical composition is also undoubted advantage of the new technology. In this case, cast iron, high-speed, tool and stainless steels and heat-resistant nickel alloys, etc., including also materials which cannot be subjected neither to cold nor to hot deformation and, respectively, not used for the traditional surfacing technologies, where the consumable electrodes in the form of wire, bars, plates, etc. are required.

2.4 Industrial application

The comprehensive examination of technological parameters of ESS LM and metal quality of composite billets produced by this method could recommend it for the industrial application to produce the billets of composite rolls of the new generation, and, in particular, rolls with a working layer from the high-speed steels. Within the scope of joint works with Novokramatorsk Machine-Building Works (NKMZ) the project of adaptation of the ESR furnace at the NKMZ was fulfilled and the first in the world industrial installation of the electroslag cladding with a liquid metal (Figure 7) designed for production of composite billets of mill rolls of up to 1000 mm diameter, up to 2500 mm length of the barrel and up to 20000 kg mass using the ESS LM, was created. The usage of these rolls in a wide-strip mill showed that the life of rolls produced by the technology ESS LM is 4 - 4.5 times higher than that of standard cast iron rolls used earlier.

Fig. 7 Appearance of ESS LM unit at "NKMZ".

172

3. Electroslag Remelting of Consumable Electrodes using Two-Circuit Diagram

As was shown above, the application of the ESR diagram in the current-carrying mould using the liquid metal guarantees the creation of optimum conditions (separation of stages of melting and solidification of metal, periphery heating and averaged cooling of the solidifying ingot, significant decrease in volume of the simultaneously solidifying molten metal) for melting and solidifying of the ESR ingots. However, the application of the filler metal in electroslag process for the complexly-alloyed steels and alloys having the easily-oxidizing elements in their composition, for example, Ti, Al can present in separate cases a certain problem, i.e. preserving these elements in the period of soaking and pouring of the molten metal is a rather complicated problem. Taking this into account and also conditions of a routine production of consumable electrodes for the ESR and absence of feasibility of preparation, soaking and dosing the filler liquid metal, a new diagram of remelting of a solid billet in the slag pool of the current-carrying mould, the so-called two-circuit diagram of the electroslag remelting of the consumable electrode was suggested (Figure 2). This diagram was called two-circuit, as the electric supply of the consumable electrode and the mould is performed in different circuits (electrical circuits) from separate power sources. The diagram is featured by the fact that two electrodes are available simultaneously in the slag pool: one is consumable in the form of a steel billet being remelted and another is non-consumable in the form of a current-carrying mould. If the non-consumable electrodes are heating the slag pool periphery, then the melting metal electrode located usually in the slag pool center is heating in this spot not only the slag pool but also brings a large part of heat with molten metal drops to a central part of the metal pool.

It is evident that such mode of the heat supply brings favourable changes in the conditions of electrode melting and formation of the metal pool. In addition, as the preliminary experiments showed, when the ESR TC is used the specific peculiarities of interrelation of temperature parameters of the process and the metal melting speed are manifested.
To study the above-mentioned peculiarities special investigations were carried out. A special series of experimental meltings resulted in obtaining basic data on the peculiarities and relation of main technological parameters of the process of the electroslag remelting of the consumable electrode using a two-circuit diagram was made.

The results showed that the two-circuit ESR diagram of the consumable electrode can reduce greatly the dependence of the electrode melting rate on the power supplied as compared with the classical diagram. And, here, the attention should be paid to the fact that the temperature conditions of the process in ESR TC can be controlled independently of its efficiency. This provides in turn the feasibility of control of volume and shape of the metal pool within wide ranges and thus influencing the quality of the ingot metal. In addition, the results of investigation of main features of the ESR TC showed that this diagram is rather promising for the solution of two-purpose problem : a good ingot formation at any low speed of its melting and, respectively, minimum depth of the metal pool.

The problem of control of depth and shape of the metal pool is especially acute in production of the ESR ingots of a large diameter (above 500 mm), produced from high-alloyed alloys, in particular superalloys on the nickel base. As the numerous investigations showed, the producing of ingots, free from the defects of a liquation origin, from these materials is possible only in case of the metal pool, being minimum in depth and flat in contour. The appropriate investigations were carried out on the effect of main technological parameters on the depth and shape of the metal pool in ESR TC. It was established, that it is possible to provide the preset contour and depth of the metal pool by changing the shape, size and arrangement of electrodes and also by a suitable redistribution of powers in "mould-bottom plate" and "electrode-bottom plate" circuits (Figure 8).

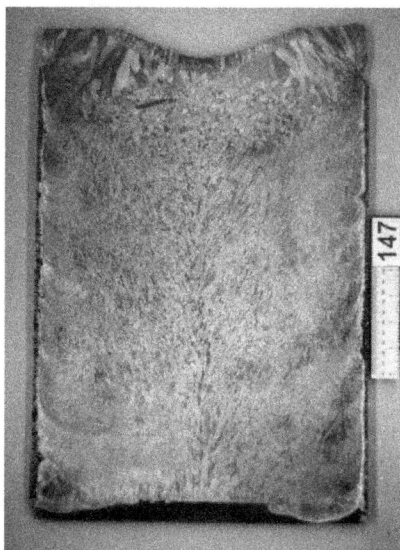

Fig. 8 Macrostructure of ESR TC ingot.

The experimental-industrial testing of ESR TC in production of ingots from complexly-alloyed steels and alloys confirmed the wide technological capabilities of this method and could recommend it as the new technological process for their industrial production.

174

To provide a wide industrial implementation of the new processes the specialized equipment was designed and the new system of control of the technological process was developed

4. Up-to Date Invertigations of ESR of Titanium

The application of technological ESR processes in the current-carrying mould opens up the new opportunity for remelting of titanium and its alloys [36-38]. Thus, the current-carrying mould can be successfully used in realizing the titanium sponge remelting. Even the first results of experiments showed that it is possible to produce inexpensive consumable electrodes for the next remelting using this method.

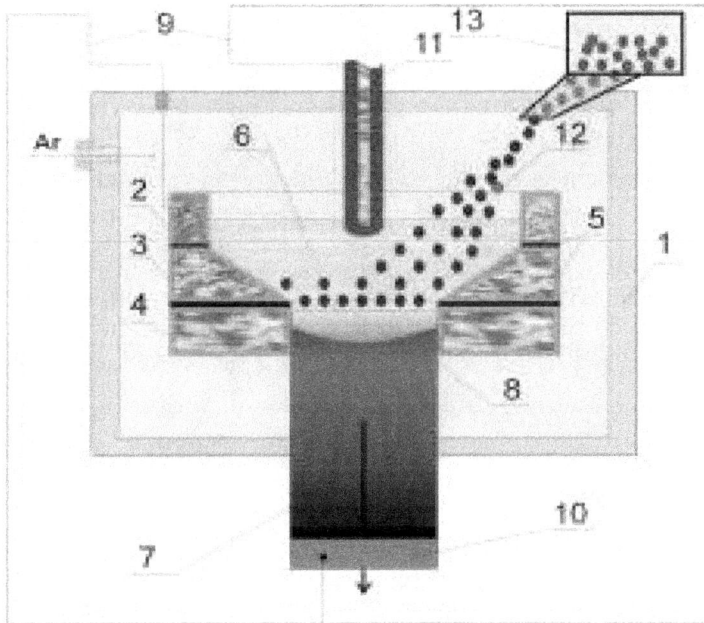

Fig. 9 Diagram of ESR of titanium sponge.

The diagram of the electroslag melting of titanium sponge in a chamber furnace created by the company "Elmet-Roll - Medovar Group" is shown in detail in Figure 9. The process of melting ingot 7 is started with filling of the chamber furnace 1 with argon. Then, the molten slag 6 is poured into a melting space, confined by upper 2 and middle 3 sections of the mould and bottom plate, mounted in a forming section 4 of the mould. Then, the power source 9 is switched on. The slag pool is electrically conductive and closes the electrical circuit of the furnace: the current-carrying section of the mould with a titanium protection-bottom plate-ingot and non- consumable electrode 11-bottom plate-ingot. After the slag heating the titanium sponge *12* is supplied to the melting space with a help of a batcher *13*. Being melted, the titanium sponge forms a metal pool *8*.

It should be noted that the experiments were carried out on the basis of a detailed computer modeling and at present this diagram is tested under the semi-industrial conditions of producing titanium ingots of diameter of up to 400 - 500 mm.

On the basis of experience of using ESR TC in the current-carrying mould, the application of this diagram was suggested for the titanium refining from high-nitrogen inclusions. It should be clarified here, that we are speaking, naturally, not about the melting of such kind of inclusions, but rather about their dissolution or even refining and uniform distribution in the titanium ingot volume. During experiments a fast partial dissolution of large nitrogen inclusions was observed .The successful experiments made it possible to offer the diagram of titanium production, shown in Figure 10, using ESR before the final VAR for purification of titanium from these superhard inclusions [36, 38]. Data about the kinetics of behaviour of these inclusions indicate that the time of dissolution of high-nitrogen inclusions of sizes, observed really in practice, can be longer than the time of molten state of the metal of ingot being melted. This is also valid for the VAR and for cold-hearth electron beam of plasma-arc melting of titanium. At the same time, during ESR the rate of dissolution of high-nitrogen inclusions can be almost twice higher at certain conditions than at the VAR. First of all, we are speaking about the application of additions of metal calcium to the slag (pure calcium fluoride) in combination with use of the current-carrying mould. At present the comprehensive investigations are carried out on a mechanism of dissolution of the mentioned inclusions in ESR both under the laboratory and semi-industrial conditions.

It quite clear that two above-given trends of present investigations of the ESR of titanium do not reveal all the capabilities of this process. It is evident that the ESR can be promising technological process for the realization of an old dream of the manufacturers of the titanium products, i.e. continuous or semi-continuous casting of titanium by the type used in a steelmaking industry. Our optimism is based on the fact that ESR possesses such indispensable attribute of the continuous casting as slag, moreover, the rather positive results of ESR of titanium ingots with an ingot withdrawal by the type of the semi-continuous casting are known.

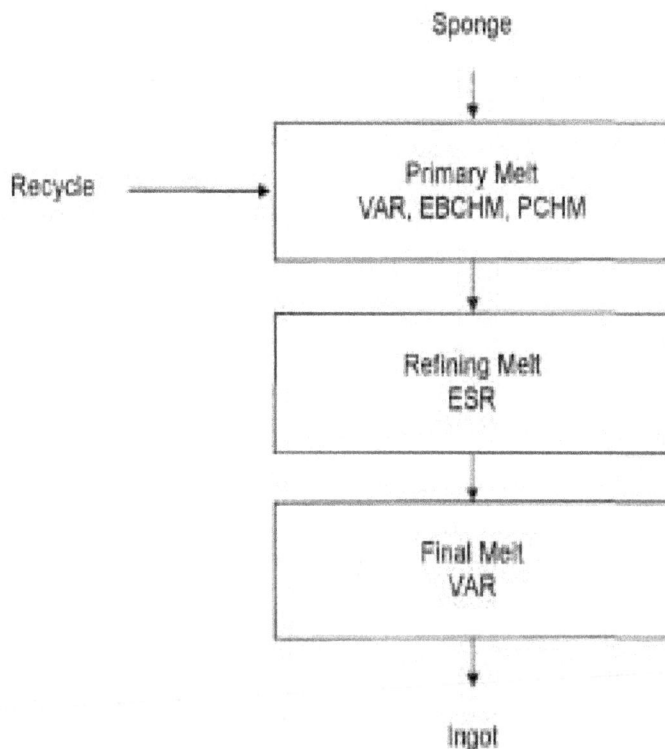

Fig. 10 Diagram of ESR application for titanium refining from high-nitrogen inclusions (by M.Benz, [36]).

5. Recent Developments

It is well know that there are several old problems in the metallurgical industry that are not yet completely solved and never could be solved due to the nature. One of such old and very important problem is the problem of segregation, especially in connection with the manufacturing of high quality forging ingots for such applications as power generating. We have to accept that it's impossible to avoid various displays of segregation as well as a undesirable structural elements formation. All these processes are going almost in parallel within time of ingot solidification. Due that, we can easy come to the conclusion about possible solution of the high quality ingot problem: low segregation big ingot should and could be made with the same volume of liquid, liquid-solid zone as small one. It's clear and well proven that, to some extend, all processes of special electrometallurgy, processes of remelting, are realizing above statement. Limitation of the standard ESR from that point of view is well known.

177

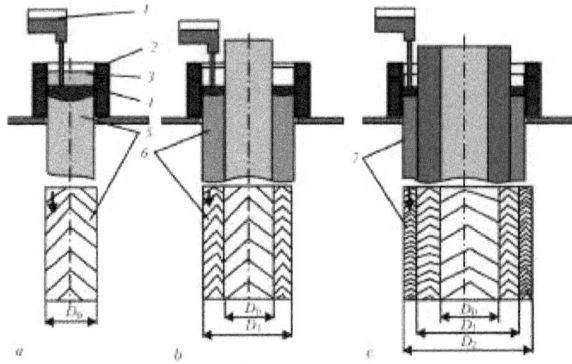

Fig. 11 Multisteps ESS LM for ingot enlargement.

a- ESS LM or standard ESR for first ingot, b-ESS LM for first one enlargement, c- final ingot manufacturing and adequate sketch of ingot structure;1 - holding-pouring furnace; 2-current conductive mold, 3-slag pool,4-metal pool, 5-ingot, 6- second step ingot, 7-third step ingot Application of the ESR TC as shown previously significantly expands capacity for controlling profile, depth of the molten metal pool and length of the double-phase zone during solidification of the ingots in order to prevent segregation processes. In spite of possibility to shallow liquid metal pool and mushy zone ESR TC can not be conducted with the enough low speed to minimized total amount of liquid metal at the metal pool good enough to stop segregation. There also a limitation from time required for whole ingot melting. In general ESR TC allows to run ESR at very low speed. But it's not practical, not realistic to melt, for example 30 ton ingot, with the speed of 300-400 kg per hour. Due above reasons it is proposed to utilize ESS LM for forging ingot enlargement [39]. Schematically it looks as presented at Figure 11.

After almost 50 years of industrial application, ESR became to be well known process with enough well developed theoretical basis. Industry demands will enforce further ESR improvements. There are several directions for such movement ahead. Traditional one of them is belong to the more and more deep investigation of the physical phenomena of the process, especially connected with various hydrodynamic effects. There are also various possibilities to increase importance of the slag influence based on active additions. Most important, per authors opinion, is increasing as much as possible ESR capabilities to control solidification. From that point of view it could be accepted, in general, that main idea to cast big ingot as a sum of several small has many positive signs. Question is how to realize them in industrial scale.

6. Conclusions

Thus, over the recent years the new technological diagrams of the electroslag process: electroslag technologies with a liquid metal (EST LM) and electroslag remelting of the consumable electrode using a two-circuit diagram (ESR TC) were suggested. The investigations showed the high level of properties of the metal of billets, produced using the new technologies and allowed these processes to be recommended for the most wide industrial application.

The analysis of works carried out now in the field of the ESR and also the presented results of works carried out by us can state that the improvement of the electroslag process both in the field of development of the technologies and in the field of creation of the new equipment is continued. We suppose that in parallel with the further improvement of the classical technology of the ESR with consumable electrodes the wider industrial application of electroslag technologies with a liquid metal, including a semi-continuous casting of a sort, the appearance of new chamber furnaces for remelting high-reaction steels and alloys, including titanium, are to be expected.

Acknowledgments

The authors will always be grateful to academician B.I. Medovar, whose ideas were used a basis of the described works, and who was an inspirer and direct participant in the present investigations and also to academician B.E. Paton for support and assistance in the fulfillment of the investigations.

References

1. B.E. Paton, B.I. Medovar, V.E. Paton, *New method of electroslag casting of ingots*, Bull. of Tech. Inform., Issue 1, 1956, p.3.

2. B.E. Paton, B.I. Medovar, Yu.V. Latash, *Electroslag melting of high-alloyed steel and alloys in a water-cooled mould*, M.: Metallurgizdat, 1957, p.7.

3. B.E. Paton, B.I. Medovar, Yu.V. Latash, *Electroslag remelting of steels and alloy in a copper water-cooled mould*, Avtomaticheskaya Svarka, No.11, 1958, p. 5.

4. B. I. Medovar, Yu.V. Latash, B. I. Maksimovich, L. M. Stupak, *Electroslag remelting*, M.: Metallurgizdat, 1963, p. 170.

5. Yu.V. Latash, B.I. Medovar, *Electroslag remelting*, M.: Metallurgizdat, 1970, p. 239.

6. *Hopkins, R.K.,* US Patents 2,191,479 (1940).

7. B.I. Medovar, L.M. Stupak, G.A. Boiko et al., *Electroslag furnaces,* Kiev: Naukova Dumka, 1976, p. 414.

8. B.I Medovar, L.M. Stupak, G.A. Boiko et al., *Electroslag metal,* Kiev: Naukova Dumka,1981, p. 680.

9. B. I. Medovar, A. K. Tsykulenko, A. G. Bogachenko, V. M. Litvinchuk, *Electroslag Technology abroad,* Kiev: Naukova Dumka, 1982, p. 320

10. B. I. Medovar, A. K. Tsykulenko, V. L. Shevtsovet al, *Metallurgy of electroslag process,* Kiev: Naukova Dumka, 198, p. 248.

11. *B. I. Medovar, V. Ya. Saenko, A. D. Chepurnoy, I. D. Nagaevsky,* Electroslag *Technology in machine-building,* Kiev: Tekhnika, 1984, p. 215.

12. B. I. Medovar, A. K. Tsykulenko, D. M. Dyachenko, *Quality of electroslag metal,* Ed. *by* Paton B. E., Medovar B.I., Kiev: Naukova Dumka, 1990, p. 312 .

13. B. E. Paton, B. I. Medovar, G. A. Boiko, *Electroslag casting,* Kiev: Naukova Dumka, 1980, p. 192 .

14. B.I. Medovar, V. Ya. Saenko, G. M. Grigorenko, Yu. M. Pomarin, V. I. Kumysh, *Arc-Slag Remelting of Steel and Alloys,* Cambridge Int. Science Pub., England, 1996, p. 160.

15. W. Holzgruber, *Recent Innovation in Electroslag Remelting,* Iron and Steel Maker. - October, 1998, p.107.

16. W. Holzgruber, *Electroslag Technologies - Overview and Outlook,* Proc. INTECO Symposium, Dusseldorf, Germany, 1999, p.1.

17. B.E. Paton, B.I. Medovar, L.B. Medovar, *ESR is 40: are there any prospects,* Stal, No.11, 1998, p.24..

18. B.I. Medovar, L.B. Medovar, V.Ya. Saenko. *Development of electroslag process in special electrometallurgy,* Avtomaticheskaya Svarka, No.9, 1999, p.7.

19. B.E. Paton, B.I. Medovar, G.S. Marinsky et al., *Electroslag crucible melting and new casting technologies developed on its base,* Electroslag remelting, Issue 9, 1987, p. 133.

20. B.I. Medovar, V.L. Shevtsov, V.M. Martynet al., *Electroslag crucible melting and pouring of metal,* Ed. by Paton B.E., Medovar B.I., Kiev: Naukova Dumka, 1988, p. 216.

21. R.J. Roberts, *Electroslag Remelting (ESR) at Consarc,* Proc. of Medovar Memorial Symposium, Kyiv, May 15 - 17, 2001.

22. H. Scholz, U. Biebricher and A. Choudhury, *Recent Development of Technology and Equipment Design of Electroslag Remelting Process,* Proc. of Medovar Memorial Symposium, Kyiv, May 15 - 17, 2001.

23. Yu. V. Latash, F. K. Biktagirov, R. G. Krutikov, B. A. Pshenichniy, *Application of non-consumable electrode process for melting,refining and casting of metals",* Proc. Of Medovar Memorial Symposium, Kyiv, May 15 -17, 2001.

24. N.I. Tarasevich, I.V. Korniets, L.B.Medovar, A.V. Chernets, V.E. Shevchenko, *Some Thermal-Physical Peculiarities of Electroslag Cladding of Rolls Using a Liquid Metal,* Proc. of Medovar Memorial Symposium, Kyiv, May 15 - 17, 2001.

25. B.E.Paton, B.I.Medovar, L.M.Stupak, *Electroslag melting of metallized pellets. State-of-the-art and prospects of application ,* Problemy Spetsalnoy Elektrometallurgii, No.1, 1986, p.43-45; No.3, 1986, p.3-9; No.4,1986, p.3.

26. A.G. Shalimov, A.E. Volkov, Al.G. Shalimovet al, *Production of quality steel using the method of electroslag melting of metalized pellets in the mould,* Stal, No.12, 1985, p.22.

27. A.G. Shalimov, N.V. Solovjev, A.A. Brodov e.a, *Economical efficiency of electroslag remelting of a metallized raw material,* Ibid, No.1, 1987, p.39.

28. A.G. Shalimov, Al.G. Shalimov, A.E. Volkov,V.A. Aleksandrov, *Production of quality steel using the method of electroslag melting of metallized pellets in the mould,* Problemy Spetsialnoy Elektrometallurgii, No.4, 1985, p.3.

29. Yu.V. Latash, V.N. Matyakh, *Advanced methods of production of superhigh-quality slabs,* Kiev: Naukova Dumka, 1987, p. 336.

30. F.K. Biktagirov, Yu.V. Latash, B.A.Pshenichniy et al, *Electroslag casting of ingot in installation UO-106,* Problemy Spetsialnoy Elektrometallurgii, 1997, No.2, p.7.

31. G.V. Ksyondzyk, V. S. Shirin, *Equipment for electroslag remelting and cladding*, Patent of the USA, No.4,185,682 (1980).

32. G.V. Ksyondzyk, Yu.M. Kuskov, V.V.Chernokozenko et al., *Optimizing of technique of start in electroslag cladding mill rolls with a shot*, Problemy Spetsialnoy Elektrometallurgii, No.1, 1986, p.18.

33. Yu.M. Kuskov, *Electroslag process without consumable electrode using a non-compact filler material*, Ibid, No.2, 1992, p.27.

34. B.I. Medovar, L.B. Medovar, A.V. Chernets et al., *ElectroSlag Surfacing by liquid Metal A new Way for HSS-Rolls Manufacturing*, 38th MWSP Conference Proceedings, Vol.XXXIV / Cleveland, Ohio, October, 1996, p.83.

35. B.I. Medovar, L.B. Medovar, A.K.Tsykulenko et al., *Electroslag Processes Without Consumable Electrodes*, Proc. of the International Symposium on Liquid Metal Processing and Casting, Santa Fe, New Mexico, February 16 - 19, 1997.

36. B.E. Paton et al., *ESR for titanium: yesterday, today, tomorrow*, Proc. of the 9th World Titanium Conference, St. Petersburg, Russia, June 7 - 11, 1999.

37. B. I.Medovar, L. B. Medovar, B. B. Fedorovsky et al., *Electroslag technology for TiAl ingots*, Proc. XITC '98, Xian, China, Sept. 15 - 18, 1998, p.123.

38. L.B. Medovar, M.G.Benz, *Electro-Conductive crucible for ESR refining of titanium alloys with independent control of slag temperature, slag rotational velocity and electrode melt rate*, Proc. XITC'98, Xian, Sept. 15 - 18, 1998, p. 12. E.O. Paton ElectricWelding Institrute, Kyiv, Ukraine

39. V. I. Mahnenko, L. B. Medovar, , V. Ya. Saenko , T. V. Korolyova, *New method of low segregationesr forging ingots production (computer simulation of the ESR ingot enlargement)*, Proceedings of IFM- 17, Santander, Spain, 2008.

Sustainable Development of Aluminium Industry

T.R. Ramachandran

Nonferrous Materials Technology Development Centre
Kanchanbagh, Hyderabad.

Abstract_*Sustainable development has been defined as "development that meets the needs of the present without compromising the ability of future generations to meet their own needs." In the context of aluminium industry, sustainable development involves the minimization of the use of raw material in both alumina and aluminium production, energy conservation at every step starting from bauxite mining to the finished product, reducing emissions, ensuring that the waste generated is disposed off in an environmentally friendly way or better still converted into useful products, carrying out efficient recycling of scrap, enlarging the applications and ensuring adequate safety for the plant personnel. The processes for alumina and aluminium production are both energy and raw material intensive. In addition, the waste products, red mud and spent pot lining and emissions of CO2, fluorides, perfluoro carbons (PFCs) and polyaromatic hydrocarbons (PAH), pose considerable challenge to efficient disposal and minimizing environmental pollution. Considerable progress has been made in energy conservation and reduction in environmental pollution in the last few decades. Concurrently trials are in progress to convert the waste material into value-added products. Light weighting and the resultant reduction in CO2 emissions is the driving force for the increasing use of aluminium in automobiles. Presently the transportation sector accounts for about 30% of the metal consumption in developed countries. Notable progress has been made in recycling aluminium scrap from automobiles and packaging sectors (mainly from UBCs, used beverage containers); however several challenges are still posed in scrap processing mainly related to metal sorting, improving the efficiency of metal recovery and energy conservation. These aspects are briefly reviewed in this presentation.*

1. Introduction

Considerable attention is paid to various aspects of sustainable development of industries in the last two decades; the Bruntland Commission of the United Nations dealing with environment and development has defined sustainable development as "development that meets the needs of the present without compromising the ability of future generations to meet their own needs." It ensures a better quality of life for everyone for now and for generations to come, combining economic, social and environmental concerns – the *Triple Bottom Line*.

In the context of aluminium industry, sustainable development involves the minimization of the use of raw material in both alumina and aluminium production, energy conservation at every step starting from bauxite mining to the finished product, reducing emissions, ensuring that the waste generated is disposed off in an environmentally friendly way or better still converted into useful products, promoting wider usage of the metal, carrying out efficient recycling of scrap and ensuring adequate safety for the plant personnel. The basic steps involved in the production of primary aluminium, Bayer process for alumina and Hall-Heroult process for the metal, are both energy and raw material intensive. In addition, the waste products, red mud and spent pot lining ande missions of $CO2$, fluorides, perfluoro carbons (PFCs) and polyaromatic hydrocarbons (PAH), pose considerable challenge to efficient disposal and minimizing environmental pollution. Considerable progress has been made in energy conservation and reduction in environmental pollution in the last few decades. Concurrently trials are on to convert the waste material into value-added products. Light weighting and the resultant reduction in $CO2$ emissions is the driving force for the increasing use of aluminium in automobiles. Presently the transportation sector accounts for about 30% of the metal consumption in developed countries. Notable progress has been made in the recycling of aluminium scrap from automobiles and packaging sectors (mainly from UBCs, used beverage containers); however several challenges are still posed in scrap processing mainly related to metal sorting, improving the efficiency of metal recovery and energy conservation. Progress achieved in meeting the goals of sustainable development of aluminium industry is briefly discussed in this presentation.

2. Aluminium Production and Consumption Pattern

World production of aluminium (both primary and secondary) over the last century is shown in Fig. 1; the present production is around 40mt; in the year 2006, scrap usage accounted for roughly 1/3 of the total. The consumption pattern for the metal (world average and advanced countries) is shown in Fig. 2 (1); transportation, packaging and building and construction sectors account for ~70% of the metal. The transportation field,particularly the automobile sector, offers good scope for expansion. The quantity of aluminium used in cars in Japan, US and Europe is shown in Fig. 3; it varies between 140-160 kg/vehicle and is showing a steady rise (2). There are opportunities for more penetration of aluminium in the automobiles market, particularly for closure panels and wheels of automobiles. Recycling is an important consideration in the expanding use of aluminium in the automobile industry with two-thirds of aluminum used in present day vehicles sourced from recycled metal. Aluminium industry and automobile manufacturers have launched several joint initiatives in the recent years to improve the recycling rate and efficiency of recovery of metal from the scrap.

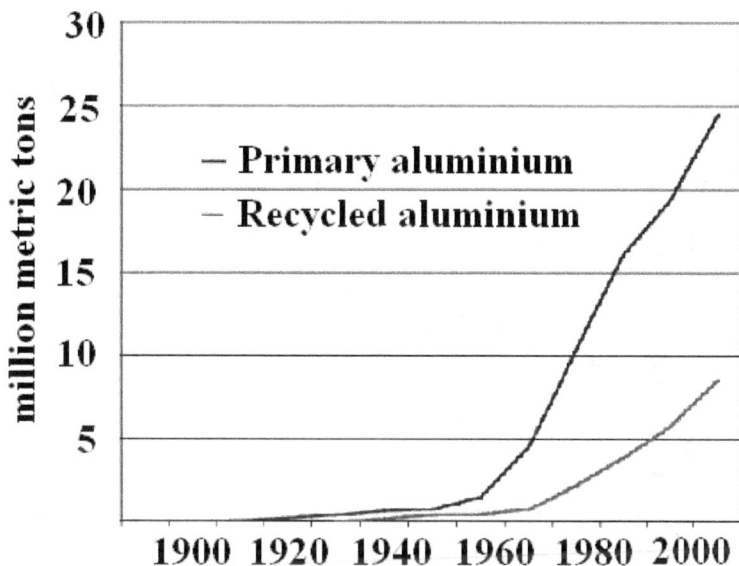

Fig. 1 World Annual Production of Aluminium in the Last Century.

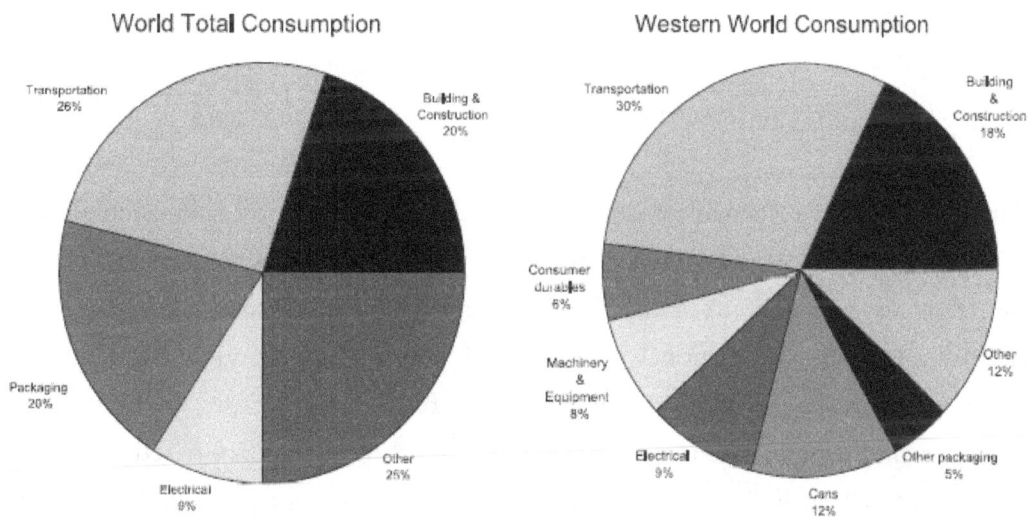

Fig. 2 Consumption Pattern for Al in the Western World and World Average.

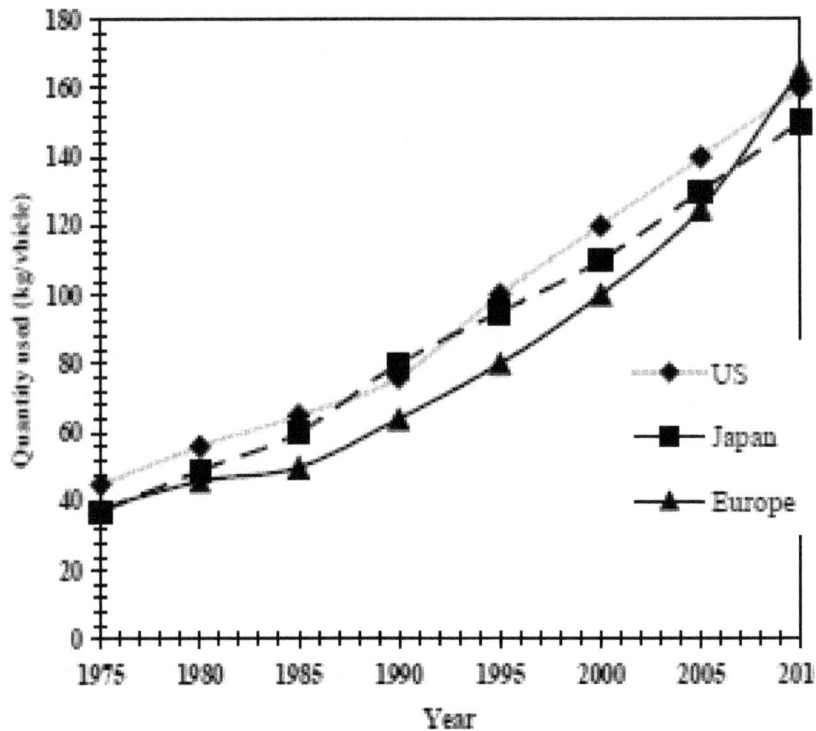

Fig. 3 Aluminium Consumption in Passenger Cars.

3. Objectives of Sustainable Development

Various aspects of sustainable development of aluminium industry have been examined in detail in the last decade and road maps and vision documents are available outlining the steps to be taken to make the industry viable environmentally, socially and economically. The International Aluminium Institute (IAI) London has launched a voluntary global undertaking with participation from ~75% of the world aluminium producers and with well defined objectives of improvement in 2010 based on the 1990 levels; the details including the present levels of achievement are given in Table I (3):

186

Table 1 Objective of Sustainable Development and Achievements in 2006.

Area	Objective	Achievements in 2006 compared to 1990
Energy conservation	10% reduction in power consumption in metal and alumina production.	16.1 kWh/kg metal to 15.2 kWh/kg in electrolysis; 12.7GJ to 12GJ/t alumina in Bayer process.
Reduction of emissions	80% reduction in PFC emissions and at least 33% reduction in fluoride emissions.	PFC reduced from 4.9 to 0.7 CO_2 equivalent/t of metal; fluoride emission reduced from 2.4 kg/t metal to 1.1 kg/t.
Waste material storage/usage	Find uses for spent pot lining (SPL); otherwise store it pending final disposal, in secure water-proof ventilated containers to maintain in a dry state and avoid build up of noxious gases.	Conversion of SPL as feedstocks for cement, steel, mineral wool, construction aggregates.
Natural resources	Increase the proportion of bauxite mining land rehabilitated annually; Reduction of fresh water consumption in the production processes	
Workers safety	50% reduction in lost time injury frequency rate and total recordable injury frequency rate by 2010 versus the corresponding figures in 2006; Implementation of management systems for environment (including ISO 14000 or equivalent certification) and /or health and safety in 95% of the plants	Lost time injury rate reduced from 4.7 to 1.6 hrs/million hours worked; recordable injury rate reduced from 22.2 to 5.2 per million hours worked..
Enhanced usage of metal	Monitoring annual aluminium semis shipments for use in transport in order to track aluminium's contribution through light weighting to reduce GHG emissions from road, rail and sea transport.	From 6.6 mt to 13.6 mt in the transportation sector
Scrap recycling	Regular reporting of global recycling performance; a mass flow model has been developed by the IAI to identify future recycling flows.	

4. Improvements in Bayer and Hall-Heroult Processes

The areas of energy and raw material conservation and handling of wastes from alumina and aluminium production have attracted extensive R &D investigations over the last three to four decades; the developments are mainly related to better understanding of the processes and improving their efficiency, large scale use of process control, use of new materials for electrodes in electrolysis and improved design of reactors and cells using mathematical modeling. Treatment of emissions (e.g. dry scrubbing in electrolytic cells) and value addition to wastes such as red mud and spent pot lining also constitute important developments. Some of these features are summarized in the following sections.

4.1 *Bayer Process:*

Bayer process for alumina production has been in use for over a century; it involves the following steps:

- digestion of bauxite with caustic soda at temperatures in the range 105°C to 260°C and at high pressures (the temperatures and pressures depend on the mineralogy of the ore)
- filtering the undissolved components which after washing is disposed off as waste (red mud)
- precipitation of alumina hydrate from the solution obtained in the digestions stage and separation of the hydrate from the spent liquor
- recirculation of the spent liquor to the digesters
- calcinations of the hydrate to anhydrous alumina.

The important characteristics of alumina for smelter use are purity, particle size distribution, amount of α-alumina phase, flowability and surface area. The main energy consuming steps are digestion, evaporation of the spent liquor before recirculation and calcination. Improvements made in some of the important stages of the Bayer process are summarized in Table II (4). Large quantities of red mud ($\sim$ 1.5 t/ t of alumina) are generated as waste in the Bayer process; the main constituents are oxides of Fe, Al, Si, Ti and Na. Problems regarding disposal are the massive amounts of waste (a 1 mt plant will produce at least 1.5mt of red mud) and the soda content (about 4 g/l). Several methods of disposal are available; the best method is dry stacking. Use of red mud as constructional material, catalysts, coagulating agents for water purification and for recovery of valuable constituents such as Ti and Fe has been tried and many trials are still in progress. The success achieved is modest. Pending large scale utilization, dry stacking appears to be the main method of storage.

4.2 *Aluminium Electrolysis*

The Hall-Heroult process involves the electrolysis of alumina dissolved in cryolite (with excess AlF3) using prebaked (or in-situ baked Soderberg) carbon anodes. The cathodes are made of carbon blocks (amorphous, semigraphitized or fully graphitized form) at about 950°. Fluoride and GHG emissions (CO_2 and PFCs) pose environmental problems. Substantial progress has been made towards improving the electrolysis parameters, reduction of emissions and utilization f spent pot lining, the waste dug up from failed cell cathodes. These improvements have been

Table 2 Improvement in Equipment/Technology in Alumina Production.

Area	Improvement
Digestion	• Double digestion / Sweetening technology for mixed bauxite and double digestion for gibbsitic bauxite with boehmite content around 5%; tube-in-tube digesters for reducing plant size and improving energy efficiency. • Hydrogarnet addition to improve digestion yield for boehmitic bauxite. • Low caustic concentration combined with tube digestion to improve liquor productivity and reduce the dissolution rate of kaolinite, thereby lowering bound soda losses by 40-50%.
Settling and Washing	• Pressure decantation and or pressure filtration to reduce hydrolysis loss even with high caustic concentration, and to higher pregnancy of the aluminate liquor. • Replacement of washers by deep thickeners / high rate thickeners and or pressure filters with or without steam wash to reduce investment costs, operating costs and also improve the overall productivity of the process.
Precipitation	• Improved liquor pregnancy combined with efficient inter-stage cooling and multiple seeding to increase liquor productivity, from 65-80 to 95-100 kg/m^3; Addition of CGM (crystal growth modifier) to improve precipitation yield; further increase in productivity possible by improving liquor purity.
Classification and Hydrate Filtration	• Improved and efficient classification for better product sizing, with the − 45 μm size in the product hydrate going closer to 5-6 %. • Steam wash of the hydrate to reduce moisture content to less than 6%, leading to reduced energy requirement during calcination and soda content of the product.
Evaporation	• Efficient handling, improved mechanical seals, reduced dilution of the system, higher caustic concentration at precipitation and lower caustic concentration at digestion to reduce evaporation requirement. • Possibility of evaporationless technology
Calcination	• Development of pressure calcination, offering substantial reduction in energy requirement (~40% less)
Alumina quality	• Better understanding of the effects of particle size, angle of repose, moisture content, attrition index, loss on ignition, BET surface area and various impurities on the smelting process and the quality of aluminium produced leading to introduction of optimum specifications for these parameters

brought about by devoting attention to cell measurements, design, materials technology, and environmental control; the following factors are important in this regard: • Measurement of magnetic field at various locations in and around the cell and compensation of the residual field leading to reduced magneto-hydrodynamic effects and consequent improvement in current efficiency

- Improved electrolyte with low ratio bath (excess aluminium fluoride), relatively low temperature of operation and low vapour pressure

- Improved process control systems leading to new cell designs and contributing to the competitiveness of many old technologies

- Point break and feed of alumina

- Multipurpose cranes for tapping, changing anodes etc.

- Improved cell hooding and alumina dry-scrubbing systems for fume treatment and recovery of the fluorides

- Air slides and dense phase systems for the delivery of alumina to cells

- Better understanding of the influence of the physical and chemical properties of the raw materials (coal tar pitch and calcined petroleum coke) and processing conditions on the properties of the anode which has led to improved anode quality; research work on development of inert anodes

- Improved cathode designs with fully graphitized cathode blocks and silicon carbide side wall lining; development of special proprietary coatings for the cathode (e.g.TINOR) to improve the cell life and the operating parameters

- Mechanization of all aspects of smelter operation

- Operator training and education.

- Fluorine balance to check the extent of fugitive emissions and taking corrective steps

Anodes made of cermets (e.g. $NiO-Fe_2O_3-Cu$) are relatively inert and contribute to oxygen emission in contrast to CO_2 release with carbon anodes. They also facilitate the use of lower anode-cathode distance in the electrolytic cell thereby bringing down the I^2R losses in the electrolyte and hence contribute to energy savings. TiB_2 coating on the cathode favours better wetting of the molten aluminium with the cathode and hence contributes to lower amount of metal bath at the bottom of the cell. The shielding of carbon by TiB_2 at the cathode and non-carbon inert anode avoid cyanide formation, an important but minor constituent of SPL. SPL is generated to the extent 20-30kg/t metal; it represents the top surface of the cathode that is dug out when an electrolysis cell has failed. The important constituents of SPL are fluorides, alumina, carbon and a small amount of sodium cyanide (~0.1%). It is necessary to recover the carbon and fluorine values and ensure that the cyanide is totally destroyed before the residue is land filled. ALCOA Australia uses the AUSMELT technology to treat SPL (5). The method involves a high temperature treatment using the carbon available in SPL for the recovery of fluoride as aluminium fluoride while destroying cyanide and other organic constituents. The remaining product removed as a slag is benign and can be used either for land filling or sent as feedstock to cement, steel, mineral wool and construction aggregates industry. In view of the fact that the typical amount of SPL generated in plant (~ 10000t/annum) is limited, the economics of the treatment may not be favorable.

Reducing PFC emissions is important as CF4 and C2F6 have three to four orders of magnitude longer life compared to CO2. These gases are emitted (in place of CO2 in normal cell operation) during the onset of anode effect in the electrolytic cell when the alumina concentration in the bath falls below ~ 2%. Environmental effects in this regard can be reduced by cutting down the frequency of occurrence of anode effect; significant efforts have gone into prediction of anode effects and in developing a protocol for measurement of these emissions (6). As pointed out in Table 1, ~ 85% reduction in PFC emission has been achieved in the last two decades.

5. Scrap Recycling

As already pointed out, scrap recycling forms an important component of sustainable aluminium industry. Recycling leads to considerable reduction in energy consumption and GHG emissions (~5% of corresponding values in electrolysis) and considerable savings in raw material, bauxite, caustic soda, carbon and fluorides. The important sources of scrap are

- Used beverage containers (UBCs), consisting of three alloys, 3004 (Al-1Mn-1Mg) for can body, 5182 (Al-4.5Mg-0.3Mn) for can end and 5042 (Al-3.5Mg-0.3Mn) for can tab
- End-of-life automobiles – cast (Al-Si with other additions from power train components, pistons, transmission case, cylinder heads and blocks and wheels), and wrought, 1xxx and 3xxx alloys (radiators and heat exchangers), 2xxx, 5xxx, 6xxx and 7xxx (from closure panels)
- Construction (1xxx and 6xxx alloys) and municipal waste.

The frequency of scrap generation varies considerably; UBCs are recycled once every six to eight weeks while automobiles and buildings and related structures have useful life of typically one to two decades. Significant advances have been made in down gauging and light weighting beverage containers in the last two decades; weight reduction achieved in cans, without sacrificing the mechanical properties, is between 25%-30%. While this development is a welcome step in material conservation, it poses problems in recycling as the light weight high surface area scrap is prone to considerable oxidation and hence to large melt loss.

5.1 *Challenges to Scrap Recycling*

Scrap from the sources mentioned above contains a number of aluminium alloys (cast and wrought), other nonferrous metals, steel, glass, plastic, organic matter and stones. The following are the challenges posed to efficient scrap recycling:

- amount of scrap collected for further processing in relation to the amount generated (collection efficiency) – the present figures are somewhat discouraging, ~ 60% for cans and ~ 80% from automobiles.

- separation of aluminium alloys from other metals and nonmetallic material

- detection and separation of alloys in the aluminium part of the scrap, either as cast and wrought or as individual alloy family (1xxx, 2xxx, 3xxx, 5xxx, 6xxx and 7xxx)

- decoating, where necessary, with minimum environmental impact

- melting the separated alloys with high recovery (low melt loss) and high energy efficiency

- melt treatment to minimize inclusions, dissolved hydrogen and demagging to adjust the magnesium content; problems in foundry due to reduced fluidity of the melt containing scrap.

- possible use alloys from scrap for conventional applications

- Conversions of waste resulting from scrap melting into useful products or develop appropriate disposal methods which would have minimum impact on the environment.

5.2 *Closed Loop Recycling:*

The concept of closed loop recycling for both UBCs and auto scrap is now well developed and implemented. The steps involved are shown in Fig. 4 for auto scrap recycling (7); they are equally applicable to the processing of UBCs. Scrapped vehicles from the consumer are passed on to the auto dismantler, where the automobile is dismantled to the economically viable extent, for the separation and storage of usable parts such as wheels and radiators. The remaining hulk is sent to the shredder for size reduction. Shredders typically operate 2000-6000 HP hammer mills, liberating individual material and helping easier separation. Dust, fluff and foam are removed by air suction from the shredded pieces and ferrous components by magnetic separation. The Non magnetic metal shredder fraction (NMSF) is subjected to the sink float process which uses water and water slurries (water-magnetite SG ~2.5, water-ferrosilicon SG ~3.5) as media for separation. Wood, paper and foam float in the first step, rubber, plastic, Mg and hollow Al float in the second. and aluminium, rock and insulated wire float in the third. As Al particles are present in both the 2.5 and 3.5 float, eddy current separation is employed to separate the aluminium from nonmetallics. The residue from the shredder (ASR) is disposed off by land filling. The resulting mixed aluminium scrap (containing cast and wrought alloys) is sold to the secondary aluminium industry for recycling into foundry alloys. Important scrap sorting technologies are based on X-ray fluorescence, imaging (colour developed after etching), laser-induced breakdown spectroscopy (LIBS) and hot crushing. NITON XRF and X-MET5000 analysers are portable instruments available for quick identification of the elements present and therefore in characterizing the scrap.

LIBS and colour etching techniques are in advanced stages of development and are promising. LIBS is a form of atomic emission spectroscopy for determination of chemical composition where output from a high power laser is focused on to the surface to be analyzed for a very short time, typically ~ 10 ns. This leads to the formation of plasma within which the ejected material is dissociated into excited atomic and ionic species. At the end of the laser pulse, the excited atoms and ions emit characteristic radiation and revert to lower energy states. Detection and spectral analysis of the characteristic radiation using a sensitive spectrograph is the basis of LIBS. Since a minute amount of material is consumed, the method is virtually nondestructive. The colour Etching technique combines chemical etching with an optical technique to sort the aluminium alloys by colour. The technique is developed by ALCOA in partnership with Pacific Northwest National Laboratory. The objective is to separate the scrap mix into five alloy families (2xxx, 3xxx, 5xxx, 6xxx and 7xxx) for recycling into higher value wrought alloy applications. Mixed scrap on a conveyor belt is passed through a hot caustic soda bath and dried; the etching process imparts colour to the pieces that are unique to the class of alloys to which they belong. Optical sensors are used to detect five different colours, black, grey, silver, tan and gold. Following detection an air jet is used to shoot the individual pieces into different bins. The hot crush system for sorting scrap is a thermo-mechanical treatment, utilizing the difference in the melting range (solidus - liquidus) of the cast and wrought alloys. Mixed cast and wrought scrap is heated to a temperature slightly below the melt temperature of the cast alloys, followed by mechanical crushing or grinding, by hammer mill or flail mill, to break down the cast alloy pieces. The wrought alloys remain solid during heating and do not fracture. The crushed mix goes through a size separation process, such as rotary trommel, which separates smaller cast fragments from the wrought alloys. The method is quite effective in separating cast from wrought components, giving a better than 96% separation.

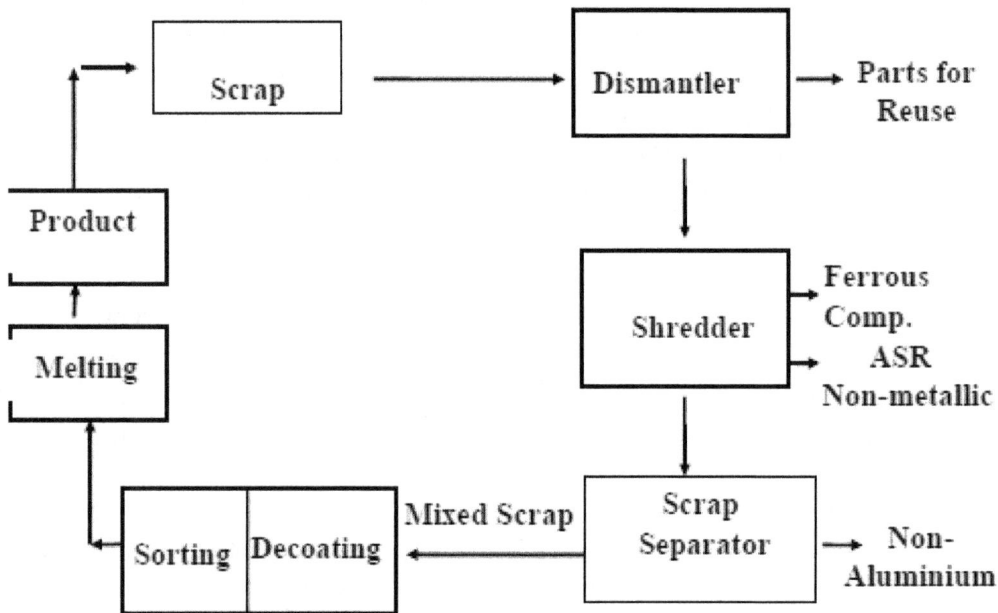

Fig. 4 Recycling of Automotive Aluminium Scrap.

Decoating can be carried out by two methods: prolonged exposure at lower temperatures (480°-520°C) which minimizes metal oxidation or rapid heating to a temperature just below the melting point (590° to 620°C) with very short exposure. Three types of furnaces have been developed for the process - rotary kiln, packed belt reactor or belt decoater and fluidized bed decoater. The hot process gas generated during decoating is used to preheat the scrap.

Several melting process are used depending on the scrap quality. Clean, uncoated scrap can be melted in a reverberatory furnace. Finely divided scrap, such as turnings, are ideally melted in a coreless induction furnace. The main advantage of rotary furnace melting is the ability to process dross and low-grade scrap which is difficult to process in other types of furnaces. The disadvantages are low efficiency, higher maintenance requirements, and considerable salt cake production which must be disposed of as a hazardous waste. Fluxless melting is well developed wherein the melting is carried out in an inert atmosphere. Molten metal fluxing and filtration technology have been developed to produce aluminium alloys of the correct quality without undue environmental impact.

Most secondary aluminium melting furnaces are fitted with fabric or ceramic filters to ensure that flue gases have a minimal dust content and other substances, regulated by the Environment Agency. Melt loss is a function of the chemistry and size of the scrap and temperature of melting. The loss can be substantially high (>10%) for thin, < 1mm samples, Mg-containing (5xxx in particular) alloys and higher temperature and agitation of the melt. Typical figures for the capacity, energy requirements and melt loss for various melting facilities are given in Table III (8)

Table 3 Typical capacity, energy consumption and melt loss in various melting furnaces

Furnace type	Typical Capacity	Energy consumption. kcal//kg	Typical dross loss
Reverberatory	1 to 150 tons	700-1400	3-% Existing, 2% New
Rotary	2 to 5 tons	> 1400	Dross feedstock 35-95% Al
Tower	10 tons/hour	650-900	< 1.5 %
Gas Crucible	< 1 ton	1100-1400	3%
Recuperated Crucible	< 1 ton	650	3%
Electric Crucible	< 1 ton	450-550 kWh/ton	2-2.5%
Electric Reverberatory	< 1 ton	400-500 kWh/ton	1%
Induction	Very small to 7.5 tons	450-550 kWh/ton	1%

Recent developments such as low turbulence submergence system (LOTUSS) minimize melt loss by making use of a metal pump to circulate the melt; the scrap is immersed in the melt rather than exposing it to the offensive atmosphere of the reverberatory furnace, ensuring better heat utilization and less oxidation (9). Similar efforts in energy conservation, reduction in gas emissions and melt loss, supported by the Department of Energy of the US include the vertical floatation melter and isothermal melting system (ITM) (10,11). The former offers simultaneous decoating, preheating and melting possibilities, with thermal efficiency of 58%. ITM uses immersion heaters to maintain the temperature of the molten metal more or less the same in all the regions of the furnace (the pumping, heating, charging and treatment bays). The requirements of the heaters are providing high heat flux and possessing chemical, thermal and mechanical robustness in the industrial molten aluminium environment. The energy requirements are about 30% of those in gas fired furnaces.

Pollutants arise from the pretreatment and melting/refining stages. These include particulates, chlorides, organics, scrubbing waste, fluorides, demagging effluents, dross and salt cake. Salt cake consists of entrained aluminium, spent salt fluxes, residual oxides. ~ 1 million tons of salt cake are land filled in USA annually. The leachable chloride content in the cake is an environmental concern. Treatment of the cake involves crushing, screening, dissolving soluble salts in water; filtration and evaporation; the method energy intensive and expensive. Investigations are concentrated on replacing the expensive evaporation stage by electrodialysis; simultaneously efforts are on to convert the nonmetallic part into more value-added products.

6. Summary

Considerable progress has been made in setting up objectives for sustainable development of aluminium industry and careful monitoring of improvements effected to achieve the objectives in the last two decades. Efficient scrap recycling is the most important step in this regard; presently secondary aluminium accounts for ~1/3 of the metal usage. Increasing the secondary metal input (which is likely to be the case with more availability of scrap from auto, packaging and construction sectors and using more recently introduced sorting and melting technologies) will reduce less usage of natural resources, significant reduction in energy consumption and GHG emissions.

References

1. Havard Bergsdal, Anders H. Stromman and Edgar G. Hertwich, The Aluminium Industry, Report No. 8/2004, Norwegian University of Science and Technology, industrial Ecology Programe Aluminium for future generations/2007 update, International Aluminium Institute London.

2. http://pc3.keikyo-unet.ocn.ne.jp/japanese/OLD/eng

3. Aluminium for future generations/2007 update, International Aluminium Institute London

4. T.R. Ramachandran and H. Mahadevan, Aluminium Industry in the New Millenium; Metallurgy in India, A retrospective, eds. P. Ramachandra Rao and N.G. Goswami, India International Publisher, 2001, p. 427.

5. Ausmlet TSL Technology, Nonferrous and Waste Processing: Technology and Commercial Experience 2005, Info@ausmlet.com.au, www.ausmlet.com.au

6. Protocol for measurement of tetrafluoromethane (CF4) and, hexafluoroethane (C2F6) emissions from primary aluminium production, US Environmental protection agency, Washington DC and International Primary Aluminium Institute, London, April 2008.

7. Aluminium Industry Road Map for the Automobile Market: Enabling Technologies and Challenges for Body Structures and Closures, The Aluminum Association, Inc.,1999.

8. http://www.energysolutionscenter.org/heattreat/metalsadvisor/aluminum/recycle_ and_scrap_melting/recycle_and_scrap_melting_equipment.htm

9. www.pyrotek.info/news.php?id=24

10. www1.enrgy.gov/industry/metalcasting/pdfs/floatation.pdf

11. www1.eere.energy.gov/industry/aluminum/pdfs/itm.pdf

Applications of High Power Lasers in Advanced Manufacturing

Shrikant V.Joshi

School of Engineering Sciences & Technology
University of Hyderabad, Hyderabad.
svjse@uohyd.ernet.in

Abstract: *By virtue of the numerous unique advantages that they afford, lasers have gradually found niche applications in many fields of science and technology, with advanced manufacturing being one of the most promising avenues. The development of CO2 lasers around 1970, with many kilowatts of power, tremendously increased the possibility of commercial development of lasers for material processing. The past decade has seen giant strides being made in the evolution of many other classes of lasers and a variety of lasers, such as pulsed CO2 lasers, solid-state lasers, excimer lasers, diode-pumped lasers and fiber lasers, have moved from the laboratory to the realm of industrial applications. The advent of these lasers has further enhanced the versatility of the laser as a processing tool of significant practical relevance. Commercial lasers available today offer an astonishingly large range of properties. These lasers span wavelengths from the ultraviolet through the infrared, pulse widths from femtoseconds to microseconds, and provide high peak powers of hundreds of joules in case of pulsed lasers and power of tens of kilowatts in continuous wave lasers.*

The considerable above development in the variety of laser sources available has led to their growing utility in recent years and numerous applications of lasers are already well-proven. The increasing interest in the use of lasers for manufacturing can be primarily attributed to the unique advantages like non-contact processing, high processing speeds, ease of automation, precise localization of treatment etc. which are generically applicable to the entire range of materials processing applications. The increasing demand for advanced difficult-to-process materials has also contributed to the growing promise of laser-based manufacturing. The more obvious laser-based operations such as laser cutting, laser welding, laser heat treatment and rapid prototyping have already found their way into actual manufacturing shop floors. In addition, the applications of lasers in surface alloying, cladding and glazing now offer the exciting possibility of producing new materials with novel properties. Apart from their selective nature, the ability of laser beams to be easily shaped into rectangular spots, lines or into more complex axisymmetric shapes for special purposes, considerably improves their versatility for surfacing operations. In recent times, the use of robots in conjunction with lasers has also served to significantly enhance productivity and quality, besides improving process flexibility. The growing adoption of lasers for materials processing is discussed in the presentation, with specific examples of applications developed in the authors laboratory.

1. Introduction

Invented as *.a solution looking for a problem.* in the 1960s, the laser has graduated to become the most versatile super-tool mankind has ever known. During the past four decades, lasers have gradually found their way into many scientific and engineering applications, which now span the entire commercial, medical and industrial scene, starting from the compact disc player at home to eye surgery at the hospital and welding in the automobile and aircraft industries. Such a diverse application spectrum has also necessitated the development of laser sources and systems with equally varied specifications. For example, the CD player uses mm size diode lasers whereas multi-kilowatt lasers for industrial processing have footprints in tens of square meters. Not surprisingly, the lasers available today also offer a wide range of energy density, varying from milliwatts/cm2 to up to 1017 watts/cm2, to adequately meet the requirements of the entire gamut of applications. Since its initial discovery when Schawlow and Towns established the theoretical concept of the laser in 1958, and Maiman first invented the working Ruby laser in 1960, a large number of laser systems have been developed. It is believed that the Ruby laser's output used to be measured in "gillettes", after the number of Gillette razor blades a given pulse of laser energy could penetrate. One could argue these were the first laser drillers and one of the early applications of the laser was to drill industrial diamond wire draw dies. The development of CO2 lasers around 1970, with many kilowatts of power, tremendously increased the possibility of commercial development of lasers for material processing. The subsequent advent of pulsed CO2 lasers, solid state lasers, excimer and diode-pumped lasers has further enhanced the versatility of the laser as a manufacturing tool of significant practical relevance.

The past four decades have seen giant strides being made in the evolution of all the above classes of lasers from the laboratory to the realm of industrial applications. Given the considerable development in the field of lasers in recent years, the already proven applications of lasers are virtually all expansive.

2. Advantages of Laser Materials Processing

The increasing demand for advanced difficult-to-process materials and the availability of high power lasers have stimulated great interest in laser-based manufacturing. The more obvious laser-based operations such as laser cutting, laser welding, laser heat treatment and rapid prototyping have already found their way into actual manufacturing shop floors. In addition, the applications of lasers in surface alloying, cladding and glazing now offer the exciting possibility of producing new materials with novel properties. The increasing interest in the use of lasers for manufacturing can be attributed to the several unique advantages listed below (along with their concomitant ramifications) which are generically applicable to the ntire range of materials processing applications:

- ✓ High processing speeds—**Increased productivity**

- ✓ Non-contact processing—**No tool related problems**

- ✓ Elimination of finishing operations—**Reduced cost**

- ✓ Ease of Automation—**Processing of complex profiles**

- ✓ Improved Product quality—**Low rejection rates**

- ✓ Precise localization of treatment—**Negligible distortion**

3. Types of Lasers

Commercial lasers available today offer with an astonishingly large range of properties. These lasers span wavelengths from the ultraviolet through the infrared, pulse widths from femtoseconds to microseconds, and provide high peak powers of hundreds of joules in case of pulsed lasers and power of tens of kilowatts in continuous wave lasers. However, the most popular currently used industrial lasers for materials processing are the CO_2 and the Nd:YAG lasers. The characteristic properties of these lasers are given in the table below:

Table 1 Comparison of CO2 and Nd:YAG Laser Characteristics

Property	CO2 Laser	Nd:YAG Lasers
Lasing medium	CO2 + N2 + He gases	Neodymium-doped Yttrium-Aluminium-Garnet
Radiation wavelength	10.6μm	1.06μm
Excitation method	Electric discharge	Flash lamps (or diode)
Consumables	CO2, N2, He, electricity	Flash lamps, electricity
Output powers	Up to 45kW	Up to 4kW
Beam transmission	Polished metal mirrors	Fiber optics or mirrors

4. Laser-Based Materials Processing

The laser beam with large amount of energy is appropriate for processing a wide range of industrial materials. In general, laser applications can be grouped as shown in the accompanying chart in Fig.1. Some of the more prominent laser materials processing operations from among those depicted in Fig.1 are briefly discussed below:

4.1 *Laser Cutting & Drilling*

A major laser material processing application, primarily seen as a complementary technology for turret punch press and now looked upon as a replacement for punch presses in industries where the product cycle is short, involves laser cutting. The laser has the special ability to cut virtually every material under the sun - even the hardest material known to man, i.e. diamond, is cut,

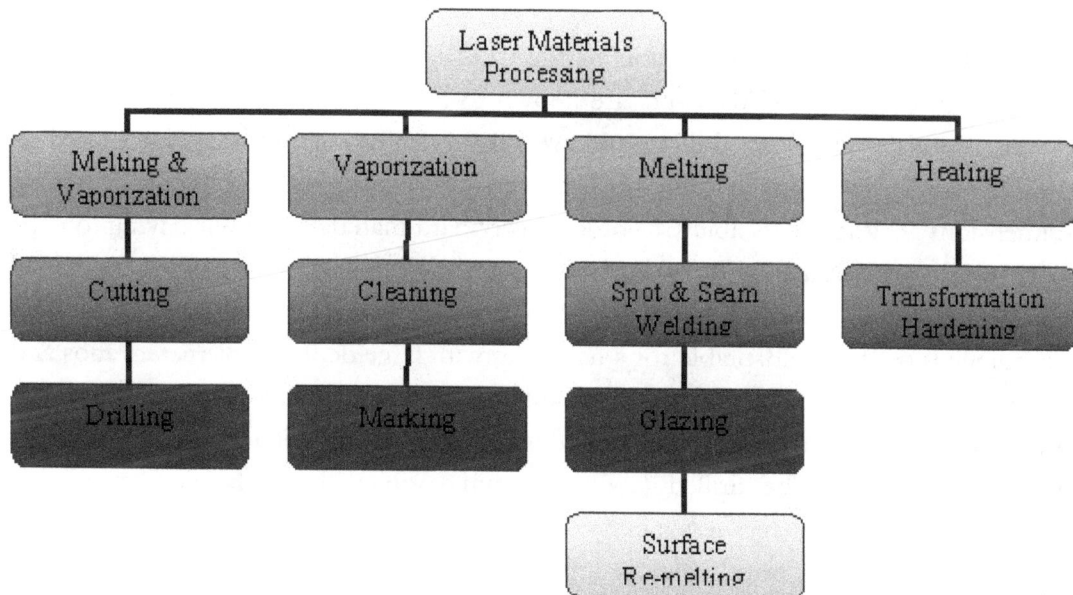

Fig. 1 Spectrum of laser-based materials processing operations.

kerfed and drilled using a laser. In the new industrial environment that demands high quality at a competitive cost, reduction of waste and rework, increased productivity and flexibility as well as superior finished products, a laser cutting system has become an absolute necessity for the sheet metal processing industries. The advantages that this technique offers as compared to conventional methods of cutting are now being increasingly appreciated by the industry.

Fig. 2 Laser drilling in action   **Fig. 3** Laser cut components

While the major current application of the CO2 lasers is for two-dimensional sheet cutting, the pulsed Nd:YAG lasers are known to be ideally suited for small precision work, especially involving drilling and welding. Nd:YAG lasers are widely used in drilling applications due to the high processing speeds, repeatability and minute dimensions achievable. Its ability to process very hard materials without having to cope with tool wear because of the non-contact nature of the process, as well the added facility of drilling at a low angle to the surface are seen as significant advantages. This technique is especially adaptable for small holes with large depth-to-diameter ratios and the operating cost is low. The above advantages far outweigh the few limitations that exist in terms of the slight taper encountered and the restriction on the material thicknesses that can be drilled (about 50 mm). While electrochemical machining (ECM), electro discharge machining (EDM) and electrochemical drilling (ECD) techniques have also been employed for fine hole drilling, laser drilling provides significant benefits and is now widely acknowledged as the state-of-the-art technique for a majority of precision drilling applications.

4.2 *Laser Welding*

Although cutting accounts for up to 80% of laser materials processing applications, laser welding is growing in the number and range of applications in the ship building, electrical/electronic, domestic products, railways, power generation and automotive industries. Laser welding is a fusion welding process where welding is performed by

202

melting part of the workpiece, normally without the addition of filler material. Some shield gas is usually applied to protect the weld from oxidization, Argon or Helium being the most common choices. The process offers significant advantages over conventional TIG and Resistance Welding techniques and also provides certain unique benefits over the sophisticated EB welding method as shown in Table 2.

Table 2 Advantages and Shortcomings of Laser Welding vis-à-vis Other Welding Processes.

Laser	Electron Beam	TIG	Resistance
Advantages			
Welding in air	Deep penetration	Thickness	Equipment cost
Equipment reliability	High weld quality		Part clamping
High throughput	High-conductivity metals		
Multiple workstations	Fast welding rates		
Weld magnetic metals	Dissimilar metals		
Simplified fixturing	Low distortion		
Multi-axis manipulation	Welds high-carbon steel		
Flexibility for production	Weld steel with P and S		
Fast Rate			
Low total heat input			
Narrow heat affected zone			
Bead width			
Narrow gap filler			
Single-pass butt welding			
Disadvantages			
Penetration limited to 1.0 "	Up time	Rate	Electrode wear
Limited success in highly reflective metals	Process in protective environment	Part clamping	Joint access
High equipment cost	Lower throughput	Inter-pass cleanup	Rate
Close part fit-up required	High equipment cost	Bead width	Part distortion
	Complex fixturing required	Thermal distortion	Dissimilar metals
	Complicated multi-axis manipulation	Penetration	Heat sensitive components
	Part cleaning		Joint efficiency
			Rectilinear shapes
			Contact process

Nowadays, increasingly, aluminium alloys are being used in the transportation industry, especially as cast components for automotive or aircraft applications, in order to reduce lighter parts. Recent innovations in high power Nd:YAG lasers are shifting the automobile manufacturers to move into making aluminium body for trucks. A major laser welding application in the automobile industry is the manufacturing of Ultra Light Steel Auto Body (USLAB). Also called laser blanking, this process is used to weld together material of different composition and thicknesses (steel and alloys) and then processed to make automobile body components that are 30 % lighter in weight at the same time have more toughness.

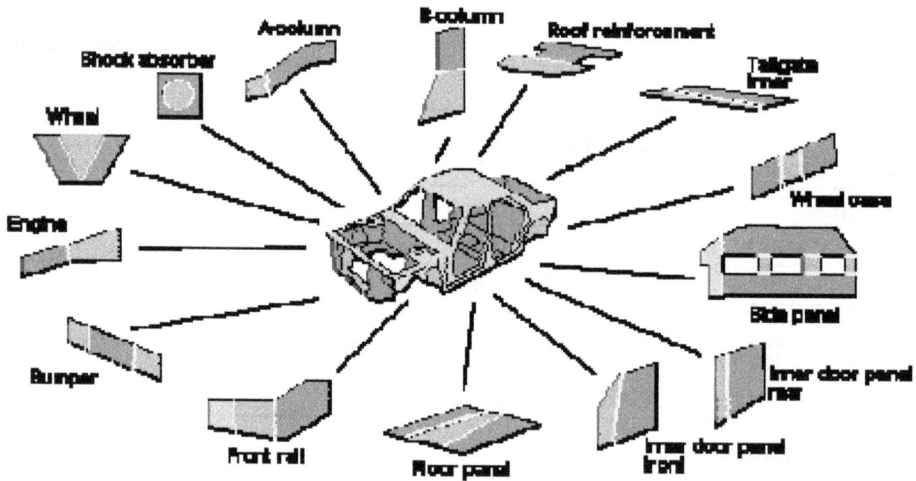

Fig. 4 Illustration of Laser welded tailored blank parts used in fabrication of a present day car.

Fig. 5 Laser welded transmission part.

Fig. 6 Laser welded solenoid valves.

4.3 Laser Surface Modification

Among the many surface treatment processes that are available today, lasers provide unique options for high quality surface modification. Their potential impact can be judged from the numerous laser assisted coating technologies that currently exist, opening up new possibilities that are both technically and economically promising. With the arrival of high power lasers, laser surface modification has gradually become a very promising alternative for producing layers that can be gainfully utilized for enhancing the performance and durability of engineering components. The process typically involves the deployment of controllable high power density available from a laser source to suitably modify the properties of a surface. While laser transformation hardening only requires the component surface to be heated above its transformation temperature, the other variants demand either melting of the surface of a component alone, or together with an extraneous pre-placed or dynamically introduced material. For example, in laser re-melting and glazing, only a thin layer on the component surface is melted to achieve improved tribological properties. By contrast, in laser alloying/cladding, a thin layer at the surface of the component is melted along with a simultaneously added alloying/cladding material. Laser alloying and cladding are extremely versatile in their ability to tailor the surface properties as demanded by any chosen application through an appropriate choice of the extraneously introduced material. The two processes are very similar except in respect of the extent of dilution by the substrate material, which is kept to a minimum in case of cladding. Laser glazing, apart from melting a thin layer of the parent material, can also be used to melt and rapidly solidify an already coated layer on the component surface. This, too, is a subject of substantial research interest in an effort to achieve pore sealing and densification to improve properties of the coated surfaces.

4.4 Laser Heat Treatment

Laser heat treatment can be broadly classified into laser transformation hardening and laser re-melting, although laser shock hardening has also been reported. It requires substantially lower laser power than alloying or cladding, usually by a factor of 2. Laser heat treatment is also selective in nature - the laser beam is defocused such that an average power density in the range 10^3-10^4 W/mm2 impinges on to the job and controlled relative motion between the work-piece and the beam results in hardening of only pre-selected areas on the surface. The most commonly applied laser heat treatment, transformation hardening, is induced by heating a material above a critical temperature and then rapidly quenching it rather than allowing equilibrium phases to form. Several advantages accrue, as summarized in Table 3. On the other hand, as the name suggests, laser re-melting operations involve melting of the surface and are carried out at relatively lower scan speeds or higher power levels. Since the high reflectivity of several metal surfaces (particularly at the 10.6 μm wavelength of CO2 lasers) can lead to poor coupling of the laser energy to the work-piece, a coating is often applied to increase absorption. Typically used coating materials are black paint, colloidal graphite or phosphates.

Table 3 Comparison of Laser Hardening with Competing Technologies.

Characteristic	Laser Hardening	Induction Hardening	Flame Hardening	Electron Beam
Distortion	Very low	High	Very high	Low
Quenchant Requirement	No	Yes	Yes	No
Process rate	High	High	Low	High
Case Depth Controllability	Easily controllable	Non-controllable	Non-controllable	Controllable
Post processing	Not required	Required	Required	Not required
Thickness Limitation	No	Depending on job profile	Yes	No
Equipment cost	Very high	High	Very low	Very high
Process flexibility	High	Low	Very low	Low
Other special Requirements	Some prior absorbent coatings	Fabrication of complex coils for specific job requirements	Not required	Vacuum
Automation	Possible	Not possible	Not possible	Possible
Selective Treatment	Possible	Not possible	Not possible	Possible

Output beams can be easily shaped into square spots, lines or into more complex axi-symmetric shapes for special purposes. This makes it easy to heat treat not only flat surfaces but also such machine parts as bearing races, gears, shafts, cylinders and camshafts. The above technique is ideal for improving the mechanical properties of above materials, with significant improvements in wear resistance and fatigue life being possible. Although typical requirements demand a 0.75-1.0 mm case depth, the depth of the hardened layer can be as high as 2.0 mm and can be controlled over a wide range depending upon the laser parameters employed. The carbon content of the steel plays a crucial role in determining the final level of surface hardness that can be achieved.

Fig. 7 Laser hardening of steam turbine blade.

Fig. 8 Laser re-melting of automobile camshaft.

4.5 *Laser Cladding & Alloying*

Laser surface alloying is a process wherein the substrate surface properties are modified as desired by first melting the surface and introducing appropriate alloying elements into the melt pool. An excellent metallurgical bond is ensured between the alloyed layer and substrate. In a sense, laser surface alloying is one extreme of laser cladding, which involves bonding a new material to the existing surface with minimal mixing of the materials. Therefore, the basic difference between laser surface alloying and cladding is that, in case of the latter, dilution from the substrate is kept to a minimum. As such, if the laser cladding operation were carried out at high power or at very low scan speeds, surface alloying would result on account of increased dilution.

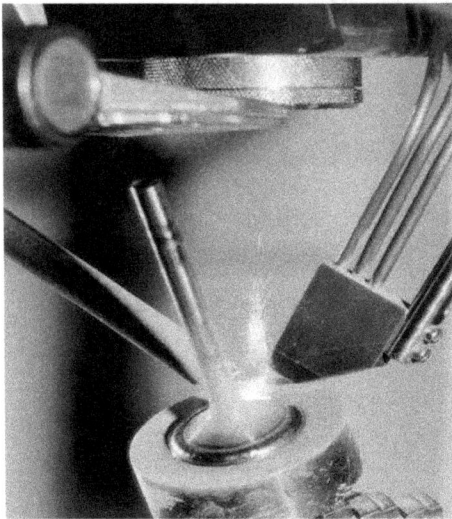

Fig. 9 Laser cladding of engine valve seat.

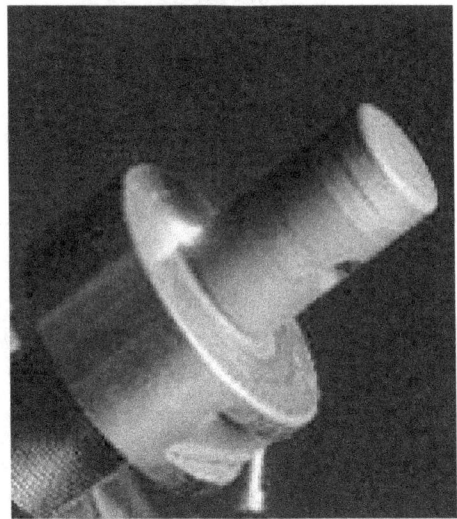

Fig. 10 Laser surface alloyed part.

Between them, the above laser surface modification techniques permit the surface properties of a component to be suitably tailored to meet virtually any functional requirement without compromising its bulk properties and, thereby, improve the parts ability to withstand aggressive environments. Typically an inexpensive substrate is alloyed or clad with a more expensive material to achieve the desired surface properties. Appropriate tailoring of the surface properties by using suitable cladding or alloying materials increases the components wear and corrosion resistance. In many instances, the properties of the clad alloy / substrate material composite are unique and cannot be obtained in any other manner regardless of cost. Laser surface modification can save costly and strategic materials by allowing the bulk of a component to be made from inexpensive materials besides making it possible to produce parts with unique properties. Cladding can also be used for refurbishing worn machine parts.

5. Laser Material Processing in India

The most popular laser for material processing, the CO2 laser, was invented by Dr. C.K.N Patel, an Indian. However, the Indian industry has remained a long way from utilizing the benefits that can potentially accrue from his invention. In spite of the proven advantages and well-documented success stories related to the industrial adoption of the laser technology in the developed countries, laser processing has not yet taken off in India. While work on the development of lasers and assessing their feasibility for industrial and medical applications was initiated long ago, most of these efforts have been restricted to the research laboratories and the use of lasers in the Indian industry is still insignificant. Probably only the diamond industry in India has fully embraced the laser technology at present. This is in stark contrast to the worldwide market for laser materials processing which currently exceeds US $ 10 billion. Only a few Indian industries use lasers in their production line. This is basically due to the fact that most high power lasers have a prohibitive capital cost. Lack of awareness of laser processing capabilities, non-availability of suitable trained manpower and absence of easily accessible laser facilities for jobbing have also been contributing factors. While there are some high power lasers set up in premier institutions for either R&D work or captive use, their accessibility to industries has been difficult. From an infrastructure standpoint, the Centre for Laser Processing of Materials (CLPM) in Hyderabad is the only comprehensive laser material processing facility in the country based on high power lasers. CLPM is a part of the International Advanced Research Centre for Powder Metallurgy & New Materials, which is a grants-in-aid research institution of the Government of India, Department of Science & Technology, and has largely focused its efforts on catalyzing the use of high power lasers by the Indian industry. The facilities in CLPM include:

Table 4 List of facilities available at CLPM

LASER	POWER	APPLICATIONS	JOB HANDLING
CW CO$_2$	9 kW	Welding, Surface HHardening, Glazing, Alloying & Cladding	4 – axis CNC Workstation
Slab CO$_2$	3.5 kW	Welding, Cutting	4 – axis CNC Workstation
Pulsed Nd:YAG	400 W avg. 20 kW peak	Precision Cutting, Drilling, Welding & Marking	4 – axis CNC Workstation
SS Diode	6 kW	Surface Modification, Welding	Robot equipped

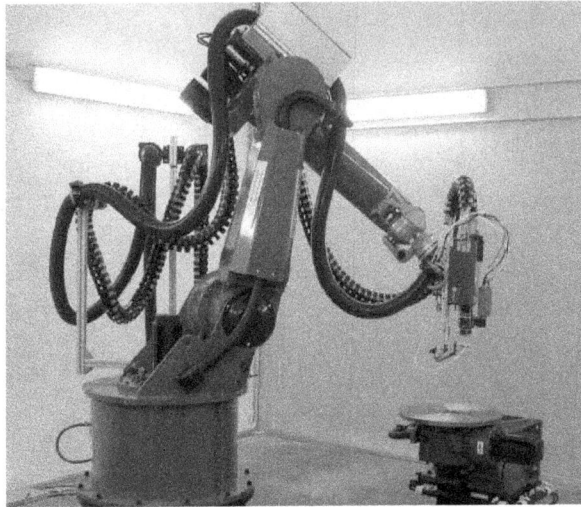

Fig. 11 Recently installed 6kW diode Laser at CLPM l.

Fig. 12 Hastelloy Combustion Chamber laser processed at CLPM.

The above high power lasers are capable of addressing most of the materials processing applications of interest in advanced manufacturing. As the only high power laser processing demonstration centre in the country, CLPM works together with institutes and industries to target research to the needs of industrial users and to develop innovative applications of high power laser processing for the benefit of the Indian industries.

6. Concluding Remarks

The escalating demands for advanced materials and the availability of high power lasers have stimulated considerable interest in research and development related to laser-based applications. By virtue of the giant strides made in the field, high-energy lasers are increasingly being inducted onto manufacturing floors. Already laser-based operations have made major inroads into materials processing techniques such as cutting, drilling and welding. These techniques are already well-studied, developed and established as reliable, efficient and cost-effective production methods. Lasers are also finding growing acceptance as convenient tools for heat treatment to enhance the tribological performance of such critical components as automotive camshafts and gas turbine blades. The promise of laser surface alloying and cladding is also now widely acknowledged. The above developments clearly point to the mounting relevance of high-power lasers in advanced manufacturing.

Modeling of Metallurgical Processes

Vijai Mohan V

Deputy General Manager
Infotech Enterprises Limited, Hyderabad, India
E-mail: vijaymohanv@infotechsw.com

Abstract_ The steel products manufacturing involves a long line of complex processes: liquid metal elaboration, solidification, hot and cold transformation by rolling surface protection by coating. The Process Control aims at improving global productivity and quality of the resulting products by optimizing each elementary process as well as management of tools or workshops interfaces. Complex processes, involving generally many variables, require for their control more or less sophisticated models. These process models are either analytical when physical and thermodynamically mechanisms are known or statistical or knowledge based, according to circumstances. In any case, it is necessary to have a reliable and precise instrumentation to adjust undetermined parameters during model development and to be able to take into account external parameters variability during current working. Complex metallurgical processes can be analyzed quickly and effectively using simulation models. Recent innovations provide a powerful tool for evaluating flow sheets, taking into account the key economic drivers over the whole operating life of a project. Mine plans, metallurgical characteristic models, equipment reliability statistics and surge capacities are all considered in an integrated plant dynamic model. The result is lower risk and greater confidence in engineering decision making. Nowadays, metallurgical process needs to stay on the cutting edge of new technologies to remain competitive in the marketplace. Computer Aided Modeling has been used by founders for several years, not only for designing new components, but also in the redesign of existing components. In order to make a powerful use of the modeling approach, foundries have to implement a numerical methodology which allows not only engineers, but all people involved in the development or redesign of the part, to use and analyze virtual results at the early stages. That way, the time between the concept stage and the production stage is drastically reduced and cost reductions are possible.

1. Introduction

Mathematical modeling has for a long time been a crucial importance for the understanding and prediction of phenomena in the heavy industries. Previously, the main source of knowledge was practical experience with operations and use of simplified, but fundamental understanding of the chemistry and materials that was involved in the processes. Lately, during the last three decades, the development of the digital computers has offered new tools to the process analyst. Today, CFD software is, at least in the larger companies, used on a daily basis in order to support new developments and troubleshoot operational problems. The cornerstone in CFD software is teh mixture of physical models 3and the numerical technologies that are used for solutions of the governing transport equations.

1.1 Ancient Shaft Furnace Model:

The course of the metallurgical process in an ancient shaft furnace may be presented in a simplified and schematic way by means of a graphical model.

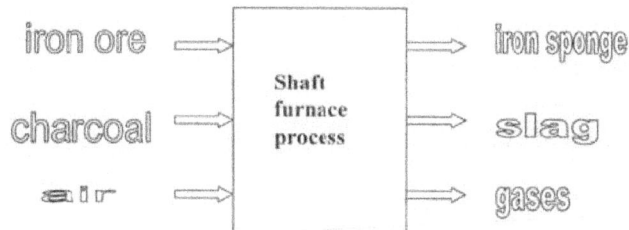

Fig. 1 Ancient iron making process model.

Figure 1 presents such a model in the form of a chemical reaction flow chart, which disregards the shape and design of the furnace and its metallurgical parameters. Modern scientific language describes such means of process presentation as the chemical reactor. Assuming that the process takes place within an enclosed space, i.e. in a chemical reactor, the following materials are supplied to the furnace: - iron ore, - charcoal, - air.

1.2 The need for innovations

Fierce competition on the world's steel markets with regard to product price and quality is moving system suppliers and steel plant operators inexorably toward highly-sophisticated automation systems capable of satisfying requirements for data acquisition, process control, logistics, and dynamic optimization. Few companies recognized early on that not only intelligent data acquisition, management, and evaluation systems would soon be needed, but innovations in the technology of the production process - metallurgy and metals making - would soon become of great importance.

For two decades, companies have been working to meet these customer needs with its own process models and those of technology partners in the process automation area. A number of reference projects underscore this experience.

2. Model concepts

The modeling concepts developed are based on the application of Newton's second law to all materials in question. In particular, each fluid is described by the well-known conservation equations for mass and momentum, expressing mass conservation by;

$$\frac{\partial}{\partial t}\left(\rho_f\right)+\frac{\partial}{\partial x_j}\left(\rho_f U_j\right)=0$$

(1)

and fluid momentum conservation by;

$$\frac{\partial}{\partial t}\left(\rho_f U_i\right)+\frac{\partial}{\partial x_j}\left(\rho_f U_j U_i\right)=-\frac{\partial}{\partial x_i}p+\frac{\partial}{\partial x_j}\tau_{ji}+\rho_f g_i$$

(2)

Where ρ_f is fluid density, U_f is fluid velocity, p is fluid pressure, g is the specific gravity and τ_{ij} is the fluid stress tensor. When the fluid stress tensor τ_{ij} is known, such as for a Newtonian fluid, the single-phase flow can be predicted by numerical solution of (1) and (2).

2.1 Dispersed flows

Extension to multiphase can be done in several ways. Let us first consider a situation when particulates (solid particles, bubbles or droplets) are dispersed in a fluid. The carrier fluid is here described by the Eulerian representation, given by the equations (1) and (2). The particle phase can now be introduced as a continuous field that can penetrate the carrier fluid. Transport equations for the phases appear from volume averages of the fluid and particles in a control volume [1]. For a gas-particle system typical transport equations that appear are for mass conservation:

$$\frac{\partial}{\partial t}\left(\rho_f \alpha_f\right)+\frac{\partial}{\partial x_j}\left(\rho_f \alpha_f U_j\right)=0$$

(3)

and for fluid momentum:

$$\frac{\partial}{\partial t}\left(\rho_f \alpha_f U_i\right) + \frac{\partial}{\partial x_j}\left(\rho_f \alpha_f U_j U_i\right) = -\alpha_f \frac{\partial}{\partial x_i} p + \alpha_f \frac{\partial}{\partial x_j} \tau_{ji} + \rho_f \alpha_f g_i + \alpha_p \beta(V_i - U_i)$$

$$(4)$$

Here we note that the volume fraction of fluid α_f and the inter-phase friction factor β (drag term) appear from the volume averages of (1) and (2).

In case of the particle mass balance we obtain:

$$\frac{\partial}{\partial t}\left(\rho_p \alpha_p\right) + \frac{\partial}{\partial x_j}\left(\rho_p \alpha_p V_j\right) = 0$$

and for particle momentum:

$$\frac{\partial}{\partial t}\left(\rho_p \alpha_p V_i\right) + \frac{\partial}{\partial x_j}\left(\rho_p \alpha_p V_j V_i\right) =$$

$$-\frac{\partial}{\partial x_i} p^s + \frac{\partial}{\partial x_j} \tau_{ji}^s + \rho_p \alpha_p g_i - \alpha_p \frac{\partial}{\partial x_i} p + \alpha_p \frac{\partial}{\partial x_j} \tau_{ji} + \alpha_p \beta(U_i - V_i)$$

$$(6)$$

Here α_p is the volume fraction of particles. Because of inter-particle collisions and momentum exchange due to collisions both the solids pressure p^s and the solid particle internal stress τ_{ji}^s is included in the equations. For more details about the origin of these equations see Laux [1]. It should be noted that in case of interphase mass transfer and more complex interphase momentum exchange mechanisms, these equations must be modified. For many metallurgical applications chemical species must be conserved within each phase. In addition the averaging procedure must be performed for the energy equation in order to yield volume-averaged equations for conservation of enthalpy.

The velocities are now averages from small control volumes and are no longer the instantaneous velocities given in the equations (1) and (2). In a swarm of large particles the fluid velocity close to the particle surface is very different from the volume-averaged velocity. However, the effects of the local variations will in practice only affect the interphase transfer terms such as drag and mass transfer.

An alternative way to include a second phase into a model is by using a Lagrangian description for the dispersed phase. In this case particle trajectories are calculated on a given flow field. The particle position is advanced in time by the momentum equation:

$$m_p \frac{d}{dt} V_i = m_p g_i + k_i + f_{ext.,i}$$

(7)

where V_i is particle velocity, m_p is particle mass, k_i is the forces caused by the surrounding fluid and $f_{ext.,i}$ is the external body forces (not gravity). The particle position X_i is calculated from:

$$\frac{d}{dt} X_i = V_i$$

(8)

By using the "Particle Source in Cell" concept due to Crowe [2] the influence of particles on the fluid's momentum can be calculated. The Lagrangian treatment of particles can be extended to deal with particle-particle interactions by the so-called Discrete Element Method (DEM) [3]. Sawley and Cleary [4] used this method to study the particle flow and inter-particle impacts in a high intensity grinding mill. See figure 1.

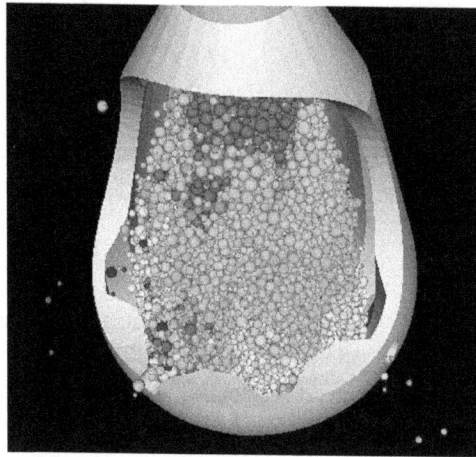

Fig. 1 From a simulation of particle milling using the Discrete Element Method (DEM) [4]. All individual particles are represented by the method.

2.2 Separated flows

When large regions of separated fluids coexist, the methods described above are hard to adopt. In this case it is possible to track the positions of the interfaces with appropriate methods and solve a single-phase momentum equation for the entire flow domain [6]. In such models the interface is a transition region where density and viscosity change abruptly over typically one single computational cell. The effects of surface tension and wetting [6] can be treated using continuous fields, associated with the interface itself. We use the method described in reference [5] to investigate free surface flow phenomena.

Wetting has recently been added to our separated flow model. At present only three wetting regimes can be reproduced. These regimes are: i) No wetting, ii) 90 ° wetting angle and iii) Complete wetting. Numerically these regimes are represented by setting the interface normal vector at the walls. In the case of a gas liquid system with no wetting, i), it is assumed that there is a microscopic gas film covering the wall when liquid contacts the wall. In the full wetting case, iii), it is assumed that all walls have a thin film of liquid. The 90° contact angle is represented by the interface normal vector that always is tangential to the wall.

We investigated the full wetting implementation on an initially cylindrically shaped droplet of water, with equivalent diameter 9.08 mm, that was placed inside a cylinder of 10 mm of height and 20 mm of diameter. The grid consisted of 40 cells in both the axial and radial direction.

Predicted interface shapes are shown in the figures 2 and 3. It is interesting to note the apparent wetting angle observed in figure 2. The Youngs contact angle imposed on the system is 180°. However, the apparent contact angle is here a result of gravity and surface tension. This tells us that:

i) The contact angle cannot be read from experiments like figure 2 without careful analyses.

ii) Numerical methods that tend to enforce a given contact angle on the interface are bound to fail. This has until recently been a popular method.

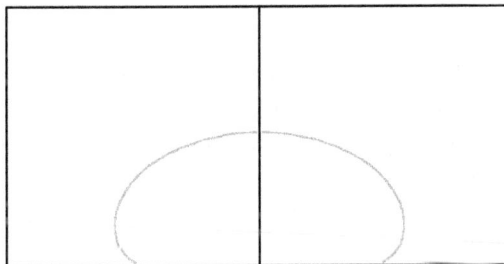

Fig. 2 Non wetted walls: Predicted equlibrium shape of 9 mm diameter water droplet.

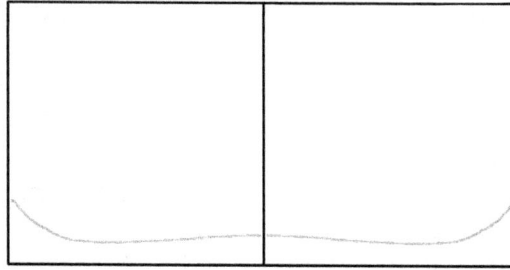

Fig. 3 Fully wetted walls: Predicted interface shape of 9 mm water droplet at time t=0.30 sec.

2.3 Turbulent flows

In most industrial flow situations the flow is turbulent. Therefore the model equations (3) to (6) must be time averaged or ensemble averaged. By using mass weighted ensemble averages for velocity, as demonstrated in Laux [1], phenomena like turbulent dispersion can be treated rigorously. The turbulence models are multiphase adoptions of the well-known k-ε model concept. In the case of dense particle suspensions we need to add model equations for the particle phase turbulent energy and dissipation rate [1].

When Lagrangian methods are applied for the dispersed phase, turbulent dispersion of particles are simulated by imposing random generated turbulent velocities [7].

In dynamic simulations like free surface flows, turbulence is most often treated by a sub-grid Large Eddy turbulence model, such as the Smagorinski model [8].

All these models need appropriate boundary conditions in order to achieve a high quality of the predictions. The most successful boundary conditions are the wall laws [7] which try to express wall fluxes as analytic expressions without solving numerically for the details in the flow boundary layers.

3. Industrial applications

In order to deal with the various multiphase flow phenomena related to the metallurgical industry a mixture of techniques may often be an advantage. In order to have a flexible tool to simulate the various phenomena we use the commercial FLUENT ^{TM} Code, which in some situations have been modified in source code or by using User Defined Subroutines (UDF).

3.1 The hierarchy of scales

In most metallurgical applications models must treat phenomena at a large number of different scales simultaneously. A typical example is a Silicon furnace, as seen in figure 4. Here the geometrical scales of the system range from micrometers (pores in carbon materials), to centimeters (quarts and iron pellets), further to meters (electrode diameter) and finally to the full diameter of the furnace that may be typically 10 m. In practical calculations the phenomena that take place on scales below, say 10 cm, must be modeled and cannot be predicted directly.

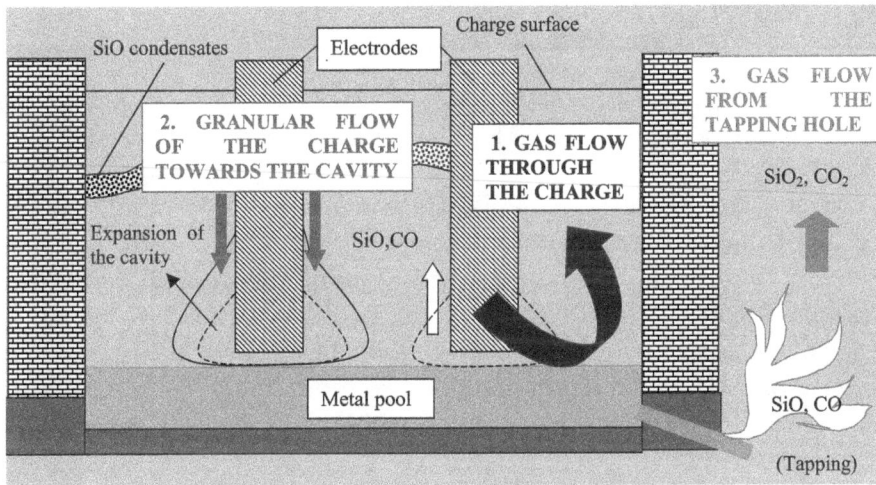

Fig. 4 Principal sketch of a Ferro Silicon furnace [9]. The raw materials are fed on the charge surface and flow slow down to the reaction zone (cavity area).

In order to model these sub-grid phenomena the flow of gas or liquid slag through a packed bed can be studied by single flow simulations as indicated in figure 5. From this type of simulation the pressure drop can be predicted and correlations regarding the permeability may be obtained for each particular system. Hence, the inter-phase friction factor β in equation (4) may be obtained even if experimental data is not available.

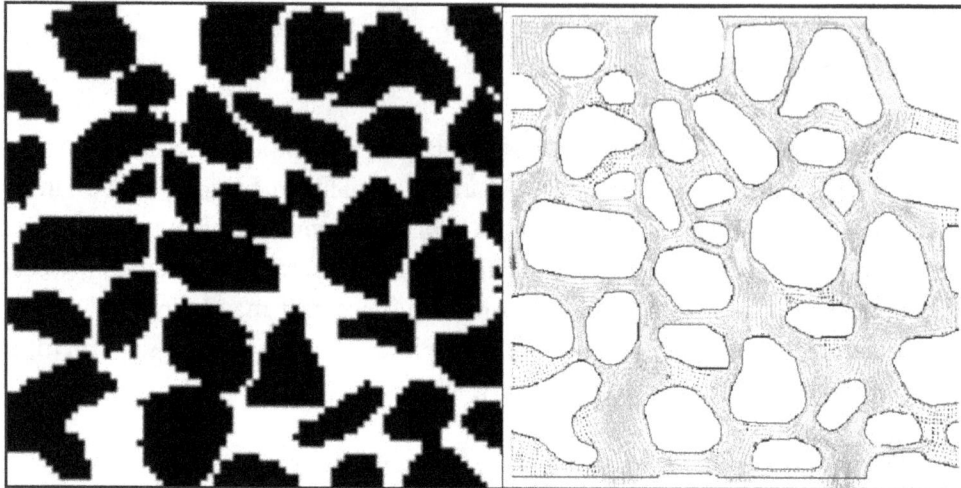

Fig. 5 To the left a digitized image of a sample of packed solid raw materials. The right hand picture shows the predicted fluid flow passing through such a bed.

We can now use the equations (3) and (4) to predict the flow of gas out from the reaction zone of a Ferro Silicon furnace. A resulting flow field of the mainly vertical gas flow is seen in figure 6. In this case the bed of solid raw materials is stagnant. The horizontal cross-section is slightly above the tip of the electrodes, as seen in figure 4. Note that, with the actual permeability for the packed bed of raw materials, the gas flow is concentrated in between the electrodes.

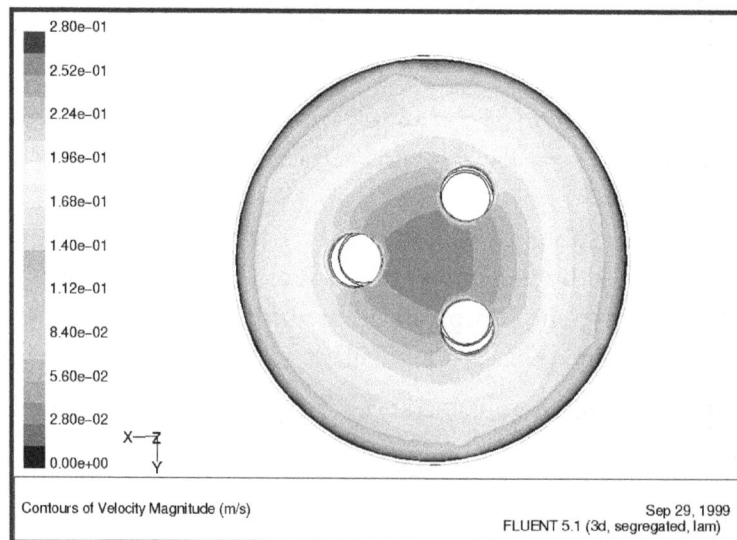

Fig. 6 The velocity distribution of CO-gas up from the bottom of a FeSi furnace [10].

By using the granular flow concept, illustrated by the equations (3)-(6), it is possible to study the flow of raw materials and gas released from the process in a two-phase simulation. In figure 7 we see the prediction of the charge surface and the velocities of raw materials and process gas. In this case the simulation is simplified to a 2D axi-symmetrical geometry with only one electrode involved. It is assumed that there is no shear stress from the raw materials on the electrode. The result is in good qualitative agreement with observation of charge movements during furnace operation [11].

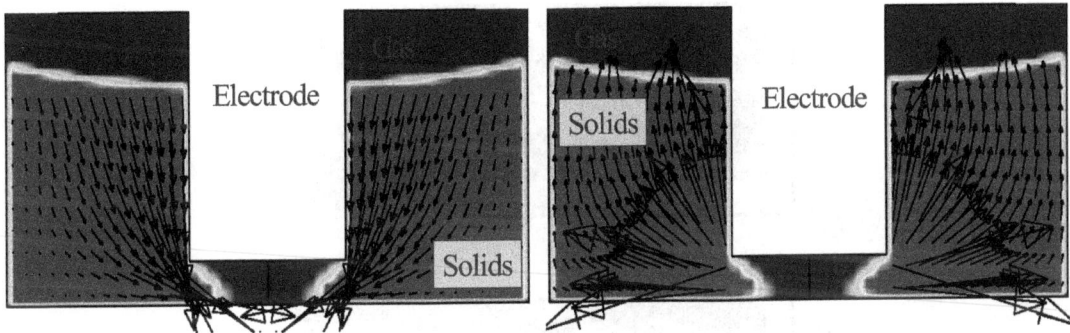

Fig. 7 Simulation of flow in FeSi-furnace using the granular flow model [1]. In the middle we se the electrode. Initially, in the computation, the bed of particles is flat after charging. Both figures show the volume fraction of solid materials after a given time. To the left we se velocity vectors for the solids. To the right the velocity vectors for the gas is displayed.

3.2 Tapping phenomena in ferro-alloy furnaces

As indicated in Figure 4 above, metal is tapped from holes in the side of these furnaces, typically ranging from 10 to 40 MW of electric power. The pool of metal indicated in Figure 4 may typically contain up to 40 tons of ferro-alloy. In the side of the furnace a hole is maintained that is opened at regular intervals. The mouth of this hole is planned to be well below the metal surface. However, immediately after opening, large amounts of combustible SiO-gas often escape from the furnace, mainly because of the large internal pressure caused by the reactions of the process itself. Flames from these "blow-outs" are a potential risk for the operators and lead to pollution of the working environment. By using a free surface flow model [5] we can explain the phenomenon in a simple manner.

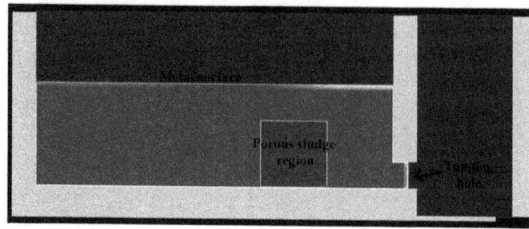

(a)

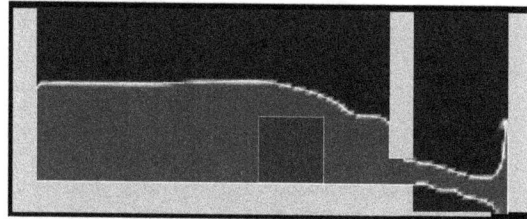

(b)

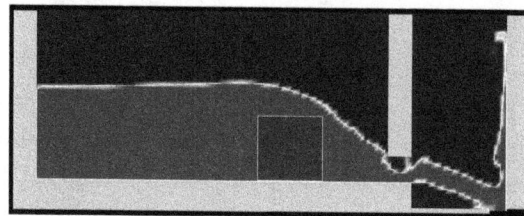

(c)

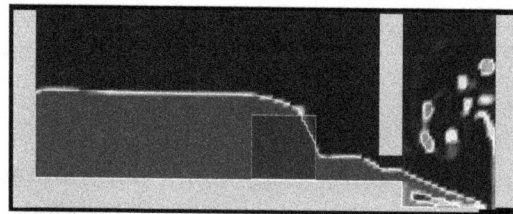

(d)

Fig. 8 The figure shows a tapping sequence in a ferro-alloy furnace. The refractory walls are colored grey. Initial metal pad depth is 30 cm. a) Initial configuration, b) Tapping is started, c) First break-through of process gas and d) Strongly reduced metal flowrate.

In Figure 8 a porous region, resembling observed regions of porous sludge, is placed inside the furnace. This porous region is in the model blocking the direct access of metal to the tapping hole. In this case the metal close to the tapping hole is rapidly drained out and process gas will escape together with metal during the complete

222

tapping period. In order to improve the situation it is important to find methods to prevent formation of the low permeability sludge region inside the metal pool.

3.3 Gas entrainment during tapping of steel into ladles

During tapping of steel from a steel converter to a ladle the fall height of the free steel jet may be more than 10 m, with a typical jet diameter of 10 cm. Two phenomena are of particular interest: i) nitrogen entrained by the plunging jet into the ladle may deteriorate the steel quality, and ii) entrained gas may influence the flow pattern in the ladle significantly and thereby effect the dissolution and mixing of alloy elements.

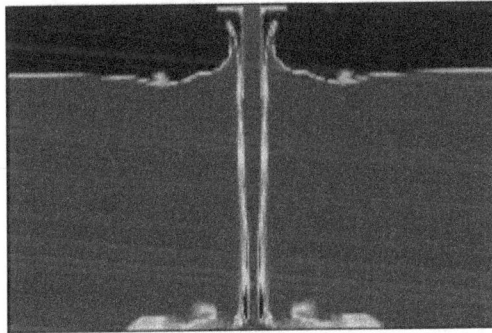

Fig. 9(a) Steel jet, plunging into a ladle at 40 % filling [21]. First bubbles arrive at the ladle bottom.

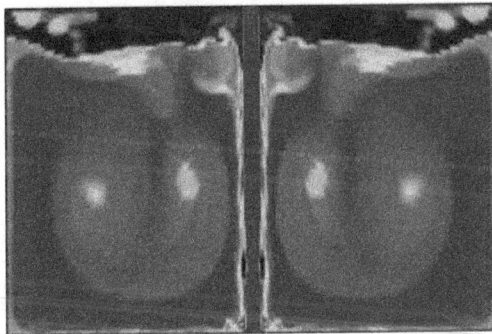

Fig. 9(b) Steel jet, plunging into a ladle at 40 % filling [21]. Surface entrainment is initiated.

In figure 9 we see the predicted gas entrainment in a 2D axisymmetric simulation. The flow is not well resolved and the bubbles are numerically diffused by the numerical technique [5]. However, the predicted entrainment flowrates are comparable to results from experimental correlations [12]. In order to predict the effect of gas entrainment on the dissolution rate of alloy materials we use an Eulerian description of the bubble phase [12], where the bubble size is predicted from a transport model [14]. The gas entrainment rate predicted by the free surface model, as seen in figure 9, is here used as a boundary condition. Using a Lagrangian description of the alloy particles [13], the effect of entrained gas on flow pattern and alloy particle dissolution have been studied [12].

In figure 10 we see the predicted bubbles sizes (background color) and the liquid steel streamlines [12]. Inside the jet region the predicted bubble sizes are typically below 5 mm in diameter. We see that the tapping jet penetrates to the bottom of the ladle and rises up along the sidewall. However, close to the jet boundary the presence of bubbles supplies sufficient buoyancy to turn the direction of the flow. Accordingly, the entrainment of gas leads to an outward flow in the surface. This surface flow direction is opposite of what is the case when gas entrainment is not present.

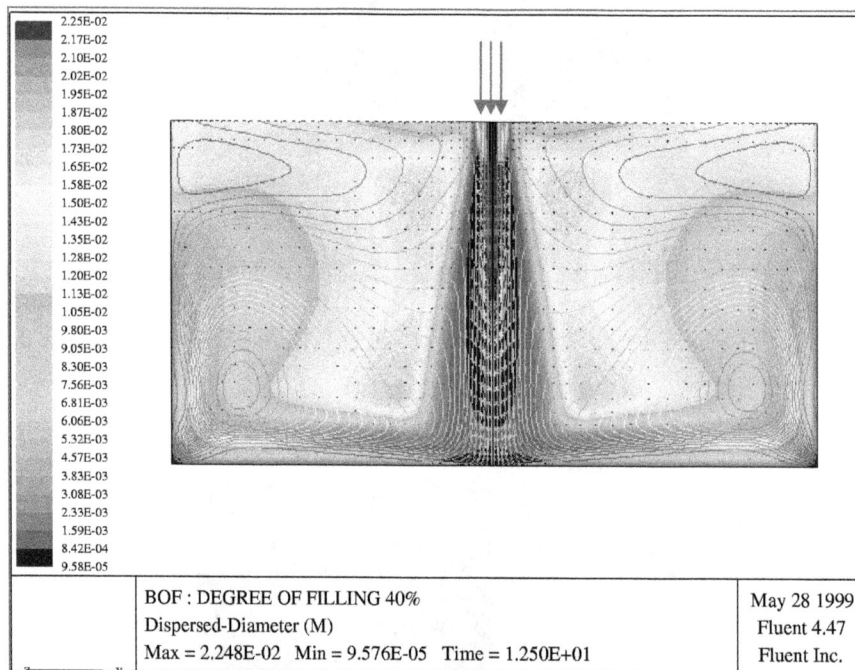

| 2.25E-02 |
| 2.17E-02 |
| 2.10E-02 |
| 2.02E-02 |
| 1.95E-02 |
| 1.87E-02 |
| 1.80E-02 |
| 1.73E-02 |
| 1.65E-02 |
| 1.58E-02 |
| 1.50E-02 |
| 1.43E-02 |
| 1.35E-02 |
| 1.28E-02 |
| 1.20E-02 |
| 1.13E-02 |
| 1.05E-02 |
| 9.80E-03 |
| 9.05E-03 |
| 8.30E-03 |
| 7.56E-03 |
| 6.81E-03 |
| 6.06E-03 |
| 5.32E-03 |
| 4.57E-03 |
| 3.83E-03 |
| 3.08E-03 |
| 2.33E-03 |
| 1.59E-03 |
| 8.42E-04 |
| 9.58E-05 |

BOF : DEGREE OF FILLING 40%	May 28 1999
Dispersed-Diameter (M)	Fluent 4.47
Max = 2.248E-02 Min = 9.576E-05 Time = 1.250E+01	Fluent Inc.

Fig. 10 The figure displays the predicted average bubble diameter distribution as well as the flow streamlines [12]. The modeling technique is the dispersed flow Eulerian description. Note that flow direction is parallel to the streamlines. The blue arrows indicate the plunging jet seen in figure 9.

4. Conclusions

By starting out from single phase flow modeling it is possible to build more complex multi-field and multiphase models. These models can be used to study complex

224

metallurgical systems with bubble break up and fragmentation as well as chemical reactions and phase transitions with heat and mass transfer.

It has further been demonstrated that also free surface modeling techniques can be used to explain a number of industrial problems. The number of possible industrial applications for such techniques is large.

The free surface techniques have a large potential in simulating detailed flow phenomena. Results from such simulations can be used to develop sub-models for studies of phenomena at larger scales. These submodels will then be used in Eulerian multi-field models that has the potential to describe very complex systems without modeling every detail directly. These macro level models will be statistically averaged, but may even so be a powerful tool in improving and designing industrial processes.

Experimental verifications at all levels are crucial for predictive CFD models. Better experimental techniques for validation of CFD models for metallurgical applications will be one important cornerstone in the future development of this field.

Carbothermic Reduction Smelting of Manganese Nodules to Recover Valuable Metals

Sanjay Agarwal*, N. S. Randhawa, K. K. Sahu & R. K. Jana

*Metal Extraction & Forming Division, National Metallurgical Laboratory,
Jamshedpur, India.*
*sanjay@nmlindia.org

Abstract : *Polymetallic sea nodule contains multiple metals like copper (1.1%), nickel (1.2%), cobalt (0.08%), manganese (24%), iron (5.4%), silica, aluminum, etc. Of these, cobalt, copper and nickel are of much importance and in great demand world over. In fact, these three metals of strategic needs are fast depleting from the earth surface. Hence a world-wide research is progressing on sea nodules as an alternative future source of these metals. India is entirely dependent on imports to meet its requirements of cobalt and nickel both of which are most strategic in nature. In this respect, India has made remarkable progress in recovering these metals from sea nodules. The recovery processes so far developed in India are based on purely hydrometallurgical routes. Alternatively, smelting of sea nodules with carbon to recover the valuable metals seems to achieve significant importance, if explored in systematic way. The present paper gives some glimpse of attempts carried out for the above work at NML, Jamshedpur. A number of experiments were carried out in 50 KVA electric arc furnace for reduction smelting of sea nodule. The smelting product contained an alloy with composition in the range of Ni: 10 -12 %, Co: 1.0 – 1.9 %, Cu: 10 – 14%, Mn: 4 - 5%, Fe: 50 - 60% and slag consisting of MnO: 42-45%, SiO2: 30-33%, FeO: 2-10%, (CaO+MgO): 5-6%, Al2O3: 6-7%. The recovery of metals in alloy were Cu: 90-95%, Ni: 90-97% and Co:80-85% with 6 % coke addition at ~1400oC.The alloy obtained may be suitably treated further to recover these metal in pure form by pyro or/ and hydrometallurgical route. The slag contained high Mn/Fe ratio and hence, may be used for ferromanganese or ferro-silico-manganese production.*

1. Introduction

Land based mineral resources are getting depleting day by day; people are looking for alternate resources of mineral available to them. One such alternate resource is sea, which contains a lot of mineral deposits in its bed. People found polymetallic nodule in the sea bed as a source of many metals like copper, nickel, cobalt, manganese, zinc, molybdenum, iron and others. The requirement of Cu, Ni, and Co by many resource-starved countries like Japan, India, China, and Korea encouraged the research organizations in these countries to develop processes, which are economical and environmental friendly. An excellent review of several processes developed during seventies was prepared (Monhemius 1980) (1) and the similarities and the differences were discussed. The nodules are rock concentrates on the sea bed sediments formed by concentric layers of iron and manganese hydroxide around a core.

Metal entities such as Cu, Ni, Co, Mo & Zn are accommodated in the complex cage of iron and manganese hydroxides. These are often called manganese nodule due to its high Mn content. So far, most of metal recovery processes developed for processing sea nodules are hydrometallurgy based, which inherit the associated problem like handling of large volume extractants, dilute leach liquor, very specific downstream processes etc. In this connection, a process to recover Cu, Ni & Co based on reduction smelting is being carried out at NML, Jamshedpur. The proposed flow-sheet for this route is shown in fig. 1.

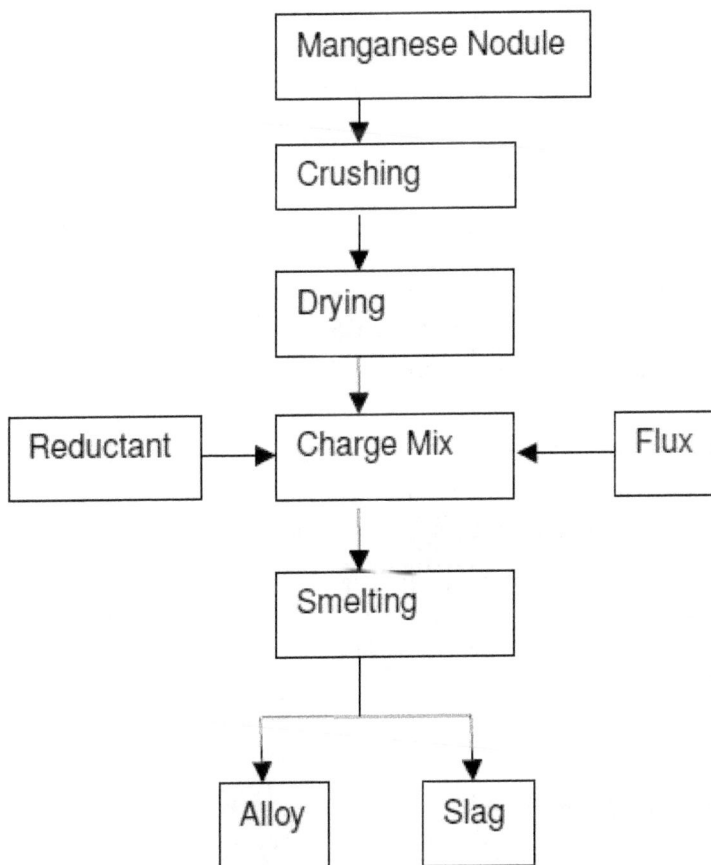

Fig. 1 Flow sheet for reduction smelting of polymetallic nodules.

After removal of Cu, Ni, Co as alloy from sea nodule smelting, the slag generated contains high manganese, which is a good starting material to produce Ferro-manganese or Ferrosilico- manganese from it. The present paper throws light on the studies carried out for the recovery of Copper, Nickel & Cobalt as alloy by Carbothermic smelting of manganese nodules.

2. Experimental Technique

Nodules were crushed down to 5-10 mm size fraction. The size of other raw materials viz. coke, quartz was taken as 5 mm. These raw materials were of commercial grade available in market. The smelting experiments were carried out with charge mix prepared by manually mixing the calculated raw materials with respect to reductant and flux requirement. Smelting was carried out on 20 Kg scale in an electric arc furnace. Furnace consisted of Mg-carbon brick-lined rectangular vessels with power rating of 50kVA and the top was lined with refractory material. There were two electrodes suspended through the top into the hearth. The electrodes were connected to the bus-bars via a water-cooled clamp connection. The electrodes and clamp formed part of a moveable electrode arm. Each mechanical arm was electrically isolated from the electrical connections (electrode and clamp) and was used to control the current and voltage ratio by adjusting the arc length i.e. moving the arm up or down. A tap hole was provided to tap out the molten mass after the completion of experiment. In a typical experiment, the furnace crucible was preheated with initial arc between electrodes on small amount of coke. The charge was added slowly in initial stage and after formation of a molten pool further addition of charge material was done. During melting period the furnace current and voltage were kept at 700 Amps and 45 volts respectively. After complete melting, another 10 minutes were provided for proper separation of slag and alloy metal. Thereafter the alloy metal and slag was poured in preheated clay bonded graphite crucible. On cooling, the slag and alloy metal was separated and ground to 100 mesh size to prepare a representative sample of slag & alloy metal. The major elements were analysed by standard wet methods and trace element analysis was done with AAS technique using Perkin Elmer Analyst 400.

3. Results and Discussion

The chemical compositions for all the raw materials are as follows

Table 1 Composition of Manganese Nodule

Const.	% (mass)	Const.	% (mass)
Cu	0.90- 1.2	S	0.2-0.4
Ni	0.95 – 1.2	P	0.007 – 0.01
Co	0.08 – 0.15	Mo	0.02 – 0.04
Mn	21-25	SiO_2	13 - 15
Fe	5-7	Na_2O	0.92 – 1.05
Zn	0.09 – 0.12	Al_2O_3	2 - 3

Table 2 Chemical composition of coke and quartz

Coke	
Fixed carbon	77.0 %
Ash	16.60 %
Volatile matter	2.35 %
Moisture	3.80 %
Quartz	
Silica	92.0 %
Al_2O_3	1.53 %
Fe (T)	3.35 %
L O I	0.37 %

The basis towards smelting of sea nodules is the well-studied Ellingham Diagram. From the diagram it is evident that the oxides of Cu, Ni & Co tends to get reduced to respective metallic state with carbon at lower temperature (about 800oC) whereas the oxides of Fe, Mn & Si needs somewhat higher temperature i.e. in the range of 1350 - 1600oC to get reduced. Therefore, it is thermodynamically favored that with little amount of coke and maintaining temperature around 14000C will result in the reduction of Cu, Ni & Co oxides along with some of iron. The resulting slag contains almost all the manganese with remaining amount of iron and silica. The basic reaction taking place during smelting in a similar system (1, 2) are described as follows:

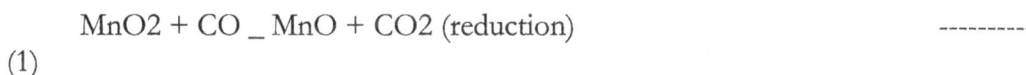

MnO2 + CO _ MnO + CO2 (reduction)   ----------
(1)

$$MnO + CO2 \_ MnCO3 \text{ (carbonation)}$$ ---------- (2)

$$Fe2O3 + CO \_ 2FeO + CO2$$ ---------- (3)

$$FeO + CO \_ Fe0 + CO2$$ ---------- (4)

$$NiO + CO \_ Ni0 + CO2$$ ---------- (5)

$$CuO + CO \_ Cu0 + CO2$$ ---------- (6)

$$Co2O3 + 3CO \_ 2 Co0 + CO2$$ ---------- (7)

$$CoO + CO \_ Co0 + CO2$$ ---------- (8)

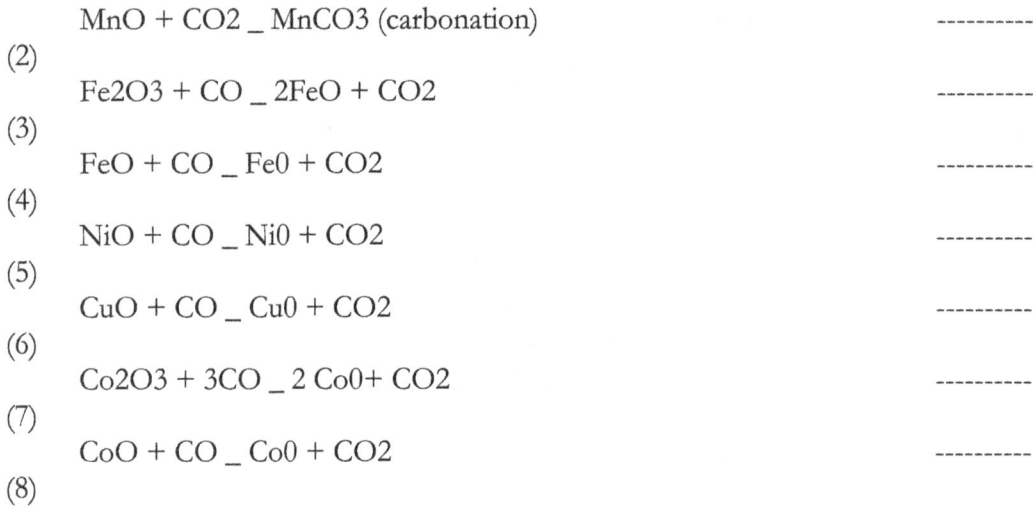

Reaction 1 is the disruption of the MnO2 structure allowing subsequent reduction of Ni, Cu, Co (reaction 5 to 8) and the carbonation of the reduced manganese (reaction 2). Iron is also reduced to either the Fe2+ state or possibly to Fe0 (reaction 3 and 4). It is desirable to reduce iron only to Fe^{2+} state because Fe0 may alloy with Ni metal. Formation of FeNi alloy retards and often prevents solubilization of the Ni values associated with the alloy formation. In this process at the temperature range of 600 – 9500C the oxides of Co; Cu & Ni are reduced to the metallic state, Co0, Cu0, Ni0 along with most of the iron present, followed by melting at around 14000C to separate alloy from slag. Under the controlled conditions, the

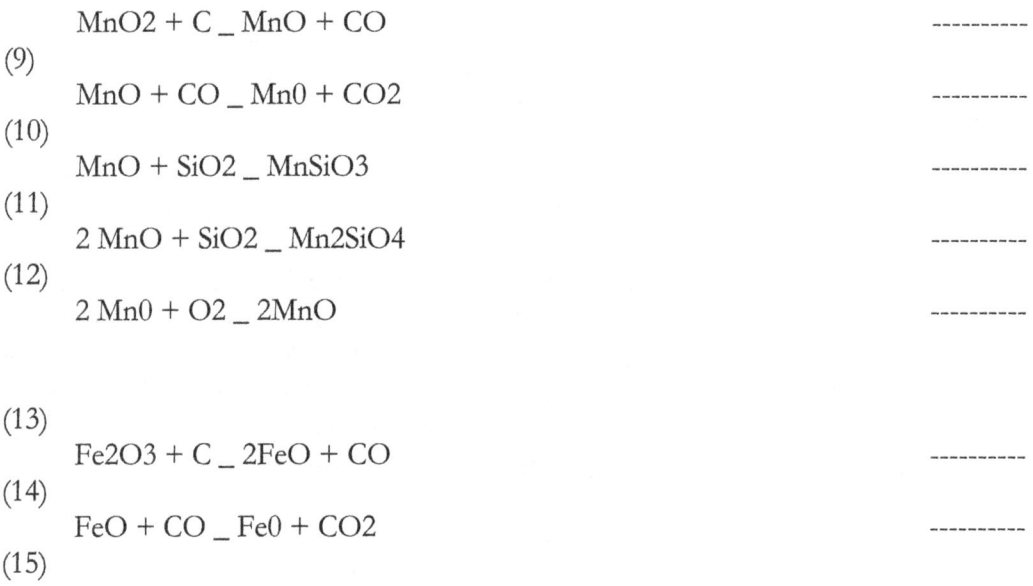

$$MnO2 + C \_ MnO + CO$$ ---------- (9)

$$MnO + CO \_ Mn0 + CO2$$ ---------- (10)

$$MnO + SiO2 \_ MnSiO3$$ ---------- (11)

$$2 MnO + SiO2 \_ Mn2SiO4$$ ---------- (12)

$$2 Mn0 + O2 \_ 2MnO$$ ---------- (13)

$$Fe2O3 + C \_ 2FeO + CO$$ ---------- (14)

$$FeO + CO \_ Fe0 + CO2$$ ---------- (15)

$$NiO + C \_ Ni0 + CO \qquad \text{----------}$$
(16)

$$CuO + C \_ Cu0 + CO \qquad \text{----------}$$
(17)

$$Co2O3 + C \_ 2\, CoO + CO \qquad \text{----------}$$
(18)

$$CoO + C \_ Co0 + CO \qquad \text{----------}$$
(19)

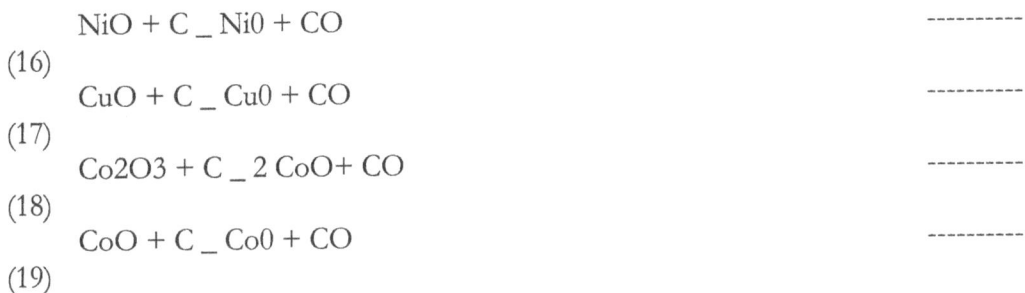

manganese is not reduced to Mn0 but remains as Mn2+ in the slag. The Co, Cu, Fe & Ni form a metal alloy that settles to the bottom of the furnace. The reduction (3, 4) is performed by using coke and slagged by using silica. The primary reactions that occur in this portion of the process involving C, CO, SiO2, Mn, Fe, Cu, Ni and Co are reactions 2, 5, 6, 8. Reaction 2, 9 10 are for the reduction of Mn4+ to Mn2+ and to Mn0. Reaction 11 & 12 are the slag formation reactions and reaction 13 is reoxidation of Mn0 to Mn2+. Reaction 5, 6, 8, and 14 to 19 shows the various steps for Fe, Cu, Ni and Co reduction to the metallic state. After complete reduction, metal alloy separates from the oxide-silicate slag by gravity. As mentioned above, the temperature & amount of reductant (coke) are two variables, which determine the degree of reduction for manganese nodule smelting. In present study the temp. was kept constant at around 14000 C and quantity of coke in charge mix was varied. Numbers of experiment were carried out varying the amount of coke addition. Graphical representations of effect of coke on reduction smelting of nodule are given in figure 2 & 3.

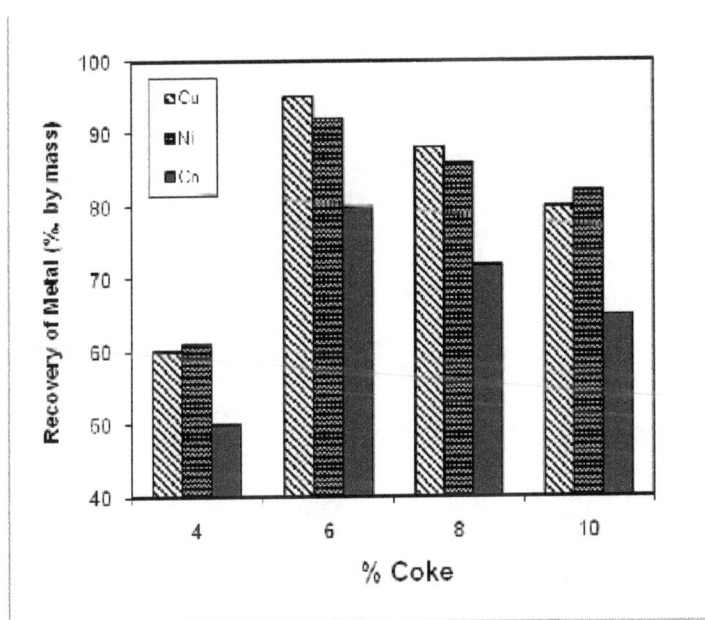

Fig. 2 % recovery of different metal with the variation of coke in charge.

231

The recoveries of Cu, Ni & Co in the alloy are depicted in fig. 2, which shows that at 4% coke addition some recovery of Cu & Ni was achieved and small amount of cobalt in alloy. Maximum recoveries of Cu, Ni & Co in alloy were obtained with addition of 6% coke.

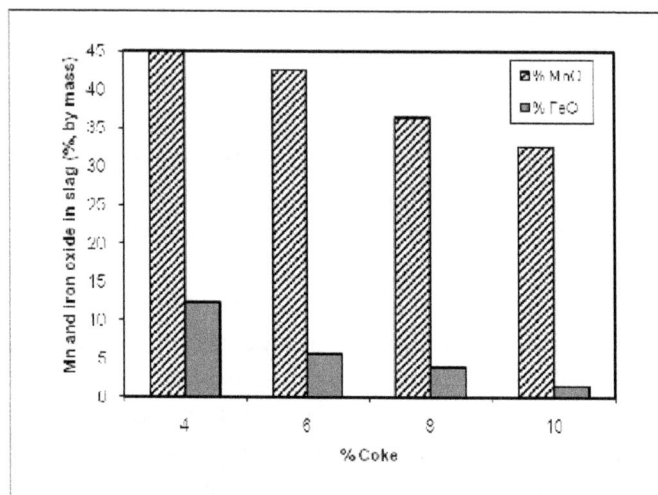

Fig. 3 % of metal oxide with the variation of coke in charge.

Thereafter, further addition of coke led to decrease the recoveries of Cu, Ni & Co. If we take a look on fig. 3, in which the FeO and MnO content of the corresponding slag is shown, it can be seen that addition of coke more than 6% bring down the MnO content of the slag. This may be due the reduction of MnO to metallic state that would bring more Mn in alloy metal & will disturb the Mn/Fe ratio in the slag. The recoveries of Cu, Ni & Co in smelting alloy metal are 95%, 92% and 80% respectively at 6% coke addition

The typical composition of alloy and slag obtained after smelting with 6% coke is given in Table 3 & 4 respectively.

Table 3 Chemical composition of alloy from sea nodules smelting

Element	%	Element	%
Cu	12.33	Mn	8.23
Ni	14.05	Fe	55.96
Co	0.75	Si	0.31

Table 4 Chemical composition of sea nodules smelting slag

Radical	%	Radical	%	Radical	%
MnO	42.61	CaO	1.01	Cu	0.10
FeO	5.66	MgO	4.41	Ni	0.05
SiO_2	33.51	Al_2O_3	6.10	Co	nf

The resulting slag contains high manganese along with iron and silica. The Mn/Fe ratio is more than 7 and hence it can be subjected to ferromanganese or silicomanganese smelting after adjusting the charge basicity by addition of flux (quartz or dolomite as required).

4. Conclusions

1. Sea nodules smelting of sea nodule with 6% coke yields maximum recoveries of Cu, Ni & Co and those are 95%, 92% & 80% respectively.
2. Smelting produces manganese rich slag, which has suitable Mn/Fe ratio for ferromanganese or silicomanganese smelting.

References

1. Monhemius, A.J., *The extractive metallurgy of deep sea manganese nodule*, In' Topics in non ferrous extractive metallurgy': R. Burkin (eds.), Society of chemical industry, London, 1980, 42.

2. Habashi F, Handbook of Extractive Metallurgy, Vol. 1, Part II, p. 421.

3. Riss M, Khodorovsky Y, Production of ferroalloys, Mir Publishers, Moscow, 1967, p. 148

4. Breg K L, Olsen S E, *Kinetics of manganese ore reduction by carbon monoxide*, Metallurgical & Materials Transactions B, Vol. 31b , No. 3, June 2000, p. 477.

Production of High Sintered Density Natural UO2 Fuel and Innovative Technique for Salvaging Pellets from Density-Quarantined Boats

D.Pramanik*, N.Rajesh, Nirmal Santra, J.V.Rajkumar and GVS Hemantha Rao

Nuclear Fuel Complex
Department of Atomic Energy, Hyderabad INDIA
* Email: dpramanik@ymail.com

Abstract_Nuclear Fuel Complex (NFC), Hyderabad manufactures high density (96-98%TD) natural UO2 fuel, highest among all reactor fuels, for the Pressurized Heavy Water Reactors (PHWRs) operating in the country. UO2 fuel pellets are produced through powder metallurgy process steps such as pressure agglomeration of UO2 powder, admixing of lubricant, die compaction of green pellets followed by high temperature (1700oC) solid state sintering of green pellets. No sintering-aid is added to UO2 powder so as to maintain the required nuclear grade purity. In order to achieve such high density fuel without any sintering-aid, stress lies on the optimization of the individual process parameters. Deviations in factors such as specific surface area of UO2 powder, particle morphology, presence of impurity elements, granule characteristics, green density, sintering profile, quantity of lubricant, etc. contribute to low sintered density or density variations in sintered pellets. Study of diametrical shrinkage vis-à-vis sintered density has been useful to sort acceptable density pellets from 5-7% boats quarantined due to density variation below specified limits, resulting in maximizing sintering throughput. The paper discusses the role of above factors and underlying mechanism for low density. It also discusses the result of the studies on effect of admixed lubricants, pressing behavior and sintering response of UO2 powder, the effect of green density, the study of sintering kinetics using high temperature dilatometer and the essence of salvage technique.*

Key Words: UO2, sintering, green pellets, admixed lubricant, dilatometer, microstructure, SEM, activation energy for diffusion

1. Introduction

Nuclear Fuel Complex, Hyderabad is an industrial unit of Department of Atomic Energy produces sintered nuclear fuel pellets in large scale (about 500TPY) through powder metallurgy process as per flow sheet given in Fig.1. Since the PHWRs operates on natural UO2 fuel having only 0.7% U235 fissile isotope, cylindrical fuel pellets are manufactured from high purity nuclear grade UO2 powder. The fuel pellets are required to be densified to 96-98% TD of UO2 i.e. to the range of 10.45g/cc to 10.75g/cc (Theoretical Density: 10.96g/cc), which is highest among all types of commercial reactor fuel. Sintering is carried out in hydrogen atmosphere so as to limit the O/U ratio to less than 2.015 in the fuel. Higher O/U ratio is not permitted as extra oxygen reduces the thermal conductivity of the fuel [1]. The fuel pellets are also required to have homogenous grain microstructures with uniform distribution of pores. The stable valencies of uranium are +6 followed by +4, because of which uranium dioxide has hyperstoichiometric stable UO2+x composition[2][3]. UO2+x exists in single phase face centered cubic structure resembling to calcium fluoride lattice structure up to its melting point (2800°C). The excess oxygen atoms occupy the interstitial sites in the lattice. The presence of excess oxygen in the lattice creates uranium ion vacancies and thus enhances the mobility of the uranium atoms, whose mobility is otherwise sluggish compared to oxygen atoms. The activation energy needed for diffusion of uranium atoms drops exponentially as the surplus oxygen increases. In order to maintain this enhanced mobility of uranium atoms for densification, the excess oxygen of the lattice is to be in equilibrium with oxygen in the sintering atmosphere. The diffusion coefficient of uranium atoms (Du) increases according to the relation Du & x^2 [2][3]. Low temperature sintering (LTS) of UO2 under controlled oxidative atmosphere (containing 500-1000ppm oxygen) is proven in R&D study at BARC [3]. But the use of oxidative sintering requires special furnaces with atmosphere safety challenges to maintain three atmospheres, oxidative atmosphere for densification, reducing atmosphere for O/U correction and an inert atmosphere in between these two atmosphere for separation. The technology is yet to be matured for continuous sintering furnaces. Obviously, this benefit of enhanced diffusivity of uranium atoms is not significant in a production sintering furnace which is maintained with hydrogen atmosphere throughout.

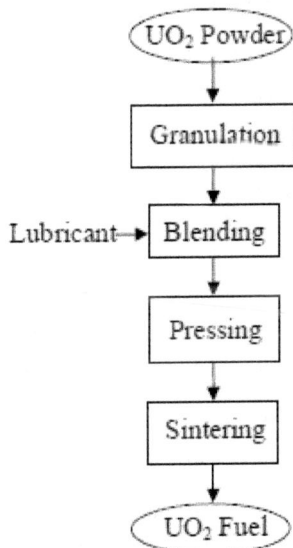

```
        ┌─────────────┐
        │  UO₂ Powder │
        └──────┬──────┘
               ↓
        ┌─────────────┐
        │ Granulation │
        └──────┬──────┘
               ↓
Lubricant→┌─────────────┐
        │  Blending   │
        └──────┬──────┘
               ↓
        ┌─────────────┐
        │  Pressing   │
        └──────┬──────┘
               ↓
        ┌─────────────┐
        │  Sintering  │
        └──────┬──────┘
               ↓
        ┌─────────────┐
        │  UO₂ Fuel   │
        └─────────────┘
```

Fig. 1 Processes for sintered UO2 pellet production.

Certain sintering aid dopants like Nb2O5, TiO2, Al2O3/SiO2, Cr2O3 added as second phase particles enhance the density of UO2 by increasing diffusion of uranium atoms in solid state or by forming partial liquid phase [4][5]. But this technique is not applicable in sintering natural UO2 fuel because of requirement of high purity in the fuel. The physical and chemical inhomogeneity of the starting UO2 powder raw material produced through a particular route greatly influence the sintered density and microstructure of UO2 fuel[6]. The physical parameters such as bulk density, particle size distribution, extent of agglomeration and specific surface area are decisive parameters. Chemical characteristics such as O/U ratio, presence of impurity elements such as sulphur Ca, Mg, Si are important for sintering performance [6]. Heterogeneous green structure may cause differential densification resulting in larger pores during sintering. Physical heterogeneity with respect to pore size and distribution in a green pellet is arisen from non-uniform particle size distribution of starting powder.

Hence, the sintering of UO2 fuel to such a high density is a tedious task. Development and optimization of processes for powder production and pellet fabrication are very essential. The interdependence nature of the parameters of powder production and pellet fabrication sometimes complicate the understanding of sintering of UO2 fuel. As per thermodynamic approach, the difference between a green pellet and a sintered pellet is that the green pellet has enormous surface energy due to particulate porous structure with contact of particles where as a sintered pellet has not because particle surfaces have been joined together and formed grain structure with pore shrinkage and reduction of pore volume. In the intermediate and final stage of sintering, the available surface energy should be utilized for diffusion of uranium and oxygen atoms from neck centre to pore-neck junction via grain boundary path (grain boundary diffusion) and/or via lattice path (volume diffusion) for neck growth and shrinkage of pores and not for diffusion of atoms from particle surface to the pore (surface diffusion). The latter mechanism at the intermediate stage will lead to premature grain growth without bringing the particles closer to each other i.e. without shrinkage.

Study of sintering kinetics of UO2 fuel pellets through high temperature dilatometer in Argon + 8% H2 atmosphere reveals important information on sintering kinetics such as temperature for onset of shrinkage, rates of shrinkage at progressive higher temperatures and effect of heating rates on shrinkage.

Eventually in production of sintered pellets, 5-7% of green pellet boats are quarantined in sintering process because some of the sample pellets have density below acceptable range of 10.45g/cc. About 40-50% pellets in a density quarantined boat containing total

840 pellets normally have acceptable range density. But since sorting of acceptable density pellets through measurement of density of each and every pellet is a tedious task, normally such quarantined boats are recycled back in powder production. Recently an innovative technique has been evolved based on obvious relation between shrinkage and sintered density. Higher sintered density is evolved from higher shrinkage, which means higher density pellets have smaller diameter and lower density pellets have higher diameter, provided green density lies in a close range and green diameter is constant. The diameter of green pellets remains constant by using wear resistant tungsten carbide dies. The technique consists of pre-grinding to a threshold diameter for all the pellets and sorting of pellets based on cut-surface appearance. The sintering throughput could be maximized by this salvage technique.

The following sections of the paper discuss the critical parameters for process optimization, study of interdependence of parameters, study of sintering kinetics, occurrences of low sintered density and the essence of salvage technique.

2. Process optimization for production of UO2 powder

Magnesium di-uranate (MDU) concentrate containing uranium equivalent to about 70wt% U_3O_8 is dissolved in nitric acid to form crude uranyl nitrate solution. The crude solution is purified through solvent extraction using tri-butyle phosphate as organic extractor. The pure solution is then subjected to precipitation of ammonium di-uranate (ADU) known as yellow cake by reacting with ammonia in aqueous form or in dry vapour form. ADU in slurry form is subjected to spray drying after which it is first calcined to U_3O_8 at about 700oC and then reduced to UO_2 at about 550oC in hydrogen atmosphere and then stabilized to UO_2+x (x<0.15). The morphology of UO_2 particle is characterized under SEM for observing shape of the crystallites, nature of packing of crystallites, particle shape and sizes, presence of intra-particle porosities, etc. The morphology combined with physical and chemical characteristis such as bulk density, average particle size, particle size distribution, specific surface area (SSA), O/U ratio, presence of elemental impurities like, S, Ca, Mg, Si govern the sinterabilty of the UO_2 powder, which is though predictable, but not fully confirmed unless a sinterability test is conducted on a sample quantity from individual UO_2 powder lots [6]. A typical SEM micrograph of UO_2 powder (morphology) is shown in Fig.2a & 2b. These powder parameters are optimized based on sinterability test experiences.

Fig. 2(a) SEM micrograph of UO2 powder crystallite

Fig. 2(b) SEM micrograph of UO2 powder particles

The powder production process is optimized to maintain the BET specific surface area (SSA) of UO2 powder expressed in m2/g, within a narrow band. Uranium concentration in pure nitrate solution, temperature of onset of precipitation and flow of ammonia are critical parameters for which control provisions are made. Reduction temperature for conversion of U3O8 to UO2 is varied to maintain the specific surface area of reduced UO2. Specific surface area is a function of particle size, particle size distribution and intra- particle porosities. As SSA is increases, sintered density of pellet increases, however the relation becomes a inverse relation when SSA increases beyond a threshold value (Fig.4).

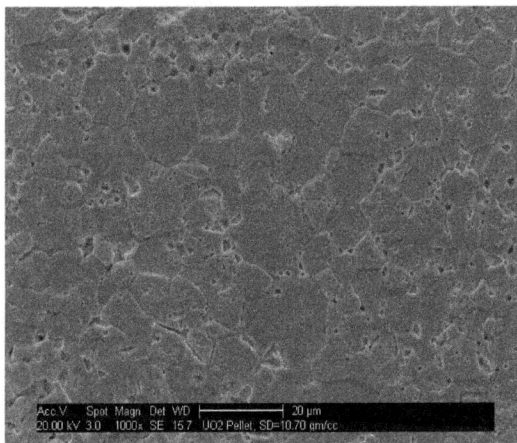

Fig. 3 SEM micrograph of high density pellet (etched sample).

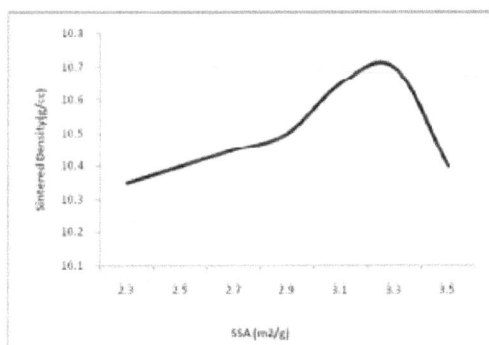

Fig. 4 Sintered Density Vs. SSA of UO_2.

3. Process optimization for pellet fabrication

Granules are produced from virgin UO2 powder by pressure agglomeration in a roll compacting press. The granules are vibro-sieved on-line using a 60 mesh. The +60 fraction goes for preparation of press feed while the minus fraction is recycled to the roll press. The press feed for final compaction is prepared by admixing powder lubricant with granules and blending by tumbling of the cylindrical granulecontainer. The lubricated UO2 granules are compacted to green pellets in multi-die cavity, double acting hydraulic presses. The density of green pellets is optimized in a narrow band of 52%TD to 54%TD (5.7g/cc to 5.9g/cc) because of its influence on sintered density (Fig.5). The pressing behavior of powder lots is compared as per the pressure/density relationship (Typical relation is shown in Fig.6.). UO2 powder is compactable to sound integrity pellets without use of any binder.

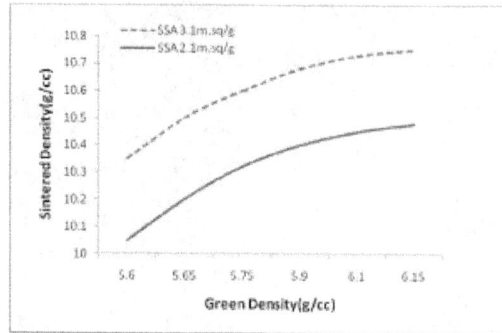

Fig. 5 Sintered Vs. Green Density.

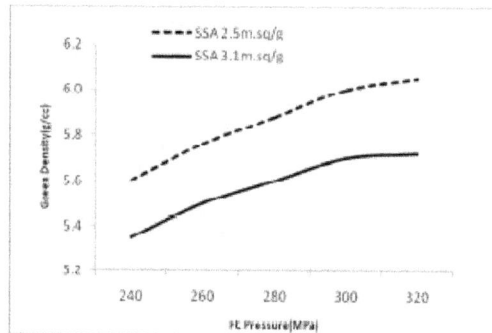

Fig. 6 Pressure Vs. Green Density.

Pressure distribution during compaction of green pellet may be non-uniform if there is a segregation of bigger agglomerates and fine fraction during die-fill. In order to avoid segregation, size distribution of granules and fine fraction is optimized (Fig.7). For a optimized size distribution the bulk density of granulated power (25-27% TD) is used as an indicator of granule softness. Higher bulk density indicates that the granules are becoming denser or harder than required. If the agglomerate is not fully disintegrated, the compact density is locally nonuniform and full densification is hard to achieve because of differential densification during sintering [7]. The granulation pressure is kept about one-third of final compacting pressure to ensure that the aggregates are destroyed uniformly during final compaction to give rise to a homogenous green microstructure. (Fig. 8).

# +60 fraction	58-65%
# -60 to +100 fraction	30-35%
# -100 to +200 fraction	3-4%
# -200 fraction	2-3%

Fig. 7 Typical size fraction of granules.

240

Fig. 8 Green microstructure.

It is generally observed that sinter density is influenced by quantity addition of lubricant admixed with granules for compaction of green pellet as shown in (Fig.9). Hence, lubricant addition is optimized in the range of 0.25-0.3wt%.

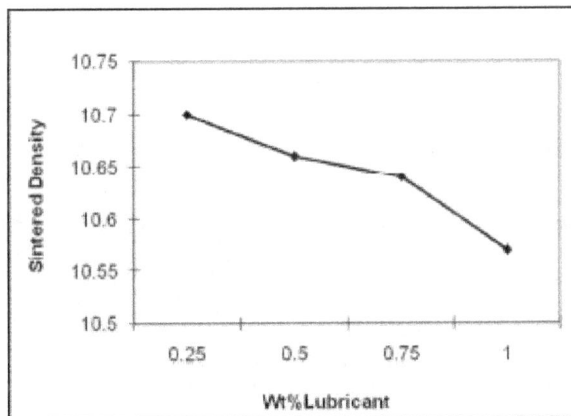

Fig. 9 Sintered Density Vs. Admixed Lubricant.

The green pellets are sintered in a multi-zone electrically heated, hydrogen atmosphere, pusher type sintering furnace. The heating zones temperature and pushing interval are so selected that the material is heated a rate of about 120oC/hr and soaked at 1700oC for about 5hrs. Such profile is decided aftersimulating the results of study of shrinkage in a high temperature dilatometer. Study of shrinkage and shrinkage rate is depicted in Fig.10 and Fig.11 respectively. Lower heating rate is found to increase the shrinkage and shrinkage rate compared to higher rate of heating.

241

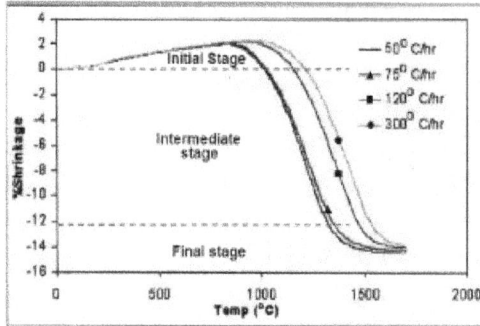

Fig.10 Shrinkage Vs heating rates.

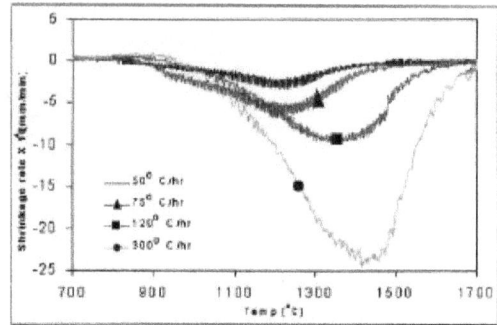

Fig.11 Shrinkage rate Vs heating rates.

4. Occurrences of low density

Degree of densification leads to micro structural homogeneity with respect to grain size and pore-size distribution. Occurrences of low density have been observed occasionally for the following types of micro structural in-homogeneity:

i)   Preferential grain growth at the pellet core with fine grains at the periphery (Coring defect) (Fig.12).
ii) Uniform large grain with large pores (Fig.13)
iii) Formation of bigger voids in large numbers (Fig.14)

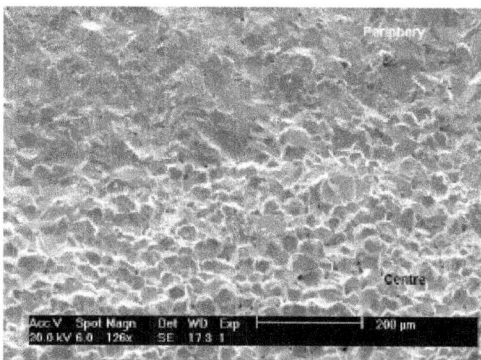

Fig.12 Coring defect (Large grain in centre and fine grain in periphery)

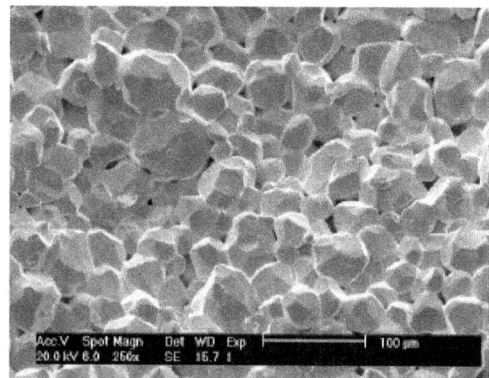

Fig.13 Large isolated grain with large pores

242

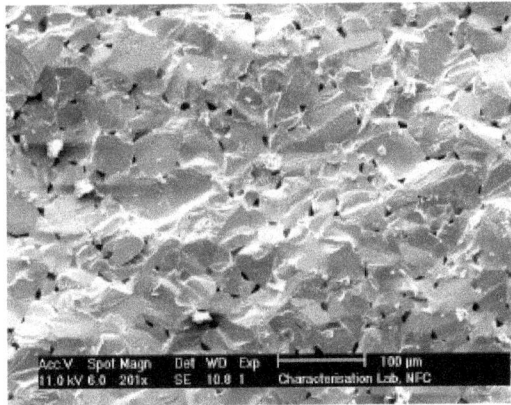

Fig. 14 Formation of bigger voids.

5. Salvaging pellets from density quarantined boats

Cent Percent sintered pellet boats are checked for density on sampling basis and only qualified boats (density >=10.45g/cc) are taken for diameter correction grinding and subsequent operations. Normally density quarantined boats (5-7%) have more than 50% OK density pellets. An innovative concept for salvaging OK density pellets has been developed which avoids checking density of each & every pellet.

Considering negligible change of mass of pellet after sintering, following relation holds good:

$$\rho_g (\pi/4) d_g^2 l_g = \rho_s (\pi/4) d_s^2 l_s$$

Where ρ_g - Green density of pellet
d_g - Diameter of green pellet
l_g - Length of green pellet
ρ_s - Sintered density of pellet
d_s - Diameter of Sintered pellet
l_s - Length of Sintered pellet

$$\therefore d_s = \left[\frac{\rho_g d_g^2 l_g}{\rho_s l_s} \right]^{1/2}$$

The length to diameter ratio remains constant after sintering. Hence the above relation can be written as,

$$\therefore d_s = \left[\frac{\rho_g}{\rho_s} \right]^{1/3} d_g$$

243

Normally sintered pellet undergoes a diametrical shrinkage of 18.5-19.5% for good densification to the acceptable range. However, the sintered diameter is also influenced by green density of the pellet. A relation curve (Fig.15) for green density versus sintered diameter is evolved accordingly, considering constant diameter of green pellet. If the pellets of known green density of a quarantined boat are ground as per the sintered diameter, a pellet after grinding have any one of the appearances as in Fig. (a) to (d). Low-density pellets are sorted out after initial grinding as per the specific appearances.

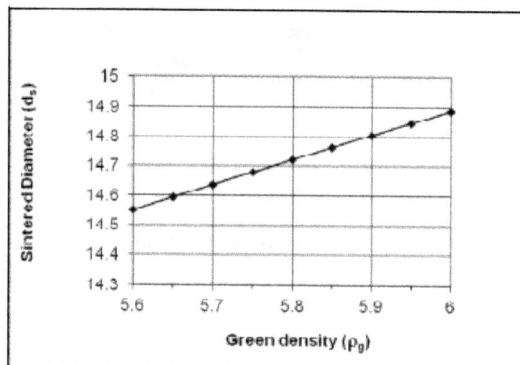

Fig. 15 Green Density Vs Sintered Diameter.

The sorted ok density appearance pellets are collected in separate trays and subjected to sample density check where four times higher sampling size (16 pellets) is used to accept such pellets. Normally about 90% pellet trays get accepted and join the main production stream.

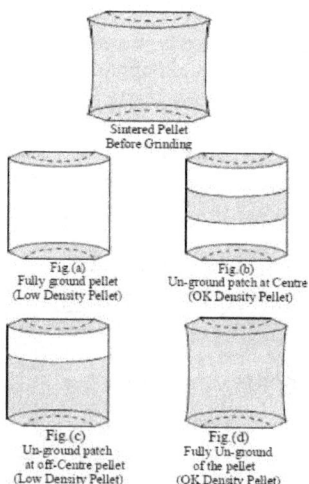

Fig. (a) to (d) Sketch on appearances of ok density and low density pellets after grinding for sorting.

6. Discussions

The performance of densification of UO2 pellet is primarily dependent on quality of the starting UO2 powder. Smaller particle size, optimum SSA, favorable particle morphology and chemical purity of powder is very important for producing high sintered density UO2 fuel. The heating rate of green pellet governs the rate of shrinkage and final density achieved. Fast heating rate delays the onset of sintering since green pellet take more time to equilibrate the temperature throughout the core. In the Initial stage of sintering, the surface diffusion becomes the active mechanism of mass transfer at lower temperature. Since the activation energy for grain boundary diffusion is lower than that of volume diffusion, grain boundary diffusion is the major mechanism of atom diffusion. Hence the rate of shrinkage is much higher at the intermediate stage of sintering [8]. At the final stage of densification, when the available surface energy is minimized, volume diffusion is the major mechanism of atom diffusion at final soaking temperature. At this stage, higher densification is attained through uniform grain growth and reduction of pore volume.

Failure to attain high density is primarily because of premature grain coarsening at the intermediate stage of sintering. Exaggerated grain growth at the pellet core (Fig.12) or throughout the pellet (Fig.13) might be originated from chemical impurity of powder. Presence of impurity like sulphur is found to play major role in such abnormal grain growth [9]. Low density due to big void in large number (Fig.14) has originated from high bulk density UO2 powder (44%higher). The close packing of crystallites and hard agglomerated nature of particles revealed in SEM morphology (Fig.16) might have produced large voids.

Fig. 16 SEM micrograph of high bulk density UO2 powder.

7. Conclusions

Densification of UO2 fuel pellets is primarily dependent on quality of starting UO2 powder. Hyperstoichiometry in UO2 does not play a significant role in conventional sintering under reducing atmosphere. Narrow particle size distribution without presence of hard agglomerates, optimized SSA, absence of specific impurities like sulpher, Ca, Mg, Si are desirable for homogenous sintered microstructure and sintered density. The pellet fabrication parameters including sintering parameters are also optimized to obtain best sintering result from a good quality powder. The innovative technique of sorting acceptable density pellets from density quarantined boats has increased the sintering yield.

Acknowledgement:

The authors are thankful to Shri R.N.Jayaraj, Chief Executive, NFC for his continual encouragement and support for study of powder characteristics and morphology vis-à-vis in-depth study of sintering process.

References

1. Masaki AMAYA1), Toshio KUBO1) and Yoshiaki KOREI1), *Thermal Conductivity Measurements on UO2+x from 300 to 1, 400K*, J. of Nuclear Science and Technology, Vol.33, No.8, 1996, p.636.

2. T.R.G. KUTTY, P.V. HEGDE, K.B. KHAN, U. BASAK, S.N. PILLAI, A.K. SENGUPTA, G.C. JAIN, S. MAJUMDAR, H.S. KAMATH, D.S.C. PURUSHOTHAM, *Densification behavriour of UO2 in six different atmosphere*", J. of Nuclear Materials, Vol. 305, 2002, p. 159.

3. C. GANGULY, *Power Metalergy of Nuclear Fuel*, Transactions of the powder Metallurgy Association of India, Vol.26, 1999.

4. J.E. LINDBACK , *Westinghouse Doped Pellet Technology*, Westinghouse Atom AB, Vasteris, Sweden.

5. KC. RADFORD AND J.M. POPE, *UO2 Fuel Pellet Microstructure Modification Through Impurity Additions*, Nuclear Materials Department, Westinghouse Research Lab, Pittsburgh, Pennsylvania 15235, USA.

6. F. GLODEANU, M SPINZI AND V. BALAN, *Correlation between UO2 powder and pellet quality in PHWR Fuel Manufacturing*, Journal of Nuclear Material vol. 153, 1988, p. 156.

7. SUK – JOOHG L. KANG, *Sintering Densification, Grain Growth & Micro-structure*, Chapter-5.5, Elsevier, Burlington, 2005, p. 66.

8. D. LAHIRI, S.V. RAMANA RAO, G.V.S. HEMANTHA RAO, R.K. SRIVASTAVA, *Study on Sintering Kinetics and Activation Energy of UO2 Pellet Using three different methods*, J.of Nuclear Materials, Vol. 357, 2006, p. 88.

9. T.W. ZAWIDZKI, P.S. APTE & M.R. HOARE, *Effect of Sulphur on Grain Growth in UO2 Pellets*, The American Ceramic Society, San Diego, California, October 26, 1978, p. 38-N-78P.

POWDER TECHNOLOGY

Innovations in Sintered Materials for Demanding Applications

Randall M. German

Associate Dean and Professor of Mechanical Engineering
College of Engineering, San Diego State University
San Diego, California, USA

Abstract: *Sintering technologies date back thousands of years, but modern sintering practice is less than 100 years old. Various traditional firing cycles are known and the devices for heating are well established. However, new and more difficult-to-process materials require innovations outside the traditional routes. For example, in systems with large polymer contents, thermal burnout leads to severe contamination of the furnace. However, in a low pressure environment it is possible to generate a plasma to ensure a clean furnace and a low contamination level. Likewise new microwave heating applications are emerging that enable rapid cooling to precipitate higher toughness microstructures. For the truly difficult to process materials, the use of a pulsed electric field has emerged as an important means to consolidate composites, such as aluminum containing carbon nano-tubes. The need for new sintering technologies propagates into the new materials that challenge traditional concepts, equipment, firing cycles, and even perceptions on performance. Consider that over the electromagnetic spectrum there exist several options for delivery of heat; yet most sintering relies on radiant heat transfer and this limits the ability to sinter new or difficult materials. This presentation covers newer ideas, ranging from the fabrication of high performance fatigue resistant ferrous alloys via sintering to novel graded composites where each two micrometer particle is a multiple layer composite useful for wire drawing applications.*

1. Introduction

Sintering is a thermal process for bonding powders into a solid structure. During heating the bonds that spontaneously grow between particles increase the component strength and reduce the surface area. Figure 1 is a picture of the sinter bonds between spherical particles. Such neck growth happens during heating when particles (wires, spheres, flakes, or other high surface area shapes) are in contact. These bonding events take place by diffusion and related atomic level events (viscous flow for amorphous materials and in some cases plastic flow) and often involve liquid formation during the

251

firing cycle. A newer variant relies on chemical reactions during heating. In hot isostatic pressing, hot pressing, and other pressure-assisted processes the slow diffusion events are accelerated by the supposition of creep or plastic flow events. Most of the properties of a powder compact are improved by sintering, and frequently density and component size change. Inherently, the atomic motion that provides sintering occurs is temperature sensitive, so sintering tends to be favored by relatively higher temperatures. The driving force for sintering is the reduction of the surface energy. Surface energy per unit volume depends on the inverse of the particle size. Thus, smaller particles with high specific surface areas have more energy and sinter faster.

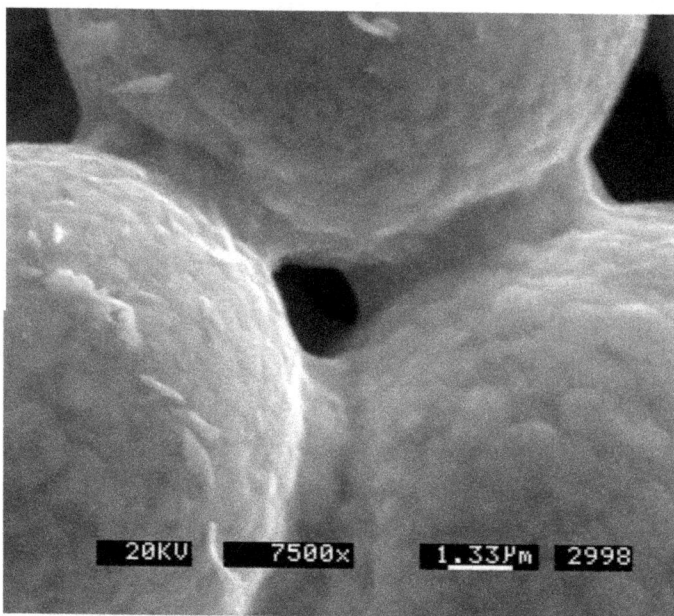

Figure 1 Sinter bonds between bronze particles.

Not all surface energy is available for sintering. For a crystalline solid, nearly every particle contact will evolve a grain boundary with a concomitant increase in grain boundary energy. So as neck growth removes surface energy, it adds grain boundary energy. Obviously, this only occurs when the decrease in surface energy is greater than the increase in grain boundary energy. Sintering temperatures vary between materials and are sensitive to particle size, but generally require higher temperatures for higher melting materials. Since sintering is a thermal softening process that induces atomic motion at high temperature, it inherently struggles with materials that are resistant to high temperatures. On the other hand, the advent of nanoscale powders has brought the sintering temperature down to nearly room temperature. For example, a 5 nm palladium powder with an initial surface area of 80 m2/g sintered to a surface area of 2 m2/g surface area after 15 minutes at 100°C.

Sintering has many different forms, with three main subdivisions - pressure-assisted, solid-state, and liquid phase sintering. Each of these then has variations depending on the process and material details, including supersolidus sintering, reactive sintering, and transient liquid phase sintering.

Sintering is associated with both bonding and densification. In the bonding situation, such as commonly used in press-sinter ferrous powder metallurgy, the pressed powder undergoes little densification or swelling during sintering. This low dimensional change is helpful in retaining component precision and in the formation of foamed or porous materials. Applications for high porosity structures include filters, flow restrictors, energy absorption materials, insulators, and capillary wicks. On the other hand, full-density sintering is used in both powder metallurgy and eramics to attain high performance via the elimination of pores. Most of the growth in both fields is associated with structures that are densified in sintering.

The demand for durability leads to requirements for higher sintering properties. This is very evident in automotive applications, where the increases use of powder metallurgy is coupled hand-in-hand with the realization of higher densities and higher properties. A good example is evident in the attached Figure 2, showing how the relative growth in ferrous powder metallurgy shifted the field to higher densities as time progressed. Densification is now a common goal in sintering, even for metallic system. To support this contention, note that since 1986 the field of powder injection molding has grown 100-fold because of the favorable shape complexity and high density attributes. During this same time, traditional press-sinter ferrous powder metallurgy expanded by a much smaller 2-fold increase, and has been declining in many areas since its peak in 2004. All of the growth since the late 1960s has been in the higher density range when 20% porosity was typical. Now, many components are shipped at 100% density. Figures 3, 4, and 5 illustrate a few examples, showing 100% dense powder metallurgy connecting rods, firearm sight, and oil well drilling tip. Indeed, the near-full-density aspect of powder injection molding is a key to its growth to become a $2 billion business.

Thus, as an early hypothesis we accept that traditional materials and traditional uses of powders are only growing at about the same rate as regional manufacturing growth (which has been negative at times), while nontraditional materials that rely on new or novel sintering technologies represent the growth opportunity.

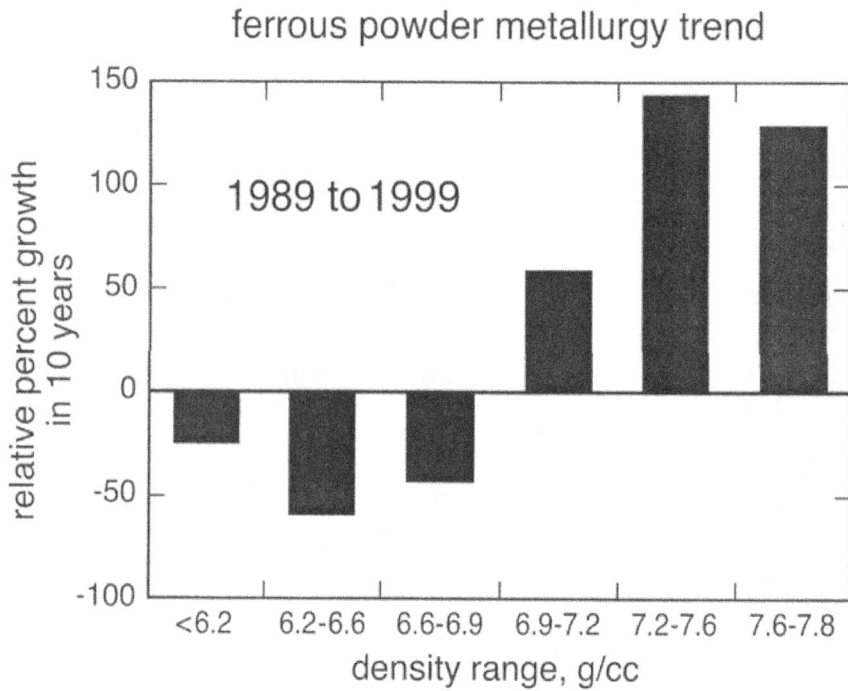

Figure 2 Trend over 10 years in the ferrous powder metallurgy field, showing relative growthonly was in the higher density range over 6.9 g/cm3.

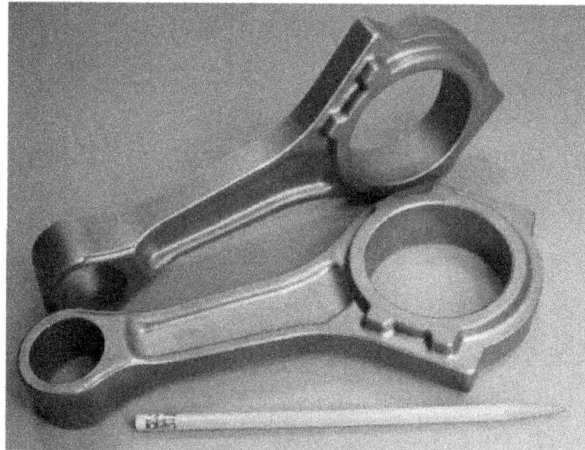

Figure 3 Powder metallurgy connecting rods, fabricated to 100% density from a mixture of iron, copper, and graphite powders.

Figure 4 Rifle sight fabricated to 100% density using sintering of an injection molded powder.

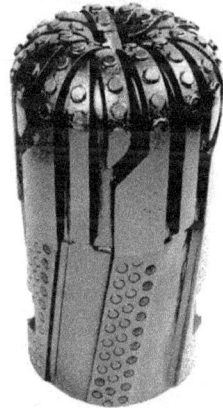

Figure 5 Oil well drilling tip fabricated by advanced sintering technologies to produce a composite that consists of diamond coated tips on a cemented carbide body, the overall diameter is 200 mm.

With respect to sintering, a few key points are important to fully understanding the opportunities. Most of the growth is in full density, and a majority of the drivers on that growth come from novel or new materials geared to high performance. We recognize that heating gives some important changes relevant to sintering:

- most materials become softer as temperature increases – termed thermal softening, so they exhibit a lower yield strength, faster recovery, and lower hardness at high temperatures
- most materials become ductile as temperature increases – even glass which is very brittle at room temperature shows a great ability to flow at high

temperatures, so this ductility couples with thermal softening to enable flow deformation at low stresses.

- at high temperatures most materials deform without hardening – which means they continuously recover during working and might even exhibit no work hardening.

As a consequence, temperature is the most important aspect of sintering. The traditional generation of temperature in a furnace is a mixture of convection, conduction, and radiation. At low temperatures it is necessary to stir the atmosphere to induce good convection, since conduction in a stagnant atmosphere is not very effective. At higher temperatures, say over 500°C, radiant heating becomes dominant. Thus, in a traditional furnace the typical means to heat is to place the work into a heated zone with stirred atmosphere to induce both convection and radiation. If sintering is in a vacuum, then low temperature heating is not effective. Many innovations are taking place in sintering, with novel applications (for example replacements for solder and silicon chips based on printed powders), novel heating techniques (laser and microwave), and property combinations (such as in computer heat dissipation devices), making sintering an exciting area for materials innovation. The attached references provide a first introduction to the activities and the following sections give a few interesting ideas.

2. Electromagnetic spectrum

The electromagnetic spectrum provides an infinite range of wavelengths and means to transport energy into a material for sintering. At the radio wavelengths (0.1 m) there is low energy and a high frequency while at the X-ray portion of the spectrum the wavelength is very short (nm) and the energy is 10-billion times higher. Gamma rays are even higher energy, but not as easy to produce. The optical range used in many sintering cycles is in the fraction of a micrometer wavelength and with an intermediate energy delivery capacity (about 4 10-19 J per quantum packet). This is relevant to understanding how we arrived at current furnaces designs. Some materials are transparent in certain wavelengths, but this is not commonly a problem for metals. Window glass is transparent to light, but steel is opaque (thus steel absorbs the energy of an intense light beam or light radiation source). At microwave wavelengths, most metals are opaque, but many ceramics are transparent. It seems strange that the visible light range dominates radiant heat transfer for metals, yet higher energy delivery efficiencies exist over a broad range of wavelengths.

Sintering furnaces are typically operating in the near white light region of the electromagnetic spectrum. At low temperatures the radiation output is low, so initial sintering heating is poor and requires a process atmosphere to transport heat, usually by

256

convection. At higher temperatures the free electrons in most metals enable direct radiation coupling This is neither optimal nor detrimental, just a fact. However, the same is true with other electromagnetic wavelengths, thus microwave, infrared, and ultraviolet heating are useful with metals, but not necessarily with all polymers and ceramics. Some of the innovations in sintering address the electromagnetic spectrum, relying on microwaves, lasers, and infrared heating as examples of new ideas.

Microwave sintering of ferrous, cermet, cemented carbide, and ceramic powders is well established. Figure 6 is a photograph of a production microwave furnace used to sinter WC-Co inserts used in oil well drilling. Rapid cooling from the sintering temperature proved beneficial to the durability of the product since the WC coming out of solution precipitated as a smaller dispersed phase in the cobalt versus the typical furnace which required hours to cool. This is a case where the heating rate was not useful (since polymer burnout was not possible), but the low thermal mass of the furnace allowed rapid cooling. Generally, microwave sintering is useful if there is no polymer present and the material can benefit from rapid heating or rapid cooling. However, often microwave sintering shows little benefit because of costs. Indeed, in several industrial trials the conclusion is that traditional, slower furnaces provide a better product with better dimensional control and similar properties. Further, the small sweet spot in most microwave furnaces makes it necessary to sinter a single component at a time. So after 25 years of demonstrations microwave sintering is still mostly a curious technology.

Figure 6 An industrial microwave sintering furnace for production of WC-Co inserts used in oil well drilling, where the rapid cooling proves beneficial to hardness and wear resistance.

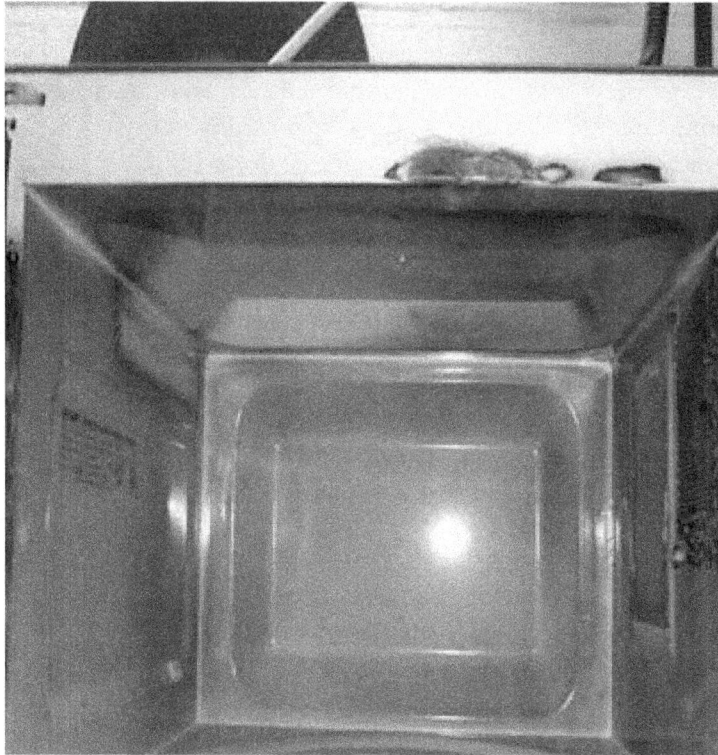

Figure 7 The interior of a home microwave oven after use for sintering powder metallurgy components.

An attraction to microwave sintering is the possible use of a home microwave. This is possible, but stray radiation usually leads to damage. Most home microwaves had nonuniform distributions to the magnetic and electric fields, so special efforts are required to properly distribute the fields inside the microwave cavity. Figure 7 is a picture of the inside of one such unit after several runs, giving evidence of blistering when used to sinter ferrous samples. The key here is to form a coffin around the component, where the walls of the coffin are microwave transparent and thermally insulating, allowing heating by the microwaves without blistering of the pain on the inside of the microwave.

On the other hand, electromagnetic fields, via electromigration, plasma activation, or similar approaches are novel ways to overcome the poor heat conduction associated with vacuum sintering. In one novel advance, a low-pressure plasma is sustained during the polymer burnout phase of the sintering cycle, to transfer heat and to ensure proper decomposition of the binders and lubricants. For the very difficult to sinter materials, there has been an explosion in plasma sintering and spark plasma sintering. These concepts have now gone commercial in the production of sputtering targets and medical implants. This shows all aspects of being very successful. Indeed, in the Sintering 08 Conference, held in La Jolla, California during November 2008, the largest topic of research were the several variants of spark plasma sintering.

3. Pressure Assisted Densification

A variety of full-density techniques have emerged that represent various combinations of stress and temperature. Pressure or stress on a powder during heating induces better densification, with the densification gain being directly proportional to the applied pressure. A variety of common approaches exist. They range from low pressure options that require high temperatures and small particles to high pressure options that require lower relative temperatures. Figure 8 plots the combination of temperature and pressure with respect to densification. Several processes are located on this plot of relative stress and relative temperature. The relative stress is the applied pressure divided by the *in situ* yield strength of the material and the relative temperature is based on the melting temperature. Categorically, full-density processes fall into a few clusters:

- low-stress routes that operate at high temperatures, such as sintering, that are dominated by diffusion controlled processes,

- intermediate-stress routes that operate at intermediate temperatures via diffusional creep processes, such as hot isostatic pressing,

- high-stress routes that operate at lower temperatures via plastic flow, such as powder forging, ultra-high-stress routes that attain full density at room temperature, such as explosive compaction.

The approximate mapping of sintering technologies with respect to pressure and temperature, showing how one or the other dominates the powder densification event.

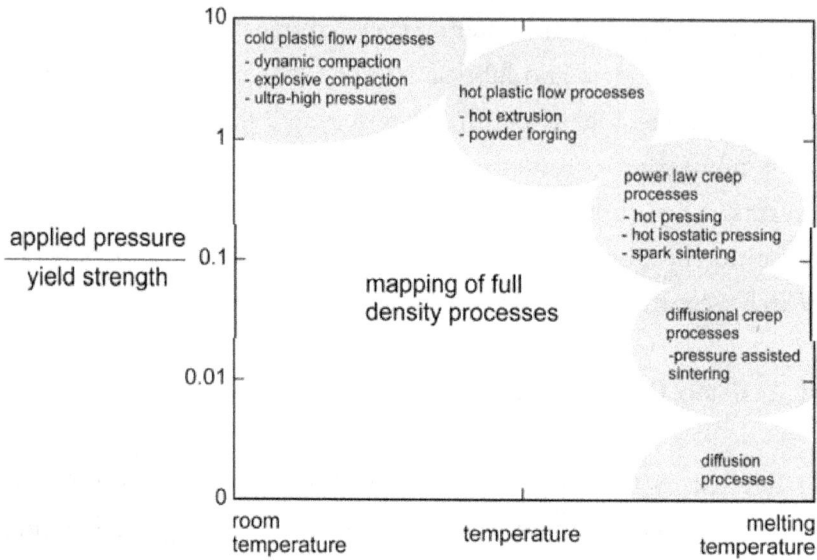

Figure 8 A mapping of sintering technologies with respect to pressure and temperature, showing how one or the other dominates the powder densification event.

Although this figure is schematic, it provides a first view of full-density options. There are many additional factors, such as particle size, green density, and component size, that separate the processes, but this simple categorization helps sort out the key parameters for densification. Performance levels can be exceptional if control is gained over defects, impurities, microstructure, and product homogeneity. That is the primary justification for full-density powder metallurgy.

The most common approach to full density is sintering at temperatures close to the melting point. Liquid phase sintering of small powders results in a composite microstructure consisting of a skeleton of high-melting-temperature phase in a matrix of solidified liquid. Applications include many technical ceramics, porcelain, and tool steels, stainless steels, W-Ni-Fe heavy alloys, Co5Sm and Fe-base magnets, WC-Co cemented carbides, TiC-Fe cermets, ruthenium, cobalt-chromium, nickel superalloys, and insoluble systems such as copper-chromium, stainless-zirconia, aluminum-graphite, and tungsten-copper.

4. Innovation Trends in Sintered Materials

To best understand the trends in sintering technologies, we need to examine the new directions in sintered materials and structures. One important trend is to expand the size range and production quantities of intered products. Most of the current products are things that you hold in your hand. The largest growth is toward small, nimble furnaces, especially for medical and electronic devices. Current needs focus on ceramics, stainless steel, and titanium. One important trend is in the production of limited quantities, such as from rapid prototyping technologies, where the green body is very fragile. One example product is shown in Figure 9

Figure 9 Limited edition sintered bronze produced using a using a novel process that involved immersion in powdered alumina to ensure good support and heat transfer.

Structures of this sort for tooling, filters, artwork, and other applications have ranged up to 20 kg, and routinely range near 2 kg.

Another area of great need in sintering is the consolidation of very hard materials, often the materials are intended for use at high temperatures, so they inherently resist sintering. These materials naturally resist sintering. For example, pure WC is very difficult to sinter yet it is desired for very abrasive applications such as water jet abrasive nozzles. Other materials include Ti(C,N) compounds, zirconia, and covalent ceramics such as titanium diboride. These almost always require pressure-assisted sintering.

Functional design materials are another challenge in sintering. They enable a change in propertics within a body. An early demonstration was the cam shaft such as shown in Figure 10, assembled using pressed powders sintered to wrought shafts. Subsequent efforts moved to functionally composites, where sintering to density without cracking proved difficult. One application is in heat sinks for computers, where one surface is high in silicon carbide and one surface high in copper, with a gradient over the body. In recent years, other areas included powder injection molding where components were combined to tailor functionality over the structure - to combine magnetic and nonmagnetic, wear and toughness, or low-cost and high cost materials

Figure 10 Sintered cam lobe that has been densified and bonded in a single vacuum sintering cycle.

Another area of great demand in sintering is the production of extremely large quantities of simple devices, such as bearings, capacitors, birdshot, and heat sinks. For example many of these devices are in production at quantities exceeding 100 million per day. Figure 11 shows some tantalum capacitors as representations of the field. How to sinter these large quantities with uniform cycles are real challenges. The novel solution is to avoid polymers in the furnace via burnout or extraction prior to sintering. Then a very small continuous furnace provides a repeatable radiant heating environment for each compact. In effect, think of how amusement rides, such as Haunted House at Disneyworld, handle large quantities of customers yet ensures the same experience for each rider. In sintering this means furnaces that heat and cool quickly, with very short hot zones, that ensure each component sees the same time-temperature history, yet to reduce floor space it is necessary the work move rapidly. Similar concepts are being considered for other high volume areas, such as the sintering of orthodontic brackets and birdshot.

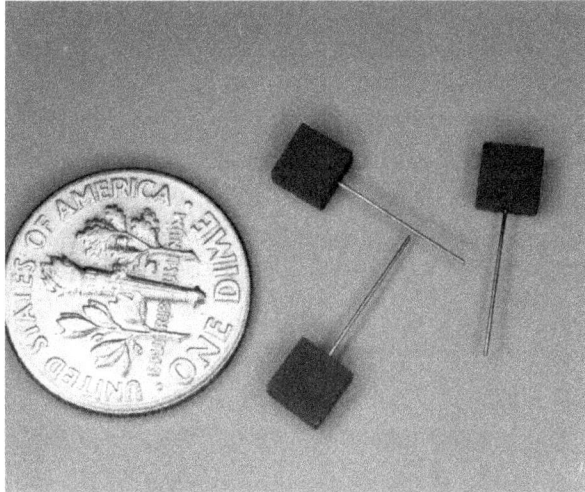

Figure 11 Picture of some pressed and sintered tantalum capacitors prior to impregnation with an electrolyte, charging, and encapsulation.

sintered precision is another challenge for powder metallurgy. The current best practices deliver standard deviations less than 0.02% of the specified dimension. A few examples are able to deliver tolerances in the 5 to 10 micrometer range. Figure 12 is an end mill fabricated to such tight tolerances. Several processing steps had to be optimized to ensure proper cycles and close dimensional control.

Figure 12 A carbide end mill formed to tolerances of 5 to 10 micrometers using advanced powder injection molding techniques and very careful sintering cycles.

Such products tax the sintering technologies, and require attention to everything. Today, advanced computational tools exist to ensure all factors that might contribute to dimensional variation are controlled or anticipated.

The drive toward minimally invasive surgical procedures creates a need for very small manipulators and devices, many with dimensions in the micrometer range. One of the first production products was for biopsy of the human heart, as shown in Figure 13. This was formed by powder injection molding out of stainless steel. Similar challenges exist in new surgical tools, orthodontic devices, and functional lab on chip devices for blood assay.

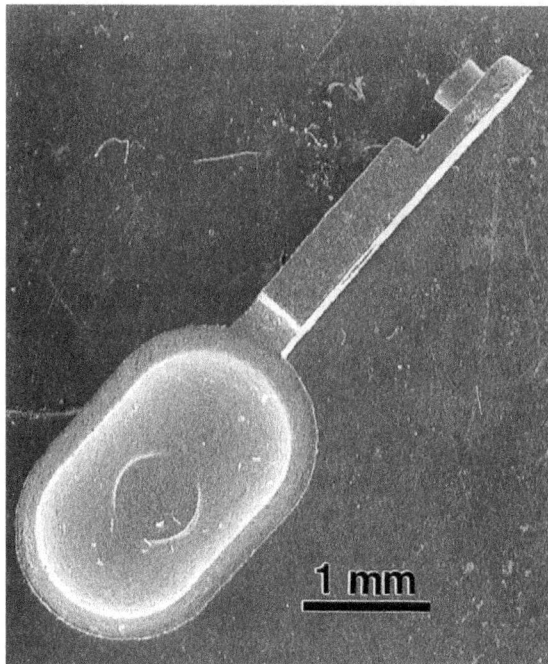

Figure 13 Heart biopsy tool formed by powder injection molding. Such small bodies challenge sintering techniques since handling and inspection are not straightforward.

Another trend is in increasing complexity. Sintered devices are pushing the limits on the number of dimensional call outs. These far exceed the traditional capabilities. A good example is the trirotor shown in Figure 14, which is a high precision and complicated assembly used in the oil fields. This tri rotor involves several parts in a hard, wear resistant composition. The tolerances, complexity, and general features simply tax sintering technologies.

Figure 14 A complex and high precision cemented carbide component used in petrochemical processing, where the design challenges the size, shape, and tolerance capabilities - it is a complex component.

5. Novel Compositions

The innovations in sintering are in means to heat materials (microwave as an example), in the designs and features (microminiature as an example), and in the materials. The field of sintered materials abounds in examples of new materials, as listed below:

o new nitrogen stabilized steels that take advantage of small powders and ability to add nitrogen from the sintering atmosphere – delivering 2-fold strength gains and 20-fold corrosion resistance gains over traditional sintered stainless steel

o carbo-nitrided tools steels that combine the strengthening of both carbides and nitrides, again based on sintering in nitrogen, giving strengths over 3000 MPa with 3% elongation to fracture

o novel titanium structures that start with titanium hydride (less explosive than titanium) and sinter in hydrogen above the hydride decomposition temperature, delivering dense, strong, and relatively clean microstructures

o coated powders provide one means to engineer at the microstructure level; one significant process relies on hard core particles and WC-Co coatings to form a dispersed composite microstructure, such as pictured in Figure 15; promises from these materials extend over a broad range of applications from cutting and wear tools, surgical devices, and even first wall materials and shape charge liners and heat dissipation devices in computers

- novel ferrous heterogeneous alloys have been developed that rely on the mixed powder route, but stabilize a multiple phase microstructure that is more resistant to fatigue; in one demonstration of this technology with a 4600 steel, the heterogeneous sintered product had a 20% gain in fatigue endurance limit

- novel materials abound via sintering and some worthy of mention are porous WC-Co materials that are infiltrated with carbon and under pressure the carbon is converted into diamond to give a novel hard composite

- reaction sintering is another option for novel materials, where for example aluminum powder is pressed to 78% density and then heated in air to form dense aluminum alumina composites; variants include formation of aluminum nitride, silicon nitride, and other compounds and composites

- composites based on mixed powders provide some novel wear resistant materials, for example 3 vol.% CaF2 addition to a tool steel (survives the sintering cycle) leads to a 30- fold reduction in dry wear rate and 3 wt.% TiN addition to a 440 stainless steel leads to a 20-fold reduction in abrasive wear rate.

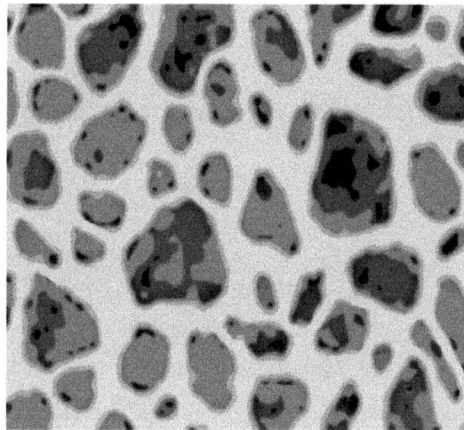

Figure 15 Consolidated hard particle cores with tungsten carbide coatings to form a hard, wear resistant composite.

6. Nanoscale Materials

Nanoscale powders represent one of the large, near-term challenges to powder metallurgy. Although there is much promise of property gains, the reality is a low ability to realize such gains. The very small powders bond and densify at lower temperatures, and thereby often do not evaporate impurities, so the properties suffer. Further, grain growth is rapid during heating, so the final product shows little gain from the high cost initial powders. Various rapid heating and pressure heating cycles have been proposed to circumvent these issues. For example, radio-frequency plasma cycles have demonstrated 100 /s heating rates and sintering times of just 10 s to attain full density.

Such short cycles mandate very clean powders, since there is no time to burnout contaminants. Pressure can be imposed on these in what is called spark plasma sintering. However, the microstructures still show large grains.

Laser sintering is another option being applied, but it too suffers from being only a surface heating route. It has found most success with depositing thin, amorphous layers and in rapid prototyping, such as in elective laser sintering. A variant has been developed that relies on electric discharge and this appears to provide better positional control for one of a kind object generation from stereo lithography ideas.

Recent progress in nanoscale powder sintering has shown that grain growth is so rapid the nanoscale powders reach a large grain size prior to attaining full density. Proposals to counter this problem include multiple step heating cycles, and even constant temperature reductions as pores are eliminated. Pressure assisted techniques are still the most fruitful. The spark plasma sintering route is useful, as demonstrated in Figure 16, but still production of nanoscale microstructures in bulk components proves elusive.

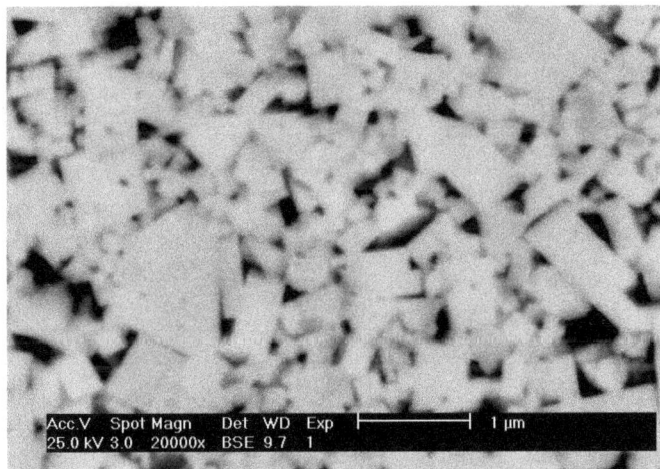

Figure 16 Spark plasma sintered nanoscale WC-Co showing the capture of the small scale microstructure via this sintering technique.

7. Computer Simulations

Final mention is required on the significant advances in computer simulations. New full threedimensional simulations enable evaluation of cycles, materials, forming processes, furnace designs, and other aspects of the technology prior to actual experimentation. In some areas the simulations are progressing ahead of practice, but in most fields the simulations are behind practice. Even so there is significant progress and soon we can expect the simulations to help remove barriers on new materials, products, and to provide a rational basis for evaluation of possible alternative designs and cycles.

Some significant advances have come from the combination of process models, such as for die compaction or injection molding, coupled to sintering models. The process model shows the green body density distribution, and when sintered the resulting nonuniform shrinkage dictates the warpage of distortion. Such ideas have been extended to hot isostatic pressing to enable prediction of the final component shape based on stiffness of the can and deformation of the powder. Figure 17 below shows the finite element predicted shape and that actual shape for a tungsten heavy alloy after sintering. Refinements in the simulation are needed, but the general attributes are seen in the simulations.

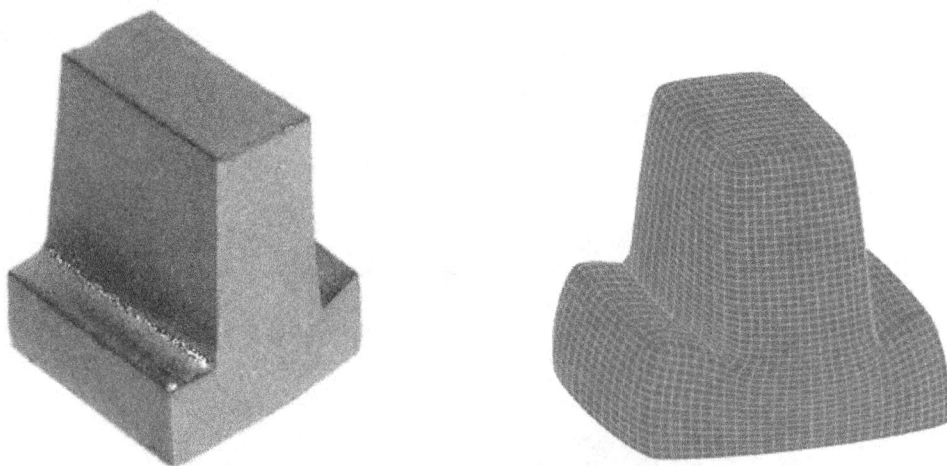

Figure 17 Actual component and finite element simulation of an inverted T shaped tungsten heavy alloy component that distorts during liquid phase sintering.

Acknowledgements

This paper is a survey of the large body of developments occurring in the field of sintering. It is impossible to properly attribute or reference all of the conferences, papers, and seminars where data used here were provided. Special mention is made of inputs from Professor Seong Jin Park, Professor Debby Blaine, Professor Hideshi Miura, Professor Eugene Olevsky, Professor Joanna Grozza, Dr. Mu-Jen Yang, and Dr. John Johnson.

References

1. R. M. German, *Basic introduction to sintering processes: Sintering Theory and Practice*, Wiley Interscience, New York, 1996.

2. E. A. Olevsky, V. Tikare, T. Garino, *Status of sintering models at multiple length scales and component levels: Multi-Scale Study of Sintering: A Review*, J. of American Ceramic Society, Vol. 89, 2006, p. 1914.

3. R. M. German, *Computer Modeling of Sintering Processes*, Int. J. of Powder Metallurgy, Vol. 38, No. 2, 2002, p. 48. J. Ma, J. F. Diehl, E. J. Johnson, K. R. Martin, N. M. Miskovsky, C. T. Smith, G. J. Weisel, B. L.

4. Weiss, D. T. Zimmerman, *Update on microwave sintering: Systematic Study of Microwave Absorption, Heating, and Microstructure Evolution of Porous Copper Powder Metal Compacts*, J. of Applied Physics, Vol. 101, 2007, p. 074906.

5. Z. A. Munir, U. Anselmi-Tamburini, M. Ohyanagi, *Review of spark plasma sintering: The Effect of Electric Field and Pressure on the Synthesis and Consolidation of Materials: A Review of the Spark Plasma Sintering Method*, J. of Materials Science, 2006, Vol. 41, p. 763.

    P. A. P. Wendhausen, J. L. R. Muzart, A. N. Klein, D. Fusao, L. A. Mendes, W. Ristow, R.

6. Machado, *plasma sintering and debinding: Plasma Assisted Debinding and Sintering: Process and Equipment*, Proceedings PM2004 Powder Metallurgy World Congress (Vienna, Austria), European Powder Metallurgy Association, Shrewsbury, UK, Vol. 2, 2004, p. 134.

7. sintering of composite hard particles:

    R. M. German, I. Smid, L. G. Campbell, J. Keane, R. Toth, *Liquid Phase Sintering of Tough*

    Coated Hard Particles, International Journal of Refractory Metals and Hard Materials,

    Vol. 23, 2005, p. 267.

8. G. Fu, N. H. Loh, S. B. Tor, B. Y. Tay, Y. Murakoshi, R. Maeda, *sintering and shaping of microminiature devices:Injection Molding, Debinding and Sintering of 316L Stainless Steel Microstructures*, Applied Physics *A*, Vol. 81, 2005, p. 495.

9. H. Ye, X. Y. Liu, H. Hong, *sintering and shaping of composites:Fabrication of Metal Matrix Composites by Metal injection Molding - A Review*, Journal of Materials Processing Technology, Vol. 200, 2008, p. 12.

10. B. Williams, *Opportunities in medical and dental devices:Powder Injection Moulding in the Medical and Dental Sectors*, Powder Injection Moulding International, Vol. 1, 2007, p. 12.

11. A. I. Taub, *Opportunities in automotive applications: Automotive Materials: Technology Trends and Challenges in the 21st Century*, Materials Research Society Bulletin, Vol. 31, 2006, p. 336.

Powder Materials for Energy Applications

R Sundaresan

NFTDC, Hyderabad

Abstract *Changes in technology are most often enabled by materials. Powder technology and powder metallurgy processing have enabled electrical energy for a long time. For example, lamp filaments and insulation materials have been processed by powder route; electrical contacts involving dissimilar metals (e.g. W-Ag, W-Cu) and dissimilar materials (e.g. Ag-Graphite) have been exclusively made by PM. In the last decades, changes in technology in generation, storage and distribution of energy have become necessary because fossil fuel production levels have peaked while energy demand continues to rise. Several alternatives are being pursued in order to fill the energy gap. These include directly accessible wind and solar energy, tidal, wave and thermal energy from the sea, and alternative fuels such as biodiesel, fuels culled from coal bed or clathrates, and hydrogen. The trend of miniaturisation of devices has also led to requirements of higher energy density in energy storage. Powder technology has continued to enable such developments, as it did the early electrical technology. Effective utilisation of hydrogen calls for major development in hydrogen synthesis, hydrogen storage and transport, and conversion to energy. Powder technology plays a major role in the storage and conversion aspects. Storage and transport of hydrogen as compressed gas or as liquid is feasible but pose economic constraints. Storage as combined with solid materials is an option that has been found very attractive because of comparatively high storage density feasible, even higher than as liquid. Such materials are often based on metals, intermetallics and complex chemicals, invariably used in powder form.*

While it is possible to convert thermal energy contained in hydrogen and other alternative fuels to mechanical energy (and, where necessary, convert further to electrical energy), hydrogen and hydrocarbons offer an opening for far superior efficiency by direct conversion to electrical energy in fuel cells. There are several classes of fuel cells, the most prominent among them being polymer electrolyte membrane (PEM) fuel cell and solid oxide fuel cell (SOFC). While PEMFC offers advantages in mobile applications such as in transport vehicles, SOFC offers distinct advantages in stationary applications, and also enables conversion of less pure hydrogen and hydrocarbons. Most of the materials of construction in SOFC are based on ceramic and metallic powders.

The paper discusses aspects of processing and application of the some powder materials used in such energy applications. Case studies on a hydrogen storage material, an interconnect material for use in SOFC and a heat transfer component in ITER are presented.

271

1. Introduction

Over the ages materials have been the enabling factor in progress of technology: so much so that eras have been named after the enabling material such as stone, bronze, and iron. The twentieth century was marked by the spurt in assimilation of science and technology, and consequently, in the consumption of energy. Increasing affluence is seen to be directly correlated with demand on energy (Fig. 1).[1] Energy has indeed become dear both because of the increased demand and because of dwindling resources. The major thrust in technology in the 21st century has therefore come to be on energy.

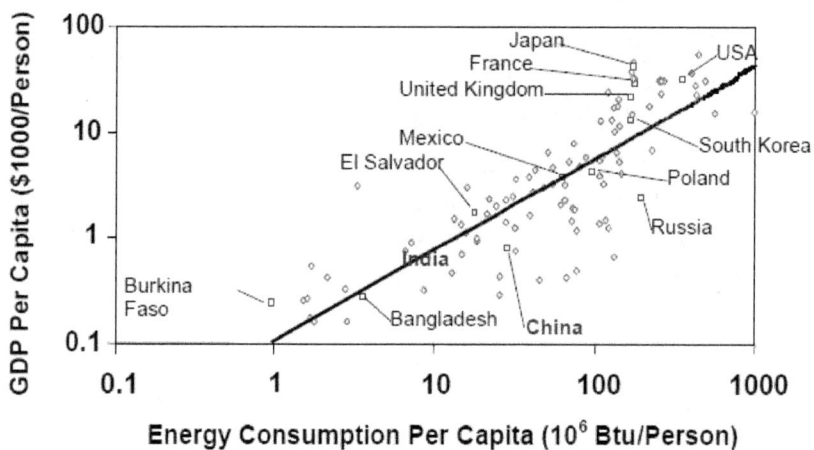

Fig. 1 Energy consumption correlates with affluence.

Over the ages, mankind has been largely dependent on fossil fuels for energy. Two major considerations in recent years dictate the need for change in the energy strategy. Recognition of the havoc on environment from CO_2 emission has inspired considerable research into energy alternatives. It is also recognised that while consumption is continuously increasing at the rate of 2% per annum, petroleum output is peaking, and the increasing gap must be met only by other energy sources. Many alternatives are being actively pursued around the world (Table 1). It is necessary that all the available alternatives should be tapped and used judiciously to augment the fossil fuel based energy, which will continue to be mainstay.

Materials will play a major role in enabling the new technologies being developed around the alternative energy schemes. We will consider contributions from powder technology in this area. Powder technology has been intimately associated with energy sector from the very beginning of electrical energy applications. The first major application of electrical energy has been lighting, and until recently thoriated tungsten filament was the mainstay of such lighting. Powder processing techniques were necessarily the route to the refractory tungsten.[2] PM processed composites containing refractory metals (W) or oxides (e.g. AgO) with metals of high conductivity (Cu/Ag) have been in use as electrical contact materials.[3] Powder technology is also extensively employed in areas such as electrical insulation ceramics and multilayer ceramic capacitors. Powder processed materials will also be widely required in enabling some of the energy alternatives considered. Among the alternatives, hydrogen has been the most prominent. Its attractiveness stems both from the abundance of hydrogen, although in combined form, and from the absolute cleanliness of hydrogen as fuel.

Table 1 Energy alternatives

No.	Energy Scheme	Options
1.	Wind	
2.	Solar	
3.	Biomass	Agrowaste
		Forestry waste
		Municipal solid waste
		Energy crops
4.	Hydrogen	From petroleum
		Steam reforming of methane
		Electrolysis of water
5.	Ocean	Tidal
		Wave
		Methane from clathrates
6.	Nuclear	Conventional fission
		Fusion
7.	Thermal	Geothermal
		Hydrothermal

2. Hydrogen

Energy is now largely transported as electrical energy over wire, rather than as raw fuels transported as mass. An electrical power grid over the nation, indeed across continents, makes it a lot easier to serve energy on demand. However, it is impossible to eliminate transport of energy sources as mass. At this time and in the near future, transport itself will need fuels. Hydrogen as the energy carrier for transport is the major scheme on which considerable work goes on across the world. The ideal scheme for energy distribution is seen as inputs into the electrical grid from various sources such as vast solar farms in deserts and even space, wind mill farms, nuclear, geothermal and ocean sources, and local inputs such as from biomass, with spare energy going into energy storage techniques such as battery, flywheel, compressed air and in generation of hydrogen which would be used as fuel for transport (Fig. 2).

Storage of hydrogen for transport is a major challenge because of the very low density of hydrogen, even when compressed to liquid state. There are other attractive options to hydrogen storage in the forms of adsorption, absorption and complexing which enable much higher hydrogen content for given volume/weight (Table 2). A storage device incorporating hydrogen stored as metal hydride is shown in Fig. 3 [4]. The hydride has to be in granular form developed through powder processing. Magnesium based materials have been developed with excellent absorption properties.

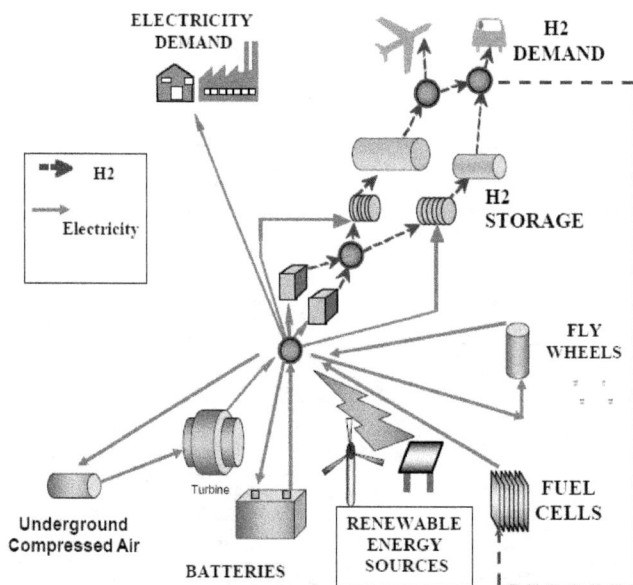

Fig. 2 Energy storage options.

Table 2 Hydrogen storage options

Storage method	Gravimetric density, mass%	Volumetric density, kg H_2/m^3	Temperature. °C	Pressure, bar
High pressure gas cylinder	13	40	RT	800
Liquid hydrogen in cryogenic tank		70	-252	1
Adsorption of hydrogen	2	20	-80	100
Absorption as interstitialcy in host metal	2	150	RT	1
Complex compounds	18	150	>100	1
Metals/complex reaction with water	40	>150	RT	1

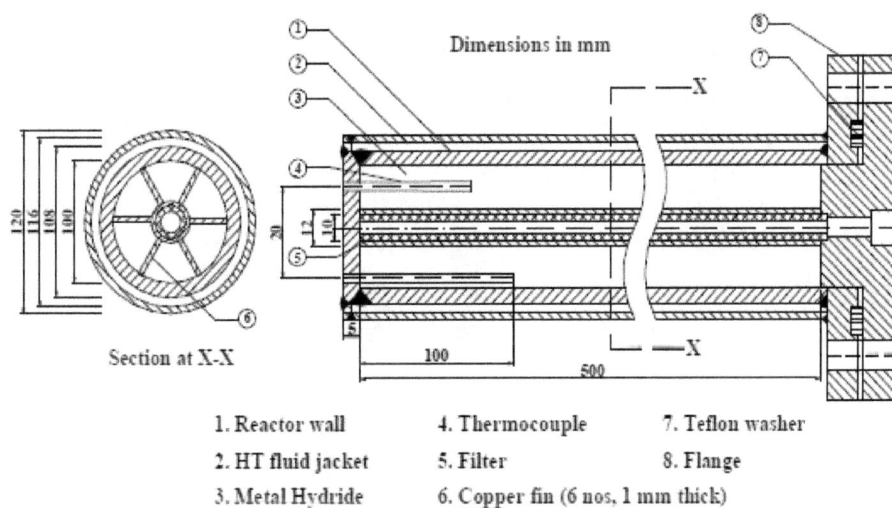

1. Reactor wall
2. HT fluid jacket
3. Metal Hydride
4. Thermocouple
5. Filter
6. Copper fin (6 nos, 1 mm thick)
7. Teflon washer
8. Flange

Fig. 3 Hydrogen storage device.

3. Magnesium with MmNi5 Dispersion

A composite powder of magnesium with mischmetal (35La-46Ce-13Nd-5Pr) based intermetallic compound MmNi5 was made by mechanical milling of the two powders in an attritor under hydrogen. Grain sizes of Mg and MmNi5 in the composites (Fig. 4) are in the nanometric regime.

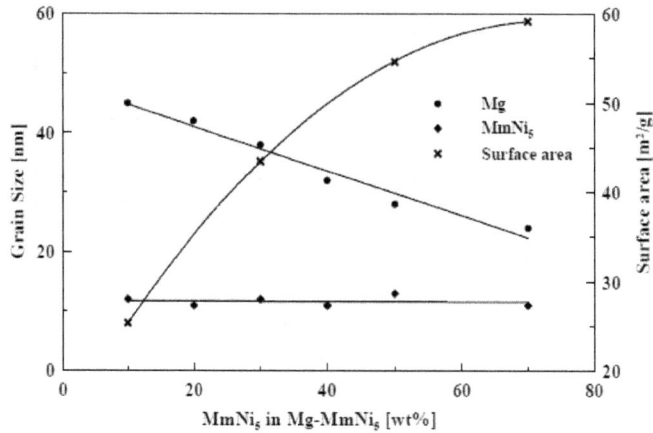

Fig. 4 Grain size and surface area in Mg-MmNi5 composites milled for 16 h.

3. 1 Hydrogen Storage Properties of Mg-MmNi5

The absorption/desorption pressure composition isotherms of Mg-MmNi5 composites were studied at 300°, 200° and 100°C. Sorption behaviour at 300°C is shown in Fig. 5. The absorption and as well as desorption plateaus for all the compositions are inclined and overlap in their respective hydrogen content range. The absorption plateaus for all the compositions begin at 0.1 MPa (1 bar) while the desorption plateaus are between 0.05 and 0.15 MPa (0.5 and 1.5 bar). At 200 and 100°C, the desorption plateaus were below 0.1 MPa (1 bar).

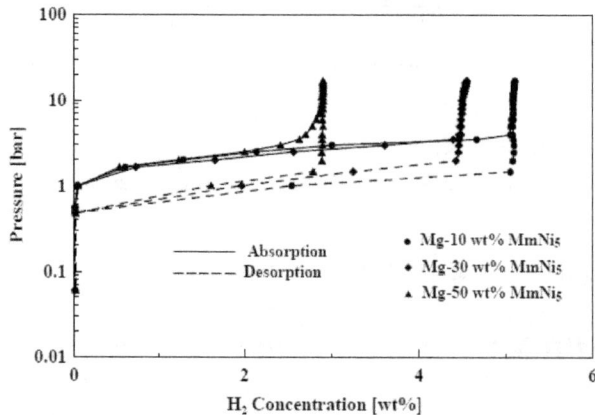

Fig. 5 PCI of Mg –MmNi5 composites at 300°C

276

The absorption kinetics are very good: the composites showed some instantaneous absorption (2 t% in less than 5 s). Even with the lower absorption rates at 100°C, the composites reached saturation in less than 400 s. The first part of absorption in the composites represents surface adsorption and grain boundary diffusion of hydrogen leading to saturation of hydrogen atoms covering surface and grain boundary area of Mg. The total quantity of hydrogen absorbed by the powders followed the respective Mg content. The extent of conversion of Mg to hydride is shown in Table 3.

Table 3 Extent of hydrogen absorption in Mg – MnNi5 composites

Composite	Pressure, MPa	Hydrogen absorption, wt%		
		Temperature, °C		
		300	200	100
Mg-10 wt% MmNi$_5$		5.1	5.1	5.0
Mg-30 wt% MmNi$_5$	3	4.6	4.4	4.3
Mg-50 wt% MmNi$_5$		2.9	2.8	2.7

MmNi5 has a profound effect on absorption of hydrogen by Mg, even though no hydride of MmNi5 was seen in the hydrided composite (Fig. 6). On the basis of analysis of the absorption kinetics data (Fig. 7), it was shown that the process of hydrogen absorption followed the following stages [5,6]:

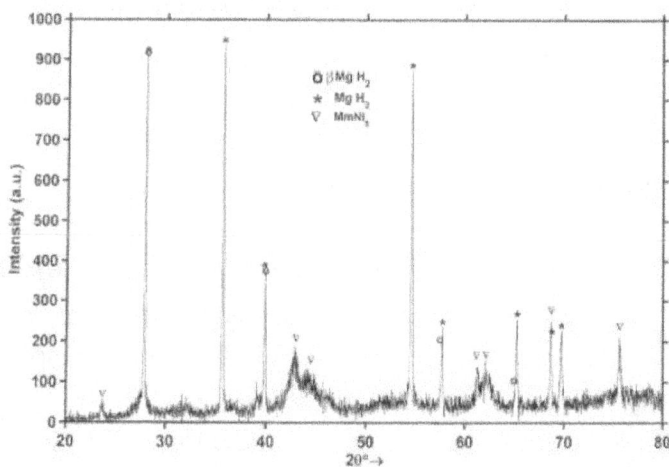

Fig. 6 XRD of hydrided Mg – 30 wt%MmNi5 composite with Mg fully hydrided and MmNi5 remaining pristine

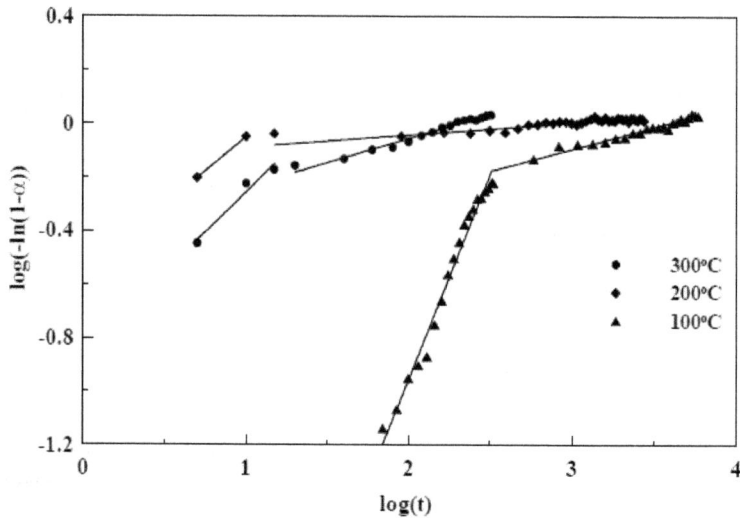

Fig. 7 Johnson-Mehl-Avrami plots of $\log(-\ln(1-a))$ vs. $\log t$ for Mg-10 wt% MmNi5 composite

(i)   MmNi5 dispersion on surfaces and grain boundaries of the magnesium particles enable break up of H2 molecule on absorption under the conditions of P and T

(ii) Lattice and grain boundary diffusion of atomic hydrogen enable rapid absorption to saturation levels on surfaces and grain boundaries

(iii) Nanometric grain sizes of Mg enable diffusion into the grains from the boundaries and formation of the hydride in short time.

(iv) Adiabatic temperature rise from the exothermic reaction enables absorption even at temperature levels as low as 100°C.

4. Solid Oxide Fuel Cells

Solid oxide fuel cells (SOFC) provide a method of generating electricity from hydrogen (or other hydrocarbon gases). They offer considerable advantages over other types of fuel cells, articularly for stationary applications. With all solid components, SOFC can be fabricated very compact, and permit the use of impure hydrogen and even hydrocarbons. With operating temperatures above 800°C, ideally SOFC would be used in conjunction with waste heat recovery. The cell configuration is shown in Fig. 8.

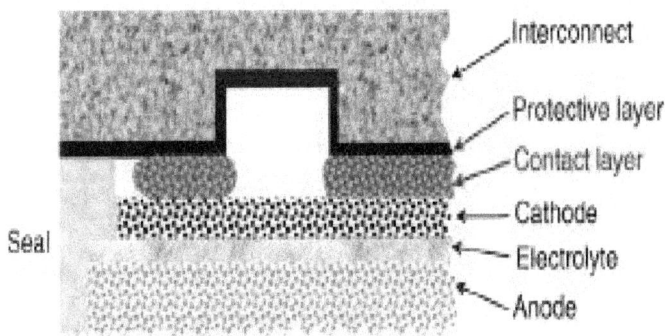

Fig. 8 SOFC configuration

The anodes, cathodes and the electrolyte in SOFC are all made through powder processing. The interconnect (bipolar layer) in SOFC has very specific functional requirements: it should be conducting at high temperature, and should retain the conductivity even on prolonged exposure to air/oxygen on one side and hydrogen/hydrocarbon on the other. Both ceramic and metallic materials have been identified for the application. One such metallic material is chromium dispersed with Y_2O_3. A cost effective process was developed in fabricating such a material with nearly full density using PM processing.

5. Chromium with Oxide Dispersion for Interconnect Application

While mechanical alloying/milling is an excellent route to uniform dispersion and nano grain sizes, the process is not universally applicable. In the case of chromium based interconnect, the oxide dispersion is provided not for strengthening but for improving the oxidation resistance. Hence the major requirement is that adequate Y_2O_3 dispersion should be available at the surfaces and interfaces. Blending ensuring very uniform dispersion on the particle surfaces, followed by a compaction and sintering process achieving over 95% sintered density would be adequate, and would keep the cost low. The objectives were realised by selection of the size distribution in raw materials, the blending process, additives and sintering parameters.

Nano Y_2O_3 (~25 nm) was generated by combustion synthesis process (Fig. 9). Chromium powder of different size ranges were blended for a suitable size distribution for maximum green density. Some metallic additives were incorporated to ensure necessary electrical and thermal properties in the component. In a process of blending and annealing of powders, ready to press powder blends were obtained with uniform distribution of yttria. Extensive check of the powder blend (Fig. 10) confirmed the uniformity of distribution of yttria as well as other elements added.

Fig. 9 Nano yttria powder synthesized.

Fig. 10 Ready to press powder blend.

The uniformity of blend at the particle level is shown in Fig. 11. The overall area of a single Cr base particle shows approximately the targeted composition. Each powder particle is seen to be stably covered with the added elements. Good distribution of nano Y_2O_3 particles as well as other elements added on the Cr particle surface can be identified.

These powders were suitably compacted and sintered. Coefficient of thermal expansion, which should match with the electrodes, was measured by dilatometry. CTE was in the range $10 - 12 \times 10\text{-}6$ /°C in the temperature range up to 1200°C meeting with the requirements. The oxidation behaviour at prolonged hold in air at 800°C showed that the weight gain stabilised within 4 h. and the maximum ranged between 0.75 and 1.4%.

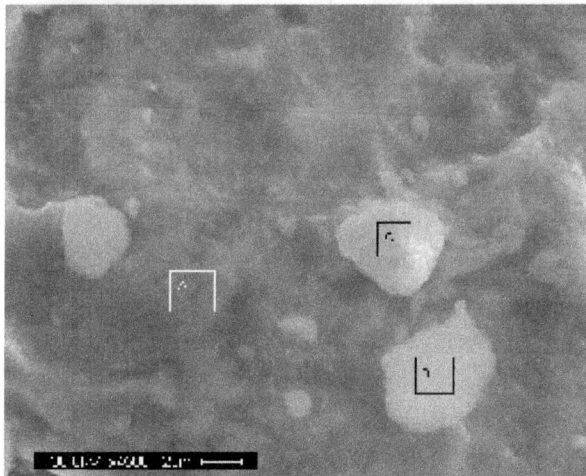

Fig. 11 Blended particle with added elements adhering.

Table 8 Coefficient of Thermal Expansion

281

Sr. No	Sample Identification	Coefficient of thermal expansion
1		30 – 250C : 10.11 x 10-6/C
		30 – 500C : 10.33 x 10-6/C
		30 – 750 C : 10.85 x 10-6/C
		30- 1000 C : 12.07 x 10-6/C

6. Nuclear Fusion for Power Generation

The international thermonuclear experimental reactor (ITER) is a joint international research and development project that aims to demonstrate the scientific and technical feasibility of fusion power. The partners in the project are the European Union, Japan, China, India, Korea, the Russian Federation and the USA. ITER is being constructed at Cadarache in the South of France.

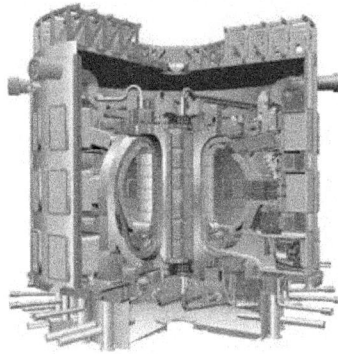

Fig. 12 The ITER device.

Fusion is the energy source of the sun and the stars. Fusion research at ITER, in a massive quest over the next few decades, is aimed at demonstrating that fusion that is the energy source of stars can be harnessed to produce electricity in a safe and environmentally benign way, with abundant fuel resources, to meet the needs of a growing world population. The fusion reaction in the ITER is Heat transfer and dissipation is one of the major engineering concerns of the reactor. NFTDC has already successfully developed one such component, and is currently engaged on the development of a plasma facing component, a divertor cassette for high flux heat removal, at the location marked in Fig. 12. The unit (Fig. 13) employs targets consisting of lanthanum oxide dispersed tungsten alloy fins on a high strength copper alloy tube (Fig. 14).[8]

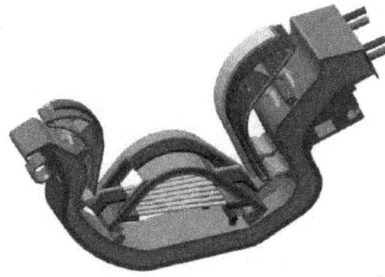

Fig. 13 Divertor Cassette.

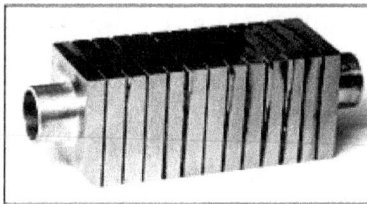

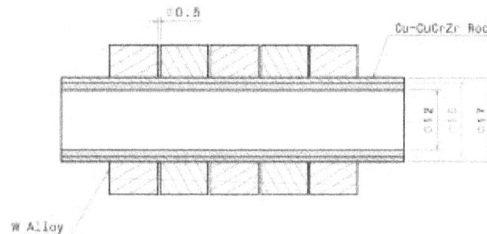

Fig. 14 Tungsten Monoblock

Doping with lanthanum oxide dispersion increases the creep strength and recrystallisation temperature of tungsten, and is also expected to improve the machinability. In this case dispersion is not achieved by mechanical alloying/milling process, but is achieved by chemical processing.

7. Conclusions

Powder technology will continue to enable energy technologies. One of the important material attributes, dispersion of particles, is easily realised by powder processing. Three case studies in such developments have been presented. Particle dispersed materials were developed for completely different application regimes in advanced energy technologies. A nanocrystalline magnesium base composite with dispersion of nanoparticles of MmNi₅ was developed for hydrogen storage. MmNi₅ particles on the Mg particle surfaces and grain boundaries dispersed by mechanical alloying provided a route for easy absorption of hydrogen by magnesium by a catalytic action of converting gaseous hydrogen to atomic hydrogen on the surface. The dispersed material showed full extent of hydrogen absorption by magnesium at considerably lower temperature and higher kinetics. Chromium with yttria dispersion was developed for application as interconnects in SOFC. In this case yttria provided enhanced oxidation resistance to the interconnect. The dispersion was achieved by blending and alloying, titanium most probably providing anchoring of yttria. Tungsten with dispersion of lanthanum oxide is used in a heat transfer device being developed for ITER. Lanthanum oxide provides both creep resistance to tungsten, but more important, raises the recrystallisation temperature enabling its application in the plasma. Dispersion in the powder in this case is achieved by chemical processing.

References

1. International Energy Annual 2000, Energy Data Reports

2. S Just and F Hanaman, Hungarian Patent 34541, 1904. (Cited in E Lassner, W-D chubert, "Tungsten", Springer, 1999.

3. Y.-S. Shen, P. Lattari, J. Gardner, and H. Wiegard, Electrical Contact Materials, ASM Metals Handbook, ASM, Vol. 7, 1998, p. 2552.

4. P. Muthukumar, M. Prakash Maiya, S. Srinivasa Murthy, R. Vijay and R.Sundaresan, J. Alloys and Compounds, No. 452, 2008, p. 456.

5. R. Vijay, R. Sundaresan, M.P. Maiya, S. Srinivasa Murthy, J. of Alloys and Compounds, No. 424, 2006, p. 289.

6. R. Vijay, R. Sundaresan, M. P. Maiya, S. Srinivasa Murthy, Y. Fu, H. -P. Klein, M. Groll, Int. J. Hydrogen Energy, Vol. 32, 2007, p. 2390.

7. R Sundaresan and S Sudhakara Sarma, unpublished work , 2007

8. M. Debata, M. Govindaraju, A. Pande, N. Panda and P K Jayakumar, NFTDC internal reports, 2009.

Physics of Pressure Transmission in Fluids and Powders

J. Viplava Kumar
Mahatma Gandhi Institute of Technology, Hyderabad, India
E-mail: j.viplavakumar@gmail.com

Abstract: *A confined liquid transmits externally applied pressure uniformly in all directions. Drawing an analogy between the atoms of a liquid and the particles of a powder and taking the indentation resistance of the material of the particles into account, the following relation is derived to calculate the load carried by the individual spherical particles of a dispersed powder at a given applied pressure during the initial stages of compaction in the absence of frictional effects:*

$$p_a = \ell_p / 2\pi a^2 = p_m \, s_c / s_o$$

where p_a is the externally applied pressure; ℓ_p is the load carried by a particle of radius a; p_m is the mean interparticle contact pressure (hardness of the material of the powder); s_c is the total contact area of the particle; and $s_o = 4\pi a^2$ is the surface area of the particle.

The relation was verified by measuring the extent of flattening of individual model-particles in a number of die-pressed compacts including 'single-sphere compacts' and a huge compact consisting of large ductile spheres of different sizes.

It was established that particle stacking and size distribution have no effect on the pressure distribution in powders and, consequently, on the progress of consolidation.

The obtained results contradict the existing views on the pressure distribution in powders and pave way to develop a well-founded theory of densification of powders during compaction, sintering, and hot isostatic pressing.

Symbols

a	radius of a particle/sphere
A	area of cross section
ℓ_p	load transferred to a particle/sphere during compaction
ℓ_a	load applied through the die-opening
ℓ_f	applied load during free-compression of a single-sphere/particle
p_a	applied pressure during die-compaction
p_f	applied pressure during free-compression
p_m	mean contact pressure (indentation hardness)

S_c total contact area of a particle/sphere

S_o original surface area of a particle/sphere $= 4\pi a^2$

S_r relative contact area of a particle/sphere $= S_c / S_o$

S_s total area of contact of a compact with punches and die-walls

r atomic radius

x radius of a contact

Z coordination number

1. Introduction

Based on experience, Pascal (1623-1662) deduced that a confined liquid (or gas) transmits externally applied pressure uniformly in all directions. Analyzing Pascal's law from the standpoint of the atomic structure of matter (Dalton, 1808) makes it possible to derive a relation between the externally applied pressure and the load carried by the individual atoms of a liquid.

Drawing an analogy between the atoms of a liquid and the particles of a powder, the derivation is extended to predict the load carried by the individual spherical particles of a dispersed powder during compaction and is verified experimentally. The obtained results pave way to develop a well-founded theory of densification of real powders during compaction and sintering.

2. Atomistic Viewpoint of Pressure Distribution in Fluids

Matter consists of atoms and voids. Evidently, voids can not transmit pressure and in a confined liquid, no atom can escape pressure. The externally applied pressure reaches each atom in the form of a set of forces through the contacts made with its neighboring atoms.

In liquids, the atomic arrangement is highly disordered. Even in a chemically simple liquid consisting entirely of mono-size atoms, the coordination number varies from atom to atom over a wide range. Consequently, the total load received by an atom ℓ is the sum of the unequal forces $\sum \ell_i$ transferred to it through the randomly oriented interatomic contacts. In spite of this apparent complexity, pressure distribution is surprisingly uniform.

Uniform pressure distribution implies equalization of pressure on all the constituent atoms (load-bearing elements) of a confined liquid. When this happens, the liquid is said to be experiencing a hydrostatic stress state.

By definition, pressure is load per unit area. For the applied pressure to be equal on all atoms, the total load carried by each atom must be directly proportional to its own surface area independent of its surroundings, i.e., independent of its coordination number, the sizes of its neighboring atoms, and its own coordinates in the confinement. This amazing self-compensating tendency of Nature can be expressed as follows:

$$p_a = \ell_a / A = p_1 = p_2 = ... = p_n = \ell_1 / 2\pi r_1^2 = \ell_2 / 2\pi r_2^2 = ... = \ell_n / 2\pi r_n^2$$

(1)

where p_a and ℓ_a are the externally applied pressure and load, respectively; A is the area of cross section of the confinement; p is the pressure acting over the surface of an atom of radius r; and ℓ is the total load carried by it. The subscripts denote individual atoms.

Note that the hemispherical surface area $2\pi r^2$ enters the denominator. Its choice is based on an analysis of the pressure acting over the surface of a spherical solid under load (§ 4).

3. Pressure Transmission in Powders

The 'particulate state' of matter is peculiar. Powders consist of solid particles yet exhibit fluidlike characteristics. They assume the shape of the vessel into which they are 'poured' and are 'compressible'.

Consolidation of powders necessarily involves pressure transmission through the interparticle contacts. There are reasons to believe that constrained powders transmit pressure much in the same way as confined liquids do.

When a powder is subjected to isostatic pressing, its linear shrinkage is equal in all directions independent of the particle shape and size distribution and the obtained compact has uniform density throughout its body. During die-compaction of a powder, the pressure-dependence of densification is close to that obtained during isostatic pressing. In the absence of pressure losses due mainly to die-wall friction, one can expect the pressure-density relation to be independent of the mode of compaction. Even under the conditions of unidirectional compaction, pores retain their shape to a large extent [1]. During sintering, a powder compact undergoes isotropic shrinkage with retention of its external shape as if it were subjected to isostatic pressing.

Thus, whether pressure is applied externally (compaction) or derived from within (sintering), it can be assumed that a hydrostatic stress state prevails during powder consolidation independent of the particle shape and size distribution.

As applied to compaction of a powder consisting of spherical particles of unequal sizes, Eq. (1) reads as follows:

$$p_a = p_{p1} = p_{p2} = ... = p_{pn} = \ell_{p1} / 2\pi a_1^2 = \ell_{p2} / 2\pi a_2^2 = ... = \ell_{pn} / 2\pi a_n^2$$

(2)

where p_a is the compaction pressure; p_p is the pressure acting over the surface of a particle of radius a; and ℓ_p is the total load transferred to the particle. Here, the subscripts denote individual particles.

If the particles are of the same size, each one of them must carry equal load independent of the number of nearest neighbors (coordination number) Z:

$$p_a = \ell_p / 2\pi a^2$$

(3)

In contrast to liquids, metal powders keep a permanent record of their response to the applied pressure due to their ability to deform plastically. During compaction of a metal powder, each particle attempts to indent its neighbors and, as a result, the interparticle contacts are converted into contact flats. At a given pressure, the extent of flattening (total contact area) of each particle depends on the inherent resistance of the material of the particles to indentation (hardness) and the total load transferred to it.

.

Thus, measuring the total contact area of the individual particles of a compacted metal powder makes physical verification of Eqs. (2) and (3) possible. We undertook this task using die-pressed model compacts [2-5] consisting of large single-spheres (§ 4) and aggregates of spheres of equal size (model of a mono-size spherical powder) and of various sizes (model of a dispersed spherical powder) (§ 5 and 6).

4. Single-Sphere Compression and Compaction: Pressure on a Solid Sphere under Load

The pressure (stress) acting on a freely compressed solid cylinder of radius a under a load ℓ_f is simply $\ell_f / \pi a^2$. However, if the cylinder is replaced with a sphere of radius a, the area over which the load acts is not quite apparent and, consequently, converting applied load into pressure poses problems. In order to solve this classic problem and, also, to establish the existence of a near-hydrostatic stress state even during die-compaction of powders, we studied the variation of the total contact area of the ductile spheres as a function of <u>load</u> during free-compression and as a function of pressure during die-compaction [5].

4.1. Free-Compression of a Ductile Sphere

A ductile sphere subjected to free-compression under a load ℓ_f undergoes deformation at the points of contact with the flats to form two circular contacts of equal radius x_f (Fig. 1). The mean contact pressure p_m is given by

$$p_m = \ell_f / \pi x_f^2 = 2\ell_f / s_c \tag{4}$$

where $s_c = 2\pi x_f^2$ is the total contact area of the sphere. Note that the mean contact pressure does not depend on the sphere size.

Neglecting the deformation-induced increase of the surface area of the sphere, Eq. (4) can be rewritten as

$$\ell_f / 2\pi a^2 = p_m \, s_r \tag{5}$$

where a is the radius of the sphere; $s_r = s_c / s_o$ is the dimensionless relative contact area of the sphere; and $s_o = 4\pi a^2$ is its original surface area.

Equation (5) suggests that the applied pressure over the surface of the freely compressed sphere $p_f = \ell_f / 2\pi a^2$. A conclusive experimental proof is given in § 4.2. Evidently, this pressure-term is independent of the material of the sphere.

On the other hand, the contact pressure p_m reflects the indentation resistance (hardness) of the material of the sphere and is, ideally, a load-independent constant.

Figure 1b shows constancy of the mean contact pressure in the $s_r = 0.10$-0.65 range for annealed nickel. It is close to (but expectedly higher than) the Brinell or the Vickers hardness of nickel (64 kg/mm^2).

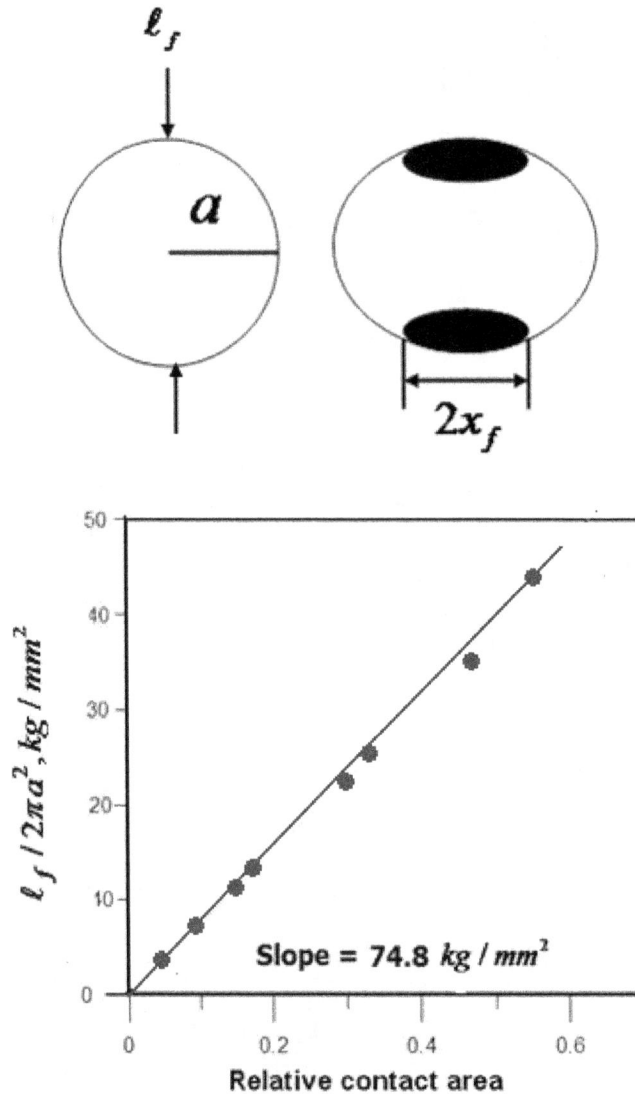

Fig. 1(a) Free compression of a ductile sphere. **(b)** Plot of Eq. (5) for a 9-mm annealed nickel sphere.

The hardness measurements carried out directly on the contact faces of the sphere [5] show that the material of the sphere undergoes significant strain hardening (HV5 $\approx$ 140 kg/mm^2 on the contact face even at $s_r \approx 0.3$). In spite of this, the mean contact pressure p_m remains unaltered at ≈ 70 kg/mm^2.

4.2 Die-Compaction of Single-Spheres

4.2.1. *Experimental*

Several annealed nickel spheres having a nominal diameter of 9.0 mm were used in this study. Spheres of this size could be slide-fitted into the specially designed dies having triangular, square, hexagonal, and circular cross sections. These dies have cross sectional areas of 102, 83, 72, and 65 mm^2, respectively. Initially, the spheres make point contacts with the die-walls and the punches ensuring coordination numbers $Z =$ 5, 6, 8, and ∞, respectively. A sphere fitted perfectly into the circular die has an infinite coordination since it makes a line-contact with the die-wall along its equator.

Compaction of the single-spheres was carried out using a 100-kN Instron machine. In each case, the applied pressure p_a was calculated from the applied load and the area of the die opening. After compacting and ejecting a sphere from the die, the linear dimensions of all its contacts were measured in two mutually perpendicular directions using a traveling microscope at 400X.

The contact areas were calculated from the dimensions and were summed up to give s_c. The radius and the original surface area s_o of each distorted sphere were calculated from its mass and the density of nickel (8.9 g/cm^3) and the relative contact area s_c / s_o was determined.

An ejected sphere can not be reintroduced into the die because of the elastic-after-effect. A new sphere is used in each compaction experiment.

4.2.2. *Results*

Figures 2-5 show the variation of relative contact area s_r of the die-compacted single-spheres of nickel as a function of the applied pressure p_a. It can be seen that the relationship is not only linear over a wide range of applied pressure but the slopes of the plots are nearly equal, independent of the number of lateral constraints confirming the existence of a hydrostatic stress state during die-compaction:

$$p_a = p_m \, s_r \qquad (6)$$

Furthermore, we note that the s_r versus $\ell_f / 2\pi a^2$ relation observed during free compression (Fig. 1) and the s_r versus p_a relation obtained during die-compaction (Fig. 3-6) are almost identical. On superimposing, they (Figs. 2-6) would all merge into a single plot. Thus, it can be concluded that the pressure acting over the surface of a freely compressed sphere is $\ell_f / 2\pi a^2$.

4.3. Load Multiplication during Die-Compaction

Combining Eqs. (5) and (6), we obtain the generalized load-pressure relation during free-compression and die-compaction of single-spheres:

$$p_f = \ell_f / 2\pi a^2 = p_a = \ell_a / k\pi a^2 = p_m \, s_r \qquad (7)$$

where the constant k depends on the number of constraints. For example, when using a square die, $k = 4 / \pi$ and in the case of a circular die, $k = 1$.

According to Eq. (7), in order to ensure equal flattening of dimensionally, chemically, and structurally identical spheres during free-compression and die-compaction, the applied pressures p_f and p_a must be equal but the applied loads differ significantly: $\ell_f = 2\ell_a / k$.

On the other hand, unless the total load transferred to the die-compacted sphere is equal to ℓ_f, it can not be flattened to the same extent as the freely compressed sphere (Eq. 4).

Thus, it can be inferred that during die-compaction of a single-sphere, the load applied through the die opening ℓ_a is multiplied due to the reaction from the die walls such that the total load transferred to the sphere is normalized to ℓ_f independent of the number of lateral constraints.

This must be true for the mono-size particles of a compacted real powder in spite of the random orientation of the constraints with respect to the direction of the applied load. At a given compaction pressure, $\ell_p = 2\pi a^2 p_a = \ell_f$, i.e., the load transferred to each of the mono-size particles in the compact is equal to the load required to ensure an equal pressure over the surface of a freely compressed identical particle.

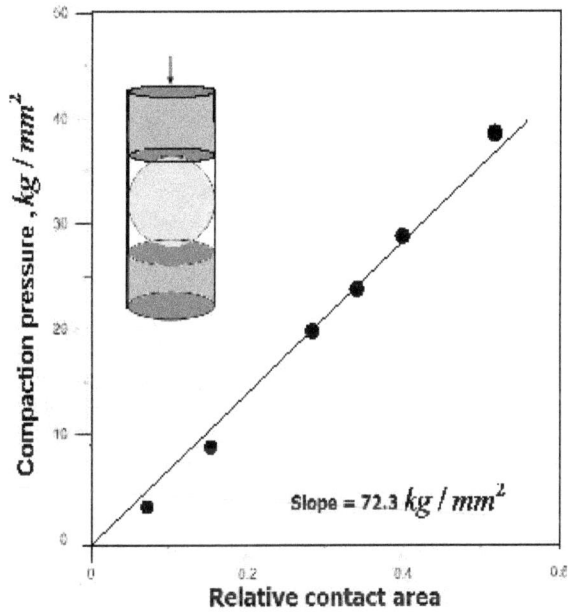

Fig. 2 Pressure dependence of relative contact area of single-spheres compacted in a **circular** die.

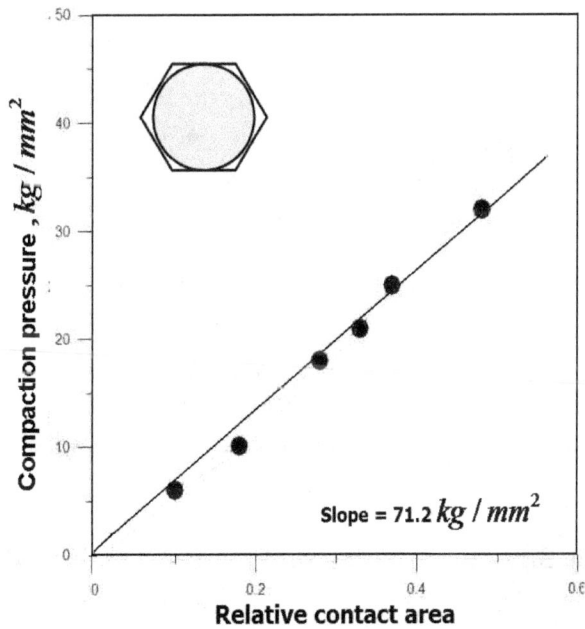

Fig. 3 Pressure dependence of relative contact area of single-spheres compacted in a **hexagonal** die.

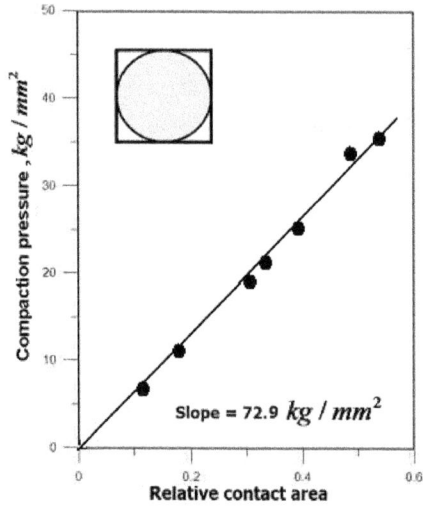

Fig. 4 Pressure dependence of relative contact area of single-spheres compacted in a **square** die.

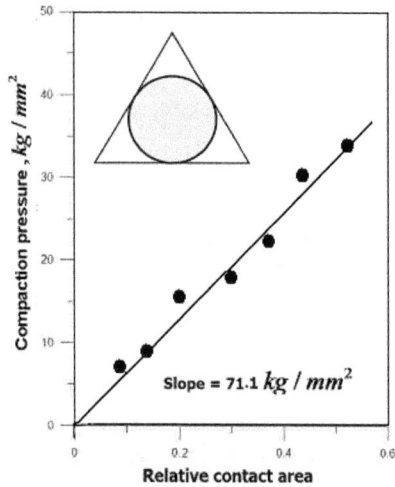

Fig. 5 Pressure dependence of relative contact area of single-spheres compacted in a **triangular** die.

5. Effect of Particle Size and Stacking on Pressure Transmission

5.1. *Experimental*

In order to establish that particle stacking and particle size have no effect on the stress state of the system, die-compaction of an annealed reference copper sphere (diameter

20.5 mm; mass 40 g) and several small mono-size randomly packed annealed copper spheres making up the same mass (40 g; 84 spheres; nominal diameter 4.68 mm) was carried out at the same applied pressure using a square die having an opening of 20.5 x 20.5 mm^2 (Fig. 6).

When placed in the die, the large reference-sphere makes point contacts with the punches and the die-walls and serves as a model of a regular simple-cubic arrangement of mono-size spherical particles. During the course of compaction, Z remains unaltered at 6. The total area of the contacts of the sphere with the punches $s_{top} + s_{bottom}$ and the four die-walls s_{walls} were determined separately and were summed up to give the total contact area s_c of the reference sphere. In the absence of frictional effects $s_{top} = s_{bottom}$. Since there are no copper-copper contacts in the single-sphere compact, the total contact area s_c is also equal to the total copper-steel contact area s_s.

The coordination number of the randomly packed small spheres varied in the 4-8 range. It was found to increase up to 5-10 as the applied pressure increases to 30 kg/mm^2 (294 MPa). After ejecting the compact from the die, the copper-steel contacts could be distinguished easily from the copper-copper contacts on the basis of the distinct shine of the contact areas. The areas of contact of these spheres with the punches $s_{top} + s_{bottom}$ and the die-walls s_{walls} were determined separately as in the previous case and were summed up to give the total copper-steel contact area s_s.

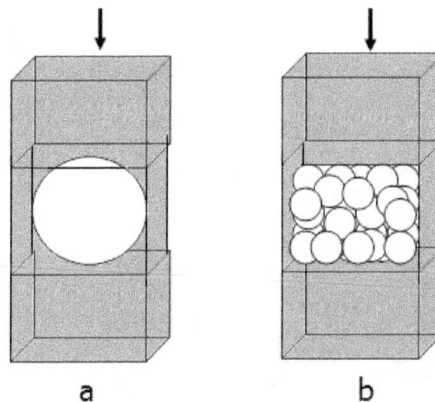

a   b

Fig. 6 Schematic of a large single-sphere placed in a square-die (a) and an equal mass of mono-size randomly packed small spheres in the same die (b).

Besides this, the total contact area s_c of each of the 84 spheres was measured in all the compacts. Using the known value of s_o (68 mm^2), s_r of each sphere was calculated and the average relative contact area of the 84 spheres was determined. The relative contact area of the large reference sphere was calculated as the ratio of its total contact area and its original surface area s_o (1320 mm^2).

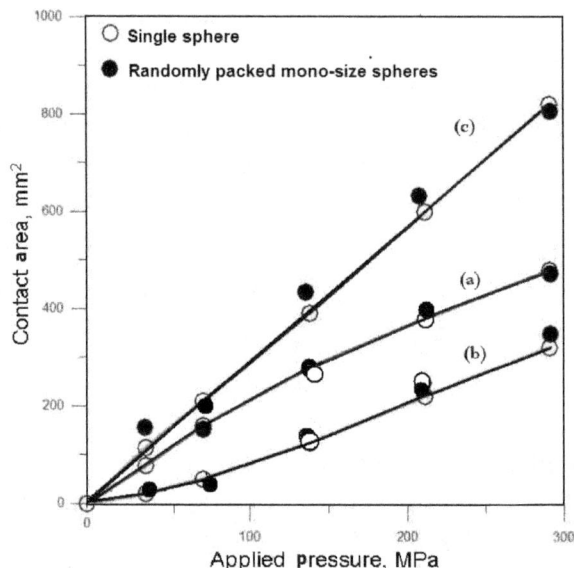

Fig. 7 Pressure dependence of the area of the copper-steel contacts formed (a) normal to the direction of applied load $S_{top} + S_{bottom}$, (b) parallel to the direction of applied load S_{walls}, and (c) their sum.

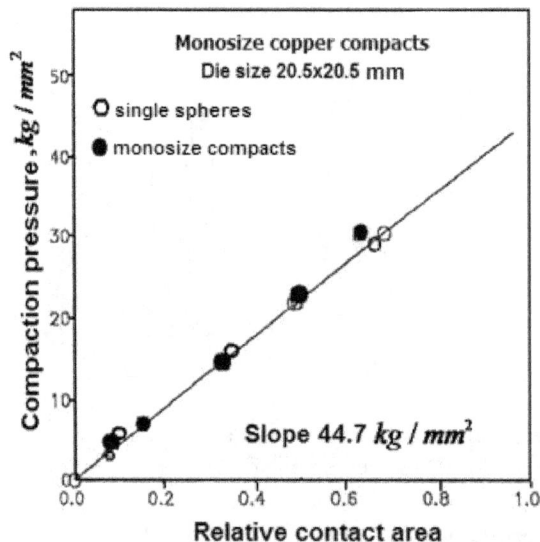

Fig. 8 Pressure dependence of the relative contact area of single-spheres s$_r$ (o) and the average relative contact area $(s_r)_{av}$ of the randomly packed mono-size spheres of copper (●).

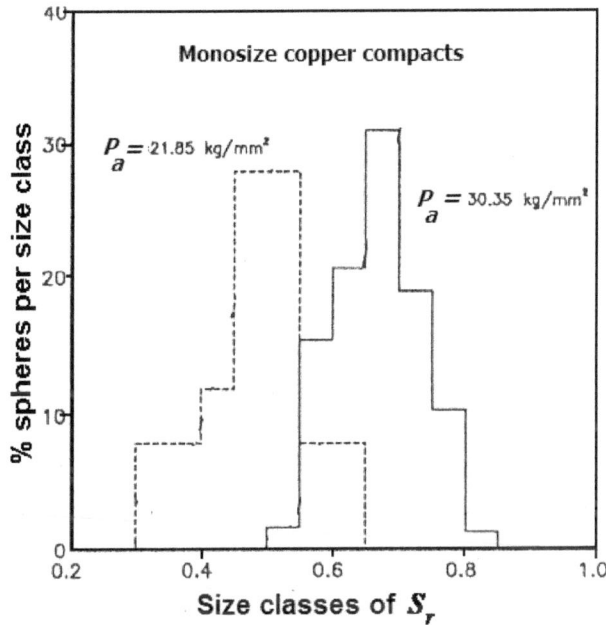

Fig. 9 Histograms of the relative contact area of the randomly packed mono-size spheres at two levels of applied pressure.

5.2. *Results*

1. At any given pressure, the copper-steel contact areas $S_{top} + S_{bottom}$ and S_{walls} measured on the compacts consisting of small spheres were found to be close to those obtained in the single-sphere experiments. Therefore, the copper-steel contact area of the entire compact $(S_{top} + S_{bottom} + S_{walls})$ is equal to the total contact area S_c of the reference sphere (Fig. 7).

2. At any given applied pressure, the average relative contact area $(s_r)_{av}$ of the 84 randomly packed small spheres is almost equal to the relative contact area of the large reference-sphere that serves as a model of a regular particle arrangement (Fig. 8).

3. The pressure-dependence of s_r of the reference sphere and the $(s_r)_{av}$ of the small spheres (Fig. 8) is not only linear but its slope is close to (but expectedly higher than) the Brinell and the Vickers hardness of annealed copper (40 kg/mm²). The plot validates Eq. (6) for real powders consisting of mono-size spherical particles.

4. The histograms of the relative contact area s_c / s_o of the small spheres (Fig. 9) show quite a narrow distribution of relative contact area at two levels of applied pressure. A few spheres do show significant deviations from the central value. Under the conditions of die compaction, attainment of a true hydrostatic stress state is hindered due to the die-wall friction.

Based on these results, it can be concluded that in the absence of frictional effects, at any given stage of compaction, all particles experience equal pressure independent of their size and stacking. Particle coordination and contact orientation have no effect on the stress state.

6. Effect of Particle Size Distribution on Pressure Transmission

6.1. *Experimental*

In order to demonstrate that particle size distribution has no effect on pressure transmission, we produced a huge compact (300x120x54 mm; mass 13.5 kg) in a split-die using a mixture of approximately 5000 annealed nickel spheres of different sizes (diameters 5-15 mm) at an applied pressure of 17.14 kg/mm² (168 MPa) on a 650-ton press (Fig. 10). On dissecting the obtained compact, it was found that the coordination number of the spheres varied in the 4-18 range. We evaluated randomly selected 288 spheres for their relative contact area. The results are shown in Figs. 11-13.

6.2. *Results*

Figure 11 shows the histogram of the relative contact area of the spheres. As in the case of mono-size copper spheres (Fig. 10), the distribution of s_r is quite narrow.

Figure 12 shows the relative contact area of the individual spheres as a function of their original surface area. The best fit of the data is a straight line parallel to the abscissa: $s_r = 0.2485$. Inserting this value into Eq. (6), we obtain a mean contact pressure p_m of 69 kg/mm² which is close to the Brinell or the Vickers hardness of nickel and to the slopes of the $p_a - s_r$ plots obtained during single-sphere compaction. It can thus be concluded that pressure transmission does not depend on particle size distribution.

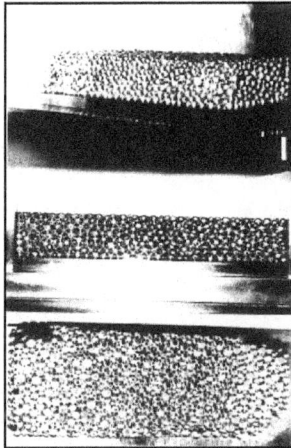

Fig. 10 Die pressed compact consisting of nickel spheres of different sizes.

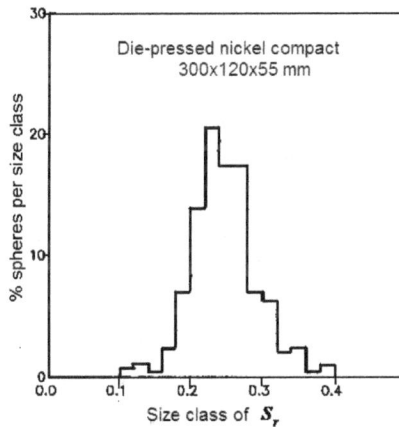

Fig. 11 Histograms of the relative contact area of the spheres.

Figure 13 shows clearly that the relative contact area is not affected be the wide variations in the coordination number of the spheres. Note that the *mean* s_r at each value of Z is close to the *average* relative contact area of all the spheres.

The observed deviations from the average value can be attributed to frictional effects but certainly not to the variation of coordination or size distribution of spheres.

Table 1 shows the results of Vickers hardness measurements carried out directly on the contact faces of a large sphere extracted from the compact. On its small contacts, measurements could not be taken except at their centers. The average hardness HV5 was found to be 143 kg/mm^2.

In spite of such a significant physical hardening in the vicinity of the contacts, the mean contact pressure p_m [according to Eq. (6)] is 62.55 kg/mm^2 which is close to the Brinell or the Vickers hardness of annealed nickel. It can thus be concluded that strain hardening of the particles does not have a significant effect on the progress of compaction, at least, up to the intermediate stages of compaction.

7. Literature Review

From the aforementioned results (§ 3-6), it is evident that even during die-compaction, the stress state of the system is essentially hydrostatic. It is to be expected that in the absence of the frictional effects (during cold isostatic pressing and sintering), equalization of pressure would occur on all the load-bearing elements just as in liquids. Consequently, the progress of the consolidation processes must be insensitive to particle coordination (initial relative density). These conclusions contradict the existing views on the pressure distribution in powders.

7.1. *Interparticle Contact Pressure: A Variable or a Constant?*

It has been considered [6-10] that an increasing particle coordination hinders the compaction process of a powder but accelerates the sintering process. The contact pressure has been treated as a variable that depends on the number of contacting neighbors per particle Z. Since the applied pressure is supported on all of them, it has been assumed that the more contacts a particle makes, the lower is the pressure at each.

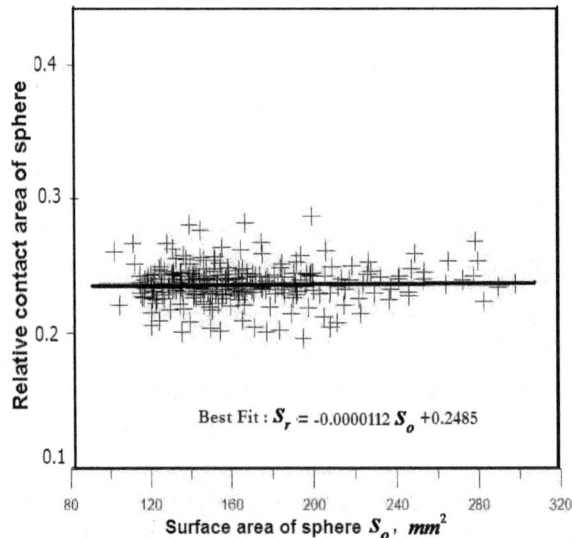

Best Fit : $S_r = -0.0000112\,S_o + 0.2485$

Fig. 12 Relative contact area of the individual spheres extracted from the huge compact as a function of their original surface area.

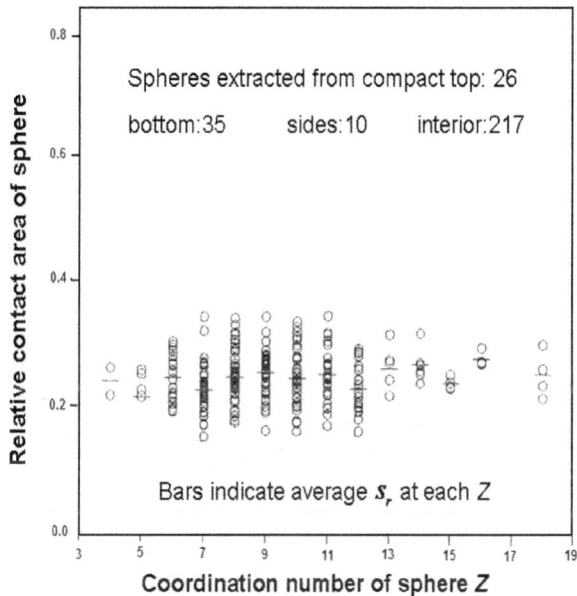

Fig. 13 Plot showing coordination independence of relative contact area of the spheres.

However, this assumption is valid only for an open system (e.g. bridge and columns). In a closed system, each particle not only supports certain load but also transfers it to its neighbors and contributes to equalization of pressure on each one of them. If a particle makes a new contact, the growth rate of some of the existing contacts is slowed down so that its total contact area is unaffected. The presence or absence of new contacts would make no difference to the progress of consolidation.

By definition [Eq. (4)], the mean contact pressure is a synonym of the indentation hardness of the material of the particles that must be treated as a material constant, at least, in the s_r = 0.10-0.65 range (Figs. 1-5 and Fig 8). In this range, it is influenced neither by an increasing particle coordination nor by the strain hardening of the material of the particles. The pressure dependence of densification is not affected either by increasing the tap density of the powder or by the increasing particle coordination during the course of compaction. The total interparticle contact area of a compact is insensitive to these factors.

It is noteworthy that some of the pioneers of the theory of powder metallurgical processes (Bal'shin [11], Kuczynskii and McQueen [12], and Kakar and Chaklader [13]) made a valid assumption that the contact areas grow to stable sizes under a constant stress but have not provided adequate experimental support.

7.2. *Geometric Responses of Particles during Isostatic Pressing and Die Compaction*

Seeking a force balance at the particle surfaces of a simple-cubic array of mono-size spherical particles subjected to isostatic compaction, Coble and Ellis [14, 15] derived exactly the same equation as Eq. (6). However, Coble himself expressed doubts [16] regarding the validity of this equation because his treatment of die-compaction [15] gave a contact pressure approximately twice that obtained in the isostatic case. Coble [15] seems to have overlooked the fact that at a given compaction pressure, the size of the individual contact flats formed on a particle during isostatic pressing and the sizes of the flats developed during die-compaction are not equal. According to the present analysis, in spite of the wide variations in the areas of the individual contacts in a mono-size powder compact, the total contact area of each particle depends on the applied pressure and not the mode of compaction - not only in regular arrays but also in randomly packed powders. The quantitative metallography studies conducted by Fischmeister, Arzt, and Olsson [17, 18] show clearly that at a given compact density, the total contact area *per* particle of a monosize powder is equal in isostatic and die compaction.

Table 1 Geometric Responses of a Nickel Sphere Extracted from the Compact and Indentation Hardness Measured on its Contact Flats

Compact dimensions	300x120x54 mm
Coordinates of the sphere (x,y,z) in the compact	(125,55,44) mm
Mass of the sphere	11.62 g
Diameter $2a$	13.58 mm
Coordination number	16
Surface area, $s_o = 4\pi a^2$	579.27 mm²
Total contact area s_c	158.55 mm²
Relative contact area $s_r = s_c / s_o$	0.274
Compaction pressure p_a	17.14 kg/mm²
Mean contact pressure p_m (Eq. 6)	62.55 kg/mm²

Contact number	Contact radius 2x, mm	Vickers hardness HV5, kg/mm² at contact		
		center	mid-radius	edge
1	6.52	161	161	161
2	6.42	130	140	150
3	3.55	135	148	
4	3.34	138	138	
5	3.32	137	138	
6	3.20	138	144	
7	3.16	128	139	
8	3.06	175		
9	3.01	153		
10	3.00	153		
11	2.80	157		
12	2.77	133		
13	2.70	118		
14	2.46	133		
15	1.90	130		
16	1.83	132		

7.3. *Effect of Particle Coordination on the Kinetics of Sintering*

Unless two particles are brought into contact with each other, they can not be sintered. However, this does not mean that the sintering process of a real system is accelerated with increasing number of contacts per particle. The new contacts that form when tapping a powder or during the course of sintering can not introduce additional driving forces. The driving force depends on the free surface area per unit mass of the system and not on the number of interparticle contacts existing or developing in the powder compact during the course of sintering. The very fact that a loose powder or a compact retains its external shape during the course of sintering implies that the system experiences a hydrostatic stress state (as if it were subjected to isostatic pressing) independent of particle shape, stacking, and size dispersion. In view of this, the kinetic equations of sintering [19, 20] must be so modified as to reflect insensitivity of specific surface reduction to particle coordination.

8. Conclusions

1. Analyzing Pascal's law from the standpoint of the atomic structure of liquids makes it possible to calculate precisely the load carried by the individual atoms of a pressurized liquid.

2. Drawing an analogy between the atoms of a liquid and the particles of a powder, the derivation was extended to predict the load carried by the individual spherical particles of a dispersed powder during compaction: $p_a = \ell_p / 2\pi a^2$ where p_a is the externally applied pressure; and ℓ_p is the load carried by a particle of radius a.

3. The relation was verified by measuring the extent of flattening of single-spheres subjected to free-compression and die-compaction as a function of applied load and applied pressure, respectively.

4. Using model compacts, it was established that during the initial stages of powder compaction, $p_a = p_m s_c / s_o$ where p_m is the mean interparticle contact pressure (hardness of the material of the powder); s_c is the total contact area of the particle; and $s_o = 4\pi a^2$ is its surface area. In spite of die-wall friction, the stress state of a die-pressed compact is essentially hydrostatic.

5. It was demonstrated that particle size, size distribution, stacking, and the changes occurring in particle coordination during the course of consolidation have no effect on the uniformity of pressure distribution in powders and, consequently, on the progress of consolidation.

6. It was shown that strain hardening of the material of the particles does not affect the constancy of the mean interparticle contact pressure significantly, at least, up to $s_r = 0.65$.

7. The conducted model studies on compaction indicate that the kinetic equations of sintering of real powders must reflect insensitivity of specific surface reduction to particle coordination.

Acknowledgements

The author acknowledges a decade-long support of his colleagues at Mishra Dhatu Nigam Limited, Hyderabad, in particular, Mr. C. Sankar Kumar and records his deep appreciation to his former students, in particular, Mr. K. Venkateswarlu, ArcelorMittal, Kolkata and Mr. D. Balakotaiah, GISH, Libya, for carrying out the experimental work meticulously.

304

References

1. B.O. Sundstrom and H. F. Fischmeister, *A continuum-mechanical model for hot and cold compaction*, Powder Metall. Int., Vol. 5, 1973, p. 171

2. J. Viplava Kumar, *Geometric responses of particulate systems during consolidation*, Sintered Metal-Ceramic Composites, G.S. Upadhyaya (ed.), Elsevier, 1984, p. 87

3. J. Viplava Kumar, *Physics of Powder Consolidation*, Technical Report No. MDN 10/1987, Mishra Dhatu Nigam Limited, Hyderabad, 1987.

4. J. Viplava Kumar, *The hypothesis of constant relative responses and its application to the sintering process of spherical powders*, Solid State Phenomena,: G.S.Upadhyaya (ed.), , Sci-Tech Publications, Switzerland, Vols. 8 & 9, 1989, p.124

5. J. Viplava Kumar and C. Sankar Kumar, *A universal pressure-densification plot based on single-particle compression and compaction models and its application to HIPing of dispersed powders*, Advances in Powder Metallurgy and Particulate Materials, MPIF, Princeton, NJ, Vols. 2, Part 5, 1995, p. 69

6. E. Arzt, *Influence of an increasing particle coordination on the densification of spherical powders*, Acta Metall., Vol. 30, 1982, p. 1883

7. F.B. Swinkels, D.S. Wilkinson, E. Arzt, and M.F. Ashby, *Mechanisms of hot isostatic pressing*, Acta Metal., Vol. 31, 1983, p. 1829

8. E. Arzt, M.F. Ashby, and K.E. Easterling, *Practical applications of hot-isostatic pressing diagrams*, Metall. Trans. A, Vol. 14A, 1983, p. 211

9. A.S. Helle, K.E. Easterling, and M.F. Ashby, *Hot-isostatic pressing diagrams – New developments*, Acta Metal., Vol. 33, No. 12, 1985, p. 2163

10. M.F. Asby, *The modeling of Hot-Isostatic pressing*, Proc. Lulea Conf. on HIPing, Lulea University Press, June, 1987.

11. M. Yu. Bal'shin, Scientific Principles of Powder Metallurgy and Fiber Metallurgy [in Russian], Metallurgiya, Moscow, 1972, p. 229

12. G.C. Kuczynskii and II.J. McQueen, *Effect of compacting on luminescence of copper activated zinc sulfide*, J. Am. Ceram. Soc., Vol. 45, No. 8, 1962, p. 399.

13. A.K. Kakar and A.C.D. Chaklader, *Deformation theory of hot pressing*, J. Appl. Phys.,Vol. 38, No. 8, 1967, p. 3223

14. R.L. Coble and J. S. Ellis, *Hot pressing of alumina: Mechanisms of mass transport*, J. Am. Ceram. Soc., Vol. 46, No. 9, 1964, p. 438

15. R.L. Coble, *Diffusion models for hot pressing with surface energy and pressure effects as driving forces*, J. Appl. Phys., Vol. 41, No. 12, 1970, p. 4798

16. R.L.Coble, Massachusetts Institute of Technology, Cambridge, MA, Private - communication, 1982.

17. H.F. Fischmeister, E. Arzt, and L.R. Olsson, *Plastic deformation and sliding during compaction of spherical powders: A study by quantitative metallography*, Powder Metal., Vol. 21, No. 4, 1978, p. 179

18. H.F. Fischmeister and E. Arzt, *Densification of powders by plastic deformation*, Powder Metal., Vol. 26, No. 2, 1983, p. 82

19. R.M. German and Z.A. Munir, *Morphology Relations during Bulk-Transport Sintering*, Met. Trans. A., No. 6, December, 1975, p. 2229

20. R.M. German and Z.A. Munir, *A kinetic model for the reduction in surface area during initial stage of sintering*, in: Sintering and Catalysis, G. C. Kuczynski (ed.), Plenum Press, New York, Vol. 6, 1975, p. 249

Effect of Alloying and Processing Conditions on Density and Coercivity of Fe-based PM Alloys

Jiten Das and Bijoy Sarma

Defence Metallurgical Research Laboratory, Hyderabad

Abstract: *Coercivity of Fe and Fe-based PM alloys are significantly affected by their density. The density of these PM parts is governed by the processing parameters (i.e. Milling condition, compaction pressure, sintering temperature and time and various post sintering operations which include repressing, rolling, swaging, HIPing etc.) and alloying additions. In the present investigation effect of various processing parameters and alloying additions (such as Ni, P, Si and Cu) on density, coercivity and hardness (in few cases) of PM Fe-based alloy parts is studied. It is in general observed that alloying additions (such as P, Si and Ni) improved densification. The higher is the green compaction pressure, sintering time and temperature, the higher is the density of the PM parts. The post sintering operations such as rolling, repressing, swaging, HIPing etc. greatly improved the density of Fe and Fe-based PM alloys and thereby lowered the coercivity values. The alloying additions such as P and Si, Ni (within their respective solubility limit in Fe) reduced the coercivity of Fe- based PM alloys significantly.*

Key words : sponge iron powder, blue dust, coercivity

1. Introduction

Sponge iron powder obtained by the reduction of Indian blue dust (fragile hematite ore) is very cheap and it is almost free from carbon. The high quality soft magnetic PM component can be produced, using these sponge iron powder, at a very low cost provided it can be densified fully. The density of the powder metallurgical component using these sponge iron powder is low (~6.5g/cc) owing to its inferior compactibility [1] and sponginess. However, the use of later powders (i.e.carbonyl iron and electrolytic iron) is limited because, the cost of the later powders is very high as compared to the cheap sponge iron powder. Therefore, effort is directed to increase the density of PM components produced using cheap sponge iron powder by improving the process parameters (i.e. milling conditions, green compaction pressure, sintering temperature and time and various post sintering operations which include repressing, rolling, swaging and HIPing) and alloying addition. Salak and Klar have clearly shown that the soft magnetic properties of pure Fe PM parts such as Saturation magnetisation, remanence, coercivity are significantly affected by its density [2,3]. They also observed that the higher the compaction pressure, sintering temperature and time, the higher the density of the PM parts. The further densification can also be achieved by special coining after sintering [4], rolling, swaging, HIPing etc after sintering operation.

Certain alloying additions also improves the density of Fe-based PM alloys. Weglinski et al [5] admixed Hoganas Fe powder with Fe3P powder(15.39%P), compacted at 600MPa pressure and sintered at 1450K to obtain series of Fe-P alloys. They found that addition of P alloying element from 0 to 1 wt% to Fe rendered increment of density from 7.2g/cc to 7.55g/cc. The increase in density is due to sintering taking place in the presence of liquid phase at the comparatively high sintering temperature employed (1450K). The advantage is taken of the local peritectic reactions : Fe3P --> Fe2P+L which is taking place at 1439K [5]. Weglinski et al also have shown that the magnetic properties such as permeability, saturation induction increases, coercivity and total losses decreased due to phosphorous addition up to ~0.8wt% and resistivity of Fe improve due to P addition up to 2wt%[5]. E. Klar [3] also has sown that P addition up to 0.8wt% increases permeability, Induction and remanent magnetisation and decreases coercivity of Fe. Silicon decreases coercivity and increases the resistivity of of iron. However, its addition beyond 4% is not recommended since it reduces saturation magnetization drastically and alloy becomes brittle with high amount of silicon. For example Fe-6.5Si alloy is brittle and its saturation magnetisation is as low as 1.3Tesla [1]. However, Anestiev et al [6] have shown that the combined influence of Si and P on the magnetic permeability and total magnetic losses of Fe is better than their individual's influence. Honejko et al [7] have shown that coercivity of Fe significantly reduces by the combined addition of P and Si. Ni on the other hand, decreases coercivity without significant loss in saturation magnetisation. Addition of 2 wt% Ni to Fe reduces its coercivity to 1.0Oe without significant reduction of its saturation magnetisation(i.e. saturation magnetization of Fe-2wt%Ni=1.5T)[1]. C is considered as impurities so far as soft magnetic properties of Fe-based alloys are concerned. C is found to decrease Induction, remanent magnetisation, maximum permeability and increase coercivity of Fe [3].

In the present investigation effect of various powder metallurgical processing parameters on density of the Fe and Fe-based PM component produced (using low carbon sponge iron obtained by reduction of Indian blue dust) has been studied. The coercivity values of these produced components were evaluated in order to assess their soft magnetic performance during service. In some cases hardness of the rolled PM product also has been evaluated in order to assess their suitability to structural applications. The effect of various alloying addition on coercivity value of Fe is also studied in this investigation.

2. Experimental Work

High density Fe-based PM parts can be obtained by choosing optimum processing conditions and suitable alloying additions, The density of the PM parts as well as alloying additions determines the coercivity of Fe-based alloys. In this investigation four different alloy systems have been chosen which are Fe-Ni-Si, Fe-P-Si and Fe-P-Si-Cu. Density of several alloys in these systems is measured after different PM processing and coercivity of the these alloys are measured thereafter. In order to understand the response of Fe PM parts to post sintering operation, a few post sintering study also undertaken for pure Fe PM parts.

2.1. Study of Fe-Ni-based system

In this system only one alloy : 97wt%Fe-2wt%Ni-1wt%Si alloy is considered for investigation. As received sponge iron powder (which are obtained by reducing Indian blue dust), Ni and Si powders are characterized in terms of particle size distribution (using Malvern particle size analyzer). These powders mixed with a ratio of Fe:Ni:Si = 97:2:1. These powder mixture is then dry milled using ball mill(capacity 25kg, stainless steel ball diameter 12mm) with a ball to charge ratio 2:1 for 15 minutes and wet milled for 15 minutes, 24 h and 48h. These milled powders are again characterized in terms of particle size distribution (using Malvern particle size analyzer). Half of these milled powders are then compacted using mechanical press (TH & J Daniel Ltd., England :Capacity: 250 tons). 30mm diameterX5mm thickness disc shaped samples are made with the help of a closed-die (material = hardened die steel, diameter = 30mm) with the application of 555 MPa pressure (40T load). The remaining iron powders are then compacted using cold isostatic press(Make : National Forge, Belgium : maximum pressure 5.2 k-bar) at 250 MPa pressure in the shape of cylindrical compacts. No lubricant is used for the powder particles, only the die wall and the punch were lubricated with zinc stearate solution (*i.e.* 2 g zinc stearate in 100 ml of benzene) during the compaction using mechanical press to reduce die wear. The green densities of these compacts are then measured by taking their weight with the help of an electronic balance (Make: Mettler, capacity: 2.4 kg, least count: 0.01 g) and calculating their volume from their dimensions. The discs and cylindrical compacts are then sintered at three different temperatures such as 1100°C, 1200°C and 1300°C in Pusher type sintering furnace (FHD Furnace Ltd, England, maximum temperature 1800°C maximum rating 100kW). Sintered density and sintering shrinkage of discs and cylinders are then calculated. Coercivity of these Fe-2Ni-1Si alloys (processed by different processing condition) is recorded along with coercivity of pure iron (processed with similar processing condition) against their respective density. Some pure Fe plates (milled for 48 hrs, compacted in mechanical press at 555 MPa pressure or at 40 T load and sintered at 1300°C) are rolled at 900°C in DEMAG Rolling Mill, Germany to reduce its thickness from 7mm to 5mm(reduction 0.5mm per pass) in order to improve the density of the PM parts. Coercivity values of these high density PM Fe and armco iron are also recorded for comparison. The microstructure of these plates along rolling direction, short transverse direction and long transvese section were taken. Hardness along these cross section is also recorded.

2.2 Study of Fe-P- based system :

In this system several Fe-P-Si based PM alloys such as Fe-0.2P-2Ni-1Cu, Fe-0.8P-1Si, Fe-0.8P-2Si, Fe-0.6P-1.5Si, Fe-0.6P-1.5Si-2Cu alloys were made by mixing Fe powder (composition given in Table I), Fe_3P powder (composition given in Table I) and Fe-Si (composition given in Table I) powder with or without elemental copper and Ni powder and compacting (using CIP at 250 MPa pressure) and sintering it. Density of the green compacted as well as sintered alloys are measured. This alloys are machined to torroid shape and coercivity is measured in as machined condition. Coercivity of some alloys were measured after annealing the machined component. One sample namely Fe- 0.6P-1.5Si alloy is swaged after sintering in order to improve density and then machined it to torroid shape to measure resistivity. This sample was also annealed to relieve residual stress due to machining and coercivity is measured again.

2.3 Study on density variation during post sintering operation :

A few study on density improvement due to various post sintering operation (i.e. repressing, HIPing) on pure iron is undertaken to understand the effect of these operation on density of PM parts. In this study a few blocks of sintered porous cylindrical shape iron piece is subjected to gradual increase of repressing load i.e. 20T, 50T, 100T) and density at each stage is measured. The repressed pieces were annealed at 800oC for one hour after every repressing operation. The pressing is done by putting grease at the contact area between the sample and the top and bottom plates in order to avoid barreling. The repressed pieces are then subjected to Hot Isostatic Pressing (1150oC/1000 bar/180 minute) in argon atmosphere to improve the density of the PM parts further. Density of the PM parts after HIPing is also recoreded.

3. Results and Discussions

3.1 Study of Fe-Ni-based system

Particle size distribution in Figure 1 shows that Fe powders are coarser than Ni and Si powder. Si is the finest powder of all the three. The finer the Si powder, faster is the dissolution in Fe. As the silicon dissolution is very slow in Fe, Si particle size is deliberately kept fine. The elemental powder mixture of Fe, Ni and Si in the ratio of Fe:Ni:Si = 97:2:1 is fast dry milled as well as wet milled (in benzene) for 15 minutes in order to see the size reduction response. We observed that wet milling is more effective in size reduction of powder mixture than the dry milling. Therefore, the powder mixture is milled for 24 hours and 48 hours. We observe that as the higher the illing time, the finer the particle size. The 48 hour milled powders are compacted using mechanical press as well as CIP. CIPed compacts show lower green density than the mechanically pressed compacts (Table II). These compacts are then sintered at 1100oC, 1200oC and 1300oC. The density data (Table II) shows that CIPed compacts (using 250 MPa pressurc) show lower sintered density than the mechanically pressed compacts (using 555 MPa pressure). Table II also shows that the higher the sintering temperature, higher is the sintered density and shrinkage (due to pore reduction) of the product. The higher the density of the PM parts, lower is the coercivity (Table III). Pure Fe processed under similar condition shows lower density than the Fe-2Ni-1Si alloy. Pure iron also shows higher coercivity than Fe-2Ni-1Si alloy processed under similar condition (Table III). This may be partly due to its lower density than Fe-2Ni-1Si alloy and partly due to absence of coercivity reducing alloying elements such as Ni, Si [1]. However, pure iron shows lower coercivity after hot rolling at 900oC (Table III). This is because of its greater part density without work hardening during hot rolling. Hardness (Table IV) data also shows that there is very little work hardening during hot rolling at 900oC.

3.2 Study of Fe-P- based system

Density (green as well as sintered) measurement of several Fe-P based alloys shows that 0.2 wt% P addition is not much effective in increasing density (Table V). However, higher P addition enhaced the densification of iron. Fe-0.6P-1.5Si alloy shows better densification than Fe-0.8P-1Si and Fe-0.8P-2Si. Therefore, the coercivity of Fe-0.6P-1.5Si is lowest among all the Fe-P based alloys developed in this investigation. Although 2wt% Cu addition improved the density of Fe-0.6P-1.5 Si alloy, it could not lower the coercivity. However, swaging of Fe-0.6P-1.5Si alloy improved its densification and significantly reduced its coercivity.

3.3 Study on density variation during post sintering operation

Study on the effect of repressing load on densification of porous sintered product shows that as the repressing load increases, density of PM parts increases. Density of PM parts improved greatly (from ~7g/cc to 7.5g/cc) with HIPing operation. This is recorded in Table VI and in Figure 2.

4. Conclusions

> Wet milling in stead of dry milling is effective in reducing particle size of elemetal mixed Fe-2Ni-1Si powder. The higher the milling time, lower is the particle size.

> Green compacts as well as as sintered density of CIPed compacts (using 250 MPa pressure) are inferior than the mechanically pressed compacts (using 555 MPa pressure).

> The higher the sintering temperature, higher is the density of Fe-2Ni-1Si alloys.

> Pure iron shows inferior density than Fe-2Ni-1Si alloy processed under similar condition, and shows also higher coercivity than Fe-2Ni-1Si alloys. However, hot rolling improves density without work hardening and thereby reduces its coercivty of Fe.

- Alloying addition such as P, Si, Ni and Cu improves density of Fe and thereby helps in reducing its coercivity. However, Cu addition is found to increase coercivity of Fe.

- Repressing after sintering improves density of sintered PM parts. The higher the repressing load, higher is the density. HIPing improves density of PM parts significantly.

Table 1 Analysed composition of iron powder, Fe3P powder and Fe-Si powder.

Elements	Fe powder(-100#)	Fe$_3$P powder(-250#)	Fe-Si powder (-250#)
Fe (wt%)	98.772	73.361	28.455
P(wt%)		25.94	0.013
C(wt%)	0.005	0.017	0.08
Si(wt%)		0.63	70.19
Al(wt%)			1.0
O$_2$(wt%)	1.0-1.2		
N$_2$(ppm)	30-33		
S(wt%)	0.023	0.028	0.018

Table 2 Density and shrinkage values of the disc and cylinders at Green comapetion and sintering.

Properties	Disc shape compact (30mm dia(D), 5mm thickness(t))						Cylidrical Shape compact (19mm dia)					
Green density (g/cc)	6.60						6.00					
Sintering temperature(°C)	1100		1200		1300		1100		1200		1300	
Sintering Shrinkage (%) at 1100, 1200, 1300°C along Dia(D), thickness (t) and Height(H)	D	t	D	t	D	t	D	H	D	H	D	H
	3.7	4.2	4.5	4.5	5.4	4.8	4.9	3.5	6.3	5.7	8.1	7.5
Sintered Density (g/cc)	6.98		7.11		7.27		6.60		6.89		6.93	

313

Table 3 Coercivity of the Fe-2Ni-1 Si alloy samples and pure iron samples processed with different conditions (powders milled for 48 h)

Material	Density				Properties		
	Green compaction Density (using Mechanical Press at 125Mpa pressure	Sintering density			Coercivity		
		Sintering at 1100C/2h/H$_2$	Sintering at 1200/2h/H$_2$	Sintering at 1300/2h/H$_2$	Sintered at 1100C/2h/H$_2$	Sintered at 1200C/2h/H$_2$	Sintered at 1300C/2h/H$_2$
Fe-2Ni-1Si alloy (PM)	6.98	6.98 g/cc	7.11 g/cc	7.27 g/cc	3.69Oe	3.274Oe	2.614Oe
Pure Fe (PM)	6.0	6.82 g/cc	6.84 g/cc	6.85 g/cc	3.959 Oe	3.276 Oe	3.067 Oe
Material	Density				Coercivity		
Pur Fe (PM) Hot rolled at 900°C	7.4 g/cc				2.966 Oe		
Armco iron	7.84				0.8 Oe		

Table 4 Hardness of the rolled plates (900 °C) along rolling direction, short transverse and long transverse direction.

Section	L1	L2	Hardness
Rolling direction	35.4	37.2	141
	38.0	38.0	128
Short Transverse direction	38.0	37.5	130
	34.2	36.2	150
	38.0	38.0	128
	38.5	39.5	122
Long Transverse direction	32.4	31.9	179
	33.0	34.2	164
	37.3	35.2	141

314

Table 5 Density and Coercivity of Fe-P-based alloys

Material	Heat Treatment	Green Density	SinteredDensity	Coercivity
Fe-0.8P-1Si	Sintered 1300°C/3h/H₂	5.5	6.8	2.39
Fe-0.8P-2Si	Sintered 1300°C/3h/H₂	5.5	6.9	2.28
Fe-0.2P-2Ni-1Cu	sintered1150°C/1.5h/H₂ machined and annealed at 1000C/1h/H₂	5.2	6.6	3.532
Fe-0.6P-1.5Si	sintered1250°C/3h/H₂ machined	5.7	7.48	1.911
Fe-0.6P-1.5Si	sintered1250°C/1h/H₂ machined and annealed at 1000°C/1h/H₂	5.7	7.48	1.520
Fe-0.6P-1.5Si-2Cu	sintered1250°C/3h/H₂	5.5	7.51	1.98

Table 6 Effect of post sintering operation on density

Post sintering operation	Density (g/cc)
As received Porous sintered Fe (cylidrical shape PM part)	4.53
Cold pressing at 20T	4.75
Cold pressing at 50T	6.20
Cold pressing at 100T	6.96
HIP (1150°C/1000bar/180 min)	7.5

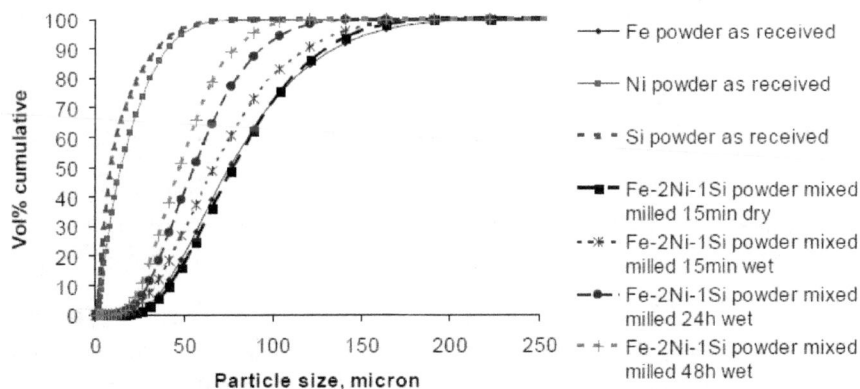

Fig. 1 Particle size distribution of different powders.

315

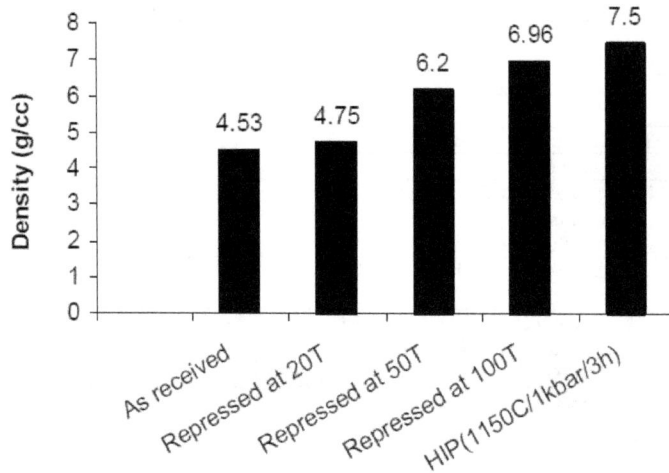

Figure 2 Density of porous sintered iron after various post sintering operations.

References

1. K. H Moyer, *Magnetic Materials and Properties for Part Applications*, ASM Hand Book on Powder Metallurgy & Application, Vol. 7, 1998, p. 1006.

2. A. Salak, *Ferrous Powder Metallurgy*, Cambridge International, 1995, p. 188.

3. E. Klar, *Powder Metallugy Applications, Advantages and Limitations*, American Society for Metals, 1983

4. P. Jones, K. B. Golder, R. Lawcock, R. Shivanath, *Densification Strategies for High Endurance P/M components*, International Journal of Powder Metallurgy, Vol. 33, No. 3, 1997, p. 37.

5. B. Weglinski and J. Kaczmar, *Effect of Fe3P additions on magnetic properties and structure of sintered iron*, Powder Metallurgy, Volume 23, No. 4, 1980, p. 210.

6. L. anestiev, M. De Wulf, L. Froyen, L. Dupre, J. Melkebeek, *Preparation of soft magnetic alloys Fe100-x-ySixPy (0<x,9, 0<y<0.6 wt%), using solid phase sintering method*, ournal of Magnetism and Magnetic Materials, Vol. 281, 2004, p 124.

7. F G Honejko, G. W Ellis, T J Hale, *Application of high performance material processing-electromagnetic products*, Presented at PM2TEC'98 : International conference on powder metallurgy & particulate materials, Las Vegas, NV USA, May 31- June 4, 1998, p 1.

316

SURFACE ENGINEERING

Cold Spraying for Coatings: Mechanisms, Properties, and Applications

Thomas Klassen, Frank Gaertner, Tobias Schmidt, and Heinrich Kreye

Helmut Schmidt University, University of the Federal Armed Forces Hamburg
Hamburg, Germany, www.hsu-hh.de

Abstract: *In thermal spraying, well-tuned amounts of thermal energy and kinetic energy are employed to form dense coatings by impacting liquid droplets or powders. The recently introduced technique of cold spraying mainly relies on the high kinetic energy of solid particles impinging the substrate surface. Since the spray material is only exposed to an inert process gas and to well tuneable temperatures below the melting point of typical feedstock powders, this method is most suitable for heat and oxidation sensitive materials.*

This presentation summarizes the current model for coating formation based on adiabatic shear instabilities, which relates critical velocities to mechanical and thermal properties of the spray material. Computer simulations visualize the bonding process and allow for extracting critical parametersthat define the "window of deposition". For selected materials, a correlation between powder characteristics, particle impact conditions and coating properties will be shown. Different examples are presented to elucidate the potential for a variety of coating and substrate materials. With respect to spray forming by cold cold spraying, microstructures and thick, further machineable structures are presented.

1. Introduction

Cold spraying was invented about twenty years ago by a group of scientists at the Institute of Theoretical and Applied Mechanics, Novosibirsk, Russia [1,2]. Meanwhile, cold spray equipment is commercially available meeting all requirements for industrial applications [3]. The principle of cold spraying is quite simple, using a highly pressurized and preheated gas, which by passing through a converging diverging DeLaval type nozzle accelerates powder particles to super-sonic velocities [1-4]. Heating of the process gas is requested to reach higher gas and thus particle velocities [4]. Even that the gas is heated, the term cold spraying is still justified since particles on their way through the nozzle and the free jet attain only temperatures below their melting temperatures, in clear contrast to thermal spraying in which at least a part of the feedstock is molten before the impact on the substrate [5].

As compared to thermal spray processes, cold spraying offers several advantages. Heating and acceleration of the feedstock material only happens in the environment of an inert gas, typically nitrogen, in some instances helium, operated at comparably low temperatures. Thus oxidation is limited down to a minimum. Moreover, the well determinable process temperatures and exposure times make the process very suitable to prevent other undesired phase transformations.

1.1 Bonding Mechanism

Whereas thermal spraying requires a certain amount of heat to melt at least a part of the feedstock material to obtain welding to the substrate or to already adhering particle layers, cold spraying just uses the kinetic energy upon impact to assure bonding [5]. For successful bonding, spray particles have to exceed a critical velocity, which is strongly dependant on the type of feedstock material, but also on the substrate properties [6]. Upon impact, the particles deform in the zones of high shear stress, i.e. at the interface between particle and substrate. This plastic deformation generates heat, which leads to thermal material softening. In turn, this results in even more highly localized deformation in these shear zones at the interface, because the heat cannot be transported into the bulk within the milliseconds of the impact event, leading to a steep temperature increase at the interface (Fig. 1). Thus, these adiabatic shear instabilities under high strain rate deformation facilitate bonding at particle-substrate or particle-particle interfaces [7,8].

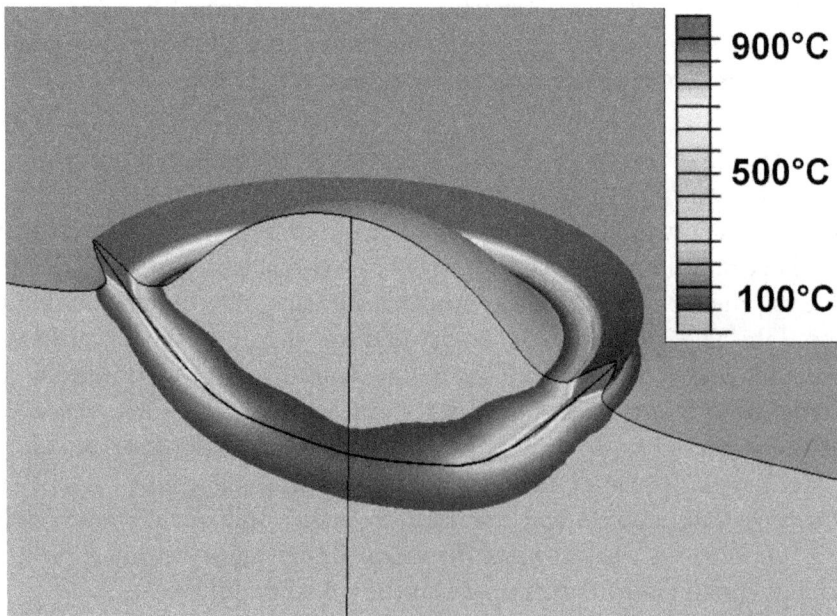

Fig. 1 Temperature rise during impact of Cu particle on Cu substrate.

320

A wide variety of pure metals, alloys, composite powders and blends, ranging from low to high melting point materials can be deposited using the cold spray process. A necessary prerequisite for bonding by shear instabilities is that at least one, but better all of the components, which interact in the particle impacts during the spray process, should have a sufficient ductility. Materials like Mg, Al, Ti, Zr, Sn, Zn, Cu, Ni, Nb, Ta, as well as alloys like steel 316L, steel 410L, Ni alloys, MCrAlY and powder blends can be sprayed successfully. Furthermore, under suitable conditions thermally sensitive materials like heat treated alloys or even very sensitive materials like plastics or polymers can be used as substrate. In special cases also ceramics and glasses can be coated.

2. Influence of Particle Impact Conditions on Coating Properties

By linking fluid dynamic modeling of individual particle impact velocities with numerical modeling of single impacts and the subsequent analyses of experimentally determined deposition efficiencies and coating microstructures, a more comprehensive understanding of the process was obtained [4,7,8].

One of the major goals within the last five years was to optimize particle acceleration, resulting in optimized nozzle geometries with respect to expansion ratios, contours and lengths of expanding sections [4,5]. Newer attempts aimed for increased particle impact temperature, supplying a well-defined amount of heat to support bonding by shear instabilities [8,9]. Since the creation of heat by deformation during impact is localized to the particle interfaces, the heat transfer to the inside of spray particles also has to be taken into account [8]. Recent investigations demonstrated that heat flux is more prominent for smaller particle sizes and increases critical velocities for bonding [8]. To obtain a high deposition efficiency or a maximum amount of bonded interfaces, the particle impact velocity should be significantly higher than the critical velocity for bonding, as illustrated in figure 2 for the example of copper particles of different sizes. Fig. 2 (c) shows that critical velocities decrease with particles sizes, which can be attributed to less effects of heat conduction. The particle impact velocities are as well highly dependant on size. Small particles are decelerated in the bow shock, means the compressed gas in front of the substrate, whereas large particles by their large momentum not reach high velocities. The intersections between both curves, for smaller and for larger particles refer to a probability that for these sizes about 50 % of spray particles would build up a coating. Between both intersections deposition, where the particle velocity is significantly higher than the critical velocity, nearly all particles should adhere to the substrate. Recent investigations demonstrated that influences of feedstock properties, process temperatures and particle sizes can be expressed by a set of easy applyable equations, supplying a guide-line to define conditions for particle bonding of a wide range of different metallic materials [8].

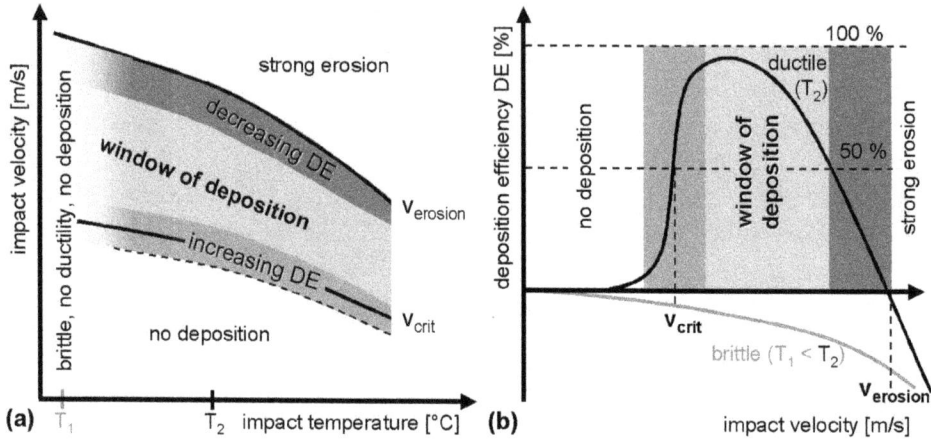

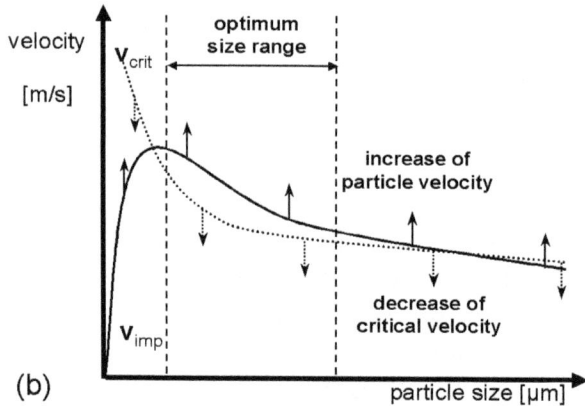

Figure 2 (a) Impact effects as a function of particle impact temperature. The window of deposition should be met in the vertical centre by particle impact conditions to obtain optimum bonding.
(b) corresponding diagram of deposition efficiency as a fuction of particle impact velocity for the Cases of brittle and ductile behaviour;
(c) Process optimisation by decreasing critical velocity, increasing particle impact velocity and by choosing an optimised size distribution.

Experimental results for deposition efficiency (DE) and coating strength in dependence of gas temperature are illustrated in Figure 3. The −38+11 μm Cu-powder was used as the spray material and nitrogen at 30 bars as the process gas. At gas temperatures up to 600°C the spray particles were injected 20 mm in front of the nozzle throat (standard particle injection) and at gas temperatures higher than 600°C they were injected 135 mm in front of the nozzle throat (elongated prechamber). The comparison demonstrates that first the DE improves steeply with increasing process gas temperature. The mechanical strength of the coatings is only slightly increased until the

322

DE reaches a saturation limit. In this parameter range, the coating strength is dominated by the particle fractions that just start to stick. Thus, the amount of bonded areas does not increase significantly. After reaching the saturation limit for the DE, the mechanical coating strength rises steeply. In this parameter range, every increase in particle impact velocity and particle impact temperature results in a significant increase of the metallurgically bonded areas. Using particle injection 135 mm before the nozzle throat leads to higher particle temperatures, resulting in a jump in the coating strength at 600°C, indicating that the particle impact temperature strongly influences the particle bonding.

The previously discussed context can also be transferred to other spray materials. In Fig. 4, the cohesive coating strengths, determined by the TCT-test, are plotted as trend lines for Al, Cu, steel 316L and Ta as a function of the process gas temperature. Of course the different properties of these four materials are playing a role in the way the cohesive strength increases, but the differences are not as strong as it could be expected. Naturally under same process conditions thermal softening of particles is more efficient for materials, having a low melting temperature (Al: Tm = 660°C), than for materials having a very high melting temperature (Ta: Tm = 2996°C). Therefore, the slope of curves in the "strength – process gas temperature" diagram is increasing with decreasing melting temperature.

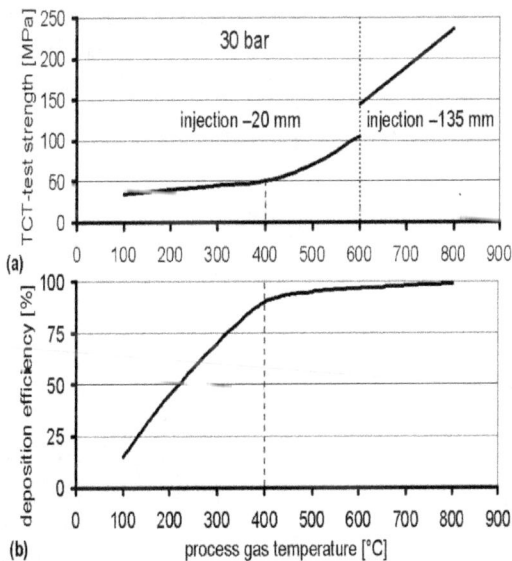

Figure 3(a) Coating strength (TCT-tests), and (b) DE as a function of the process gas temperature. A -38+11 μm Cu powder was used. The process gas was nitrogen at 30 bar. At a gas temperature of 600°C the injection point was changed from 20 mm to 135 mm in front of the nozzle throat.

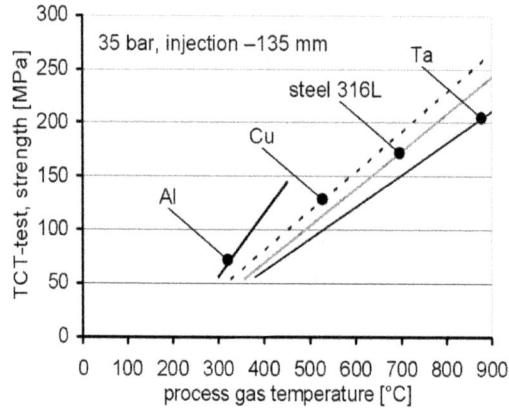

Figure 4 Trend lines of the coating strength (TCT-test), for different materials as a function of process gas temperature. The particle size distribution was around -45+15 μm

Figure 5: shows three stress-strain-curves of copper coatings measured with the MFT-test. One specimen was prepared from a coating sprayed with the standard parameter settings (Fig 5a, black line). This sample ruptured at 57 MPa after reaching an elongation of 0,1 %. The Young's modulus was determined to 71 GPa, being much lower than the reference data for copper taken from literature (125 GPa). The difference can be explained by the insufficient and inhomogeneous particle bonding and contributions by plastic deformation at the locally bonded particles. The two other samples were cut from a coating sprayed under optimised conditions, one of these was annealed. The as-sprayed sample fractured at 405 MPa at an elongation of 0,7 %, demonstrating properties close to highly deformed bulk material (Fig. 5a, grey line). In agreement, the measured Young's modulus of 117 GPa is similar to the literature value for copper. Figure 5b compares the stressstrain- curves of a coating processed under the optimised parameter in as sprayed condition (grey line) and after annealing (black line). The mechanical properties of this annealed coating are comparable to the mechanical properties of annealed bulk material.

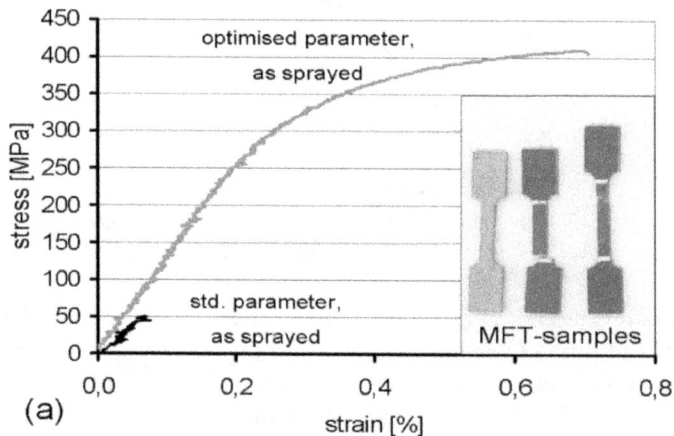

(a)

324

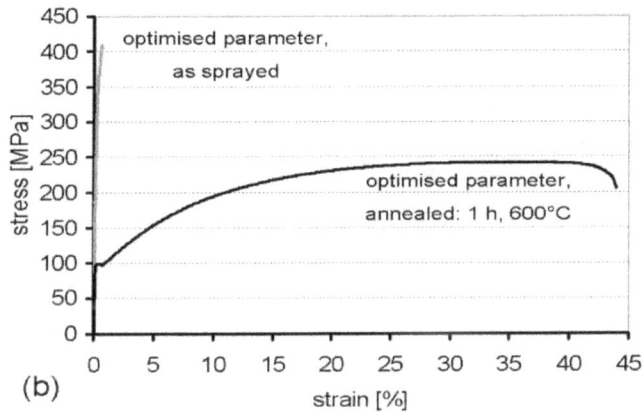

Figure 5 Stress-strain curves of cold sprayed coatings (MFT-tests). (a) using standard and optimized conditions, (b) as-sprayed and annealed conditions of coatings sprayed under optimized conditions.

3. Coating Microstructures and Applications

Figures 6 to 11 show some state of the art cold sprayed coatings. In most cases nitrogen was used as process gas.

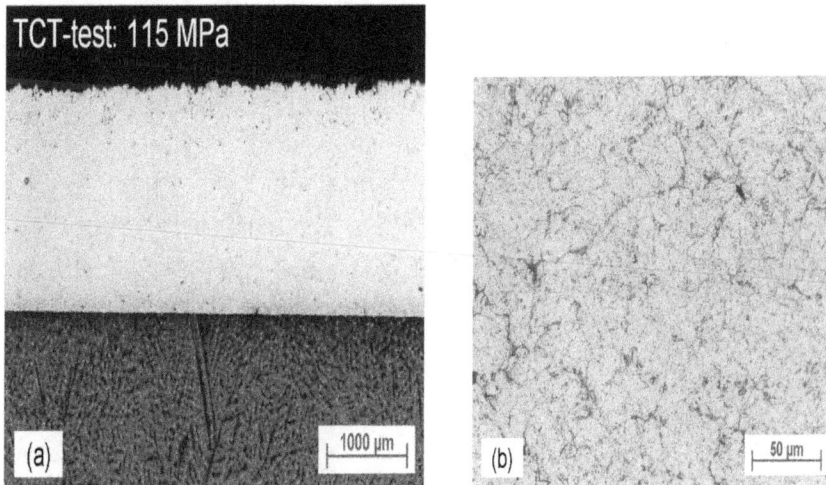

Figure 6 Al coating on Mg, nitrogen sprayed, (a) polished, (b) etched.

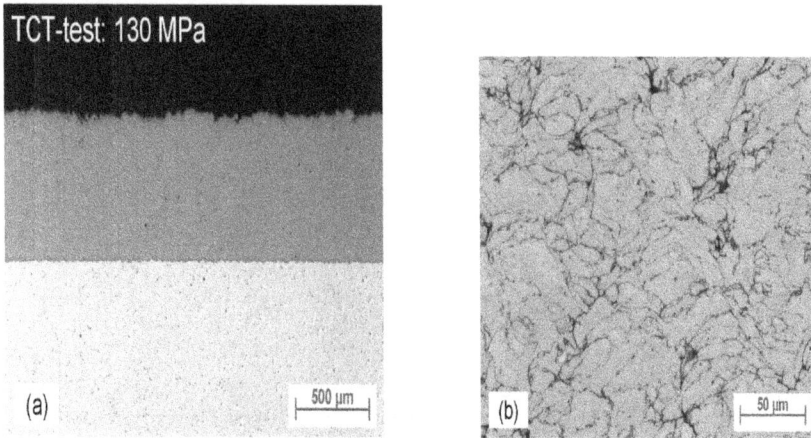

Figure 7 Ti coating on Al, nitrogen sprayed, (a) polished, (b) etched.

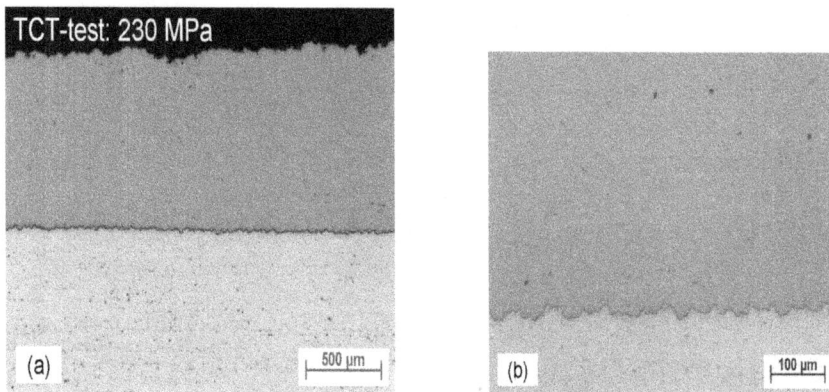

Figure 8 Steel 316L coating on Al, nitrogen sprayed, (a+b) polished.

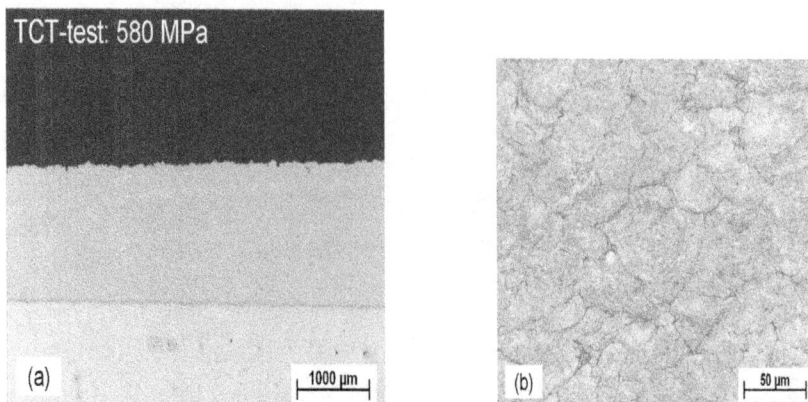

Figure 9 Steel 316L coating on Cu, helium sprayed, (a) polished, (b) etched.

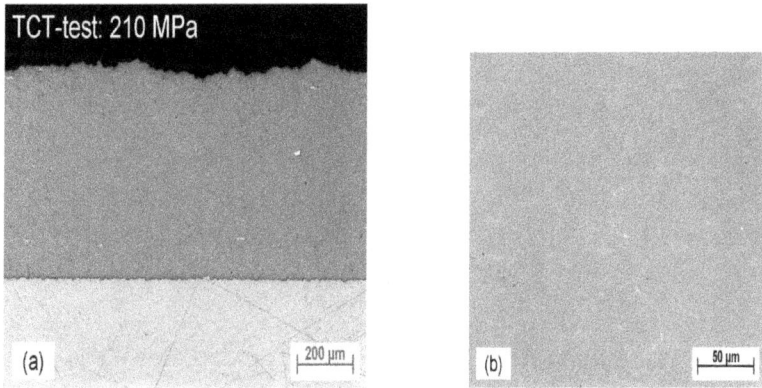

Figure 10 Ta coating on low carbon steel, nitrogen sprayed, (a+b) polished.

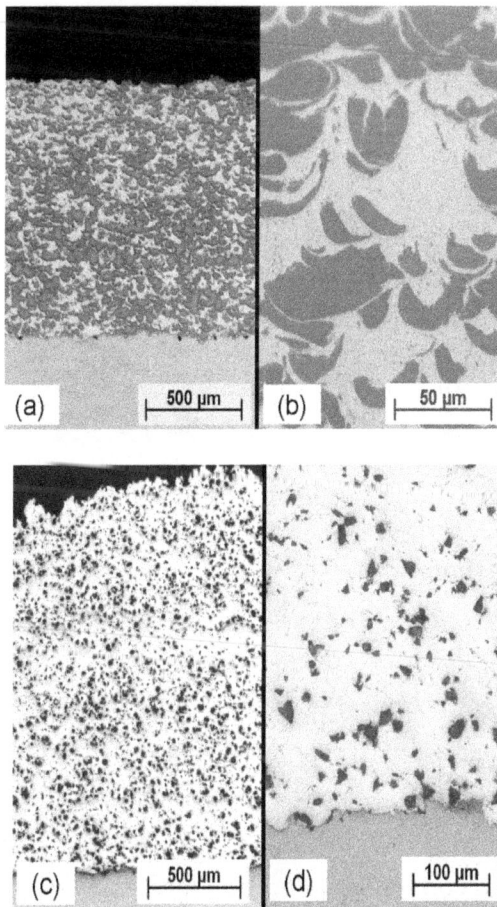

Figure 11 (a+b) Cu-Ta composite coating on carbon steel, helium sprayed, (c+d) Al-Al2O3 composite coating, nitrogen sprayed

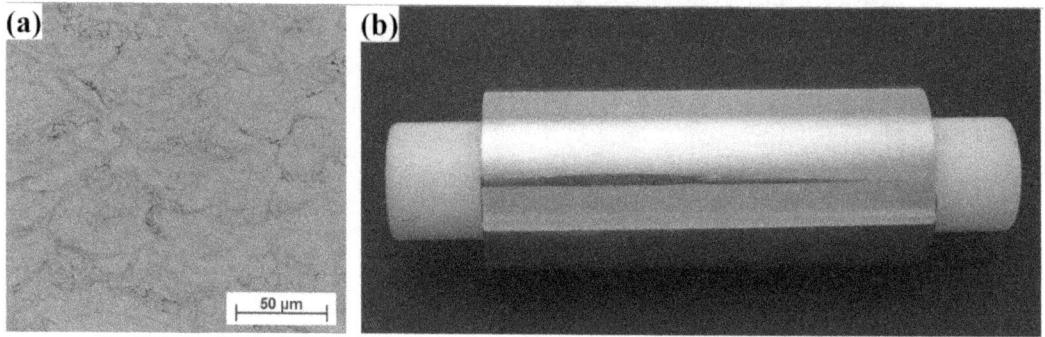

Figure 12 Etched microstructures of a cold sprayed copper coating processed with N2 as process gas (a) and an example for a print roll, produced under optimum cold spray conditions after final finish (b). The aluminum tube used as substrate for the latter has a diameter of 44 mm. After finishing, the copper coating has a thickness of 2.5 mm.

Up to present, most interests aimed to customize cold spraying for well-established engineering materials by retaining the feedstock properties. Already that could open new applications for powder spray techniques, complementary to thermal spraying. Typical cold sprayed coating thicknesses range from about less than 50 μm, corresponding to a couple of monolayers of spray particles, to some tens of millimeters. Within this range, the comparably thin coatings are interesting with respect to especially designed functional materials. Within the wide spectrum of possible applications, the present report focuses on spray forming of machinable parts. Building up thicker structures is of particular interest for repair work or rapid protyping. As compared to thermal spraying, building up thicker coatings by cold spraying is possible due to the lack of thermal tensile stresses. Morover, the compressive stresses present in the cold-sprayed coatings by the deformation upon solid-state impact can be benifcial for coating performance under fatigue conditions. Figure 12 shows the microstructure of a copper coating and the example of a thick coating sprayed on an aluminum tube, which in the polished state could be used as a print roll. Using nitrogen as process gas and recently optimized spray conditions [10], the etched coating cross section reveals no remarkable differences between grain boundaries and particle-particle interfaces.

Figure 13 As polished coating microstructure (a) and a thick, machined layer of steel 316L on a copper tube (b). The example in (b) illustrates that stainless steel 316L can be well bonded onto copper and that as prepared parts can be further finshed similar to conventional material, by drilling holes and fixing screwes in such structures. The copper tube has a diameter of 42 mm and the machined steel 316L coating in final finish has a thickness of 8 mm.

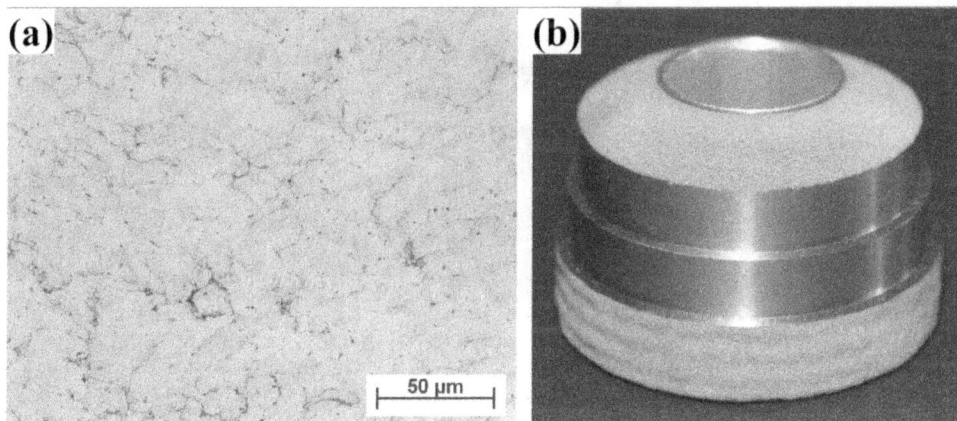

Figure 14 Etched coating microstructure (a) and a thick, machinable layer of titanium. The aluminum tube serving as substrate has a diameter of 24 mm. The titanium coating thickness was adjusted to 14 mm and machined down to thicknesses of 12.5 mm and 11 mm, respectively.

The good resistance against the chemical attack demonstrates the quality of metallurgically bonded particle-particle interfaces. The cross section also shows that grains close to the interface are much more elongated than those of the particle core. Building up a thicker copper coating on a low carbon steel tube does not restrict coating performance or adherence. The final product is well machinable and can be

329

polished to similar standards as highly cold worked copper bulk material. By using stainless steel 316L powders as feedstock material, currently attainable conditions by cold spraying result in coatings with very low porosities far less below 1 vol. % and a high strength suitable for further maching, as illustrated in figure 13. The cross section demonstrates the low porosity of the as sprayed coating. Thicker coatings or structures, as shown for the example on a copper tube, can be finished and prepared for further construction as conventional, highly deformed steel 316L.

Under currently available conditions of cold spraying with nitrogen as process gas, pure titanium coatings can be processed with porosities of less than 1 vol. % (figure 14a). As compared to steel 316L or copper as spray material, the slightly higher porosity can be attributed to the low deformability of titanium. Nevertheless, particle-particle interfaces are well bonded and thick coatings can

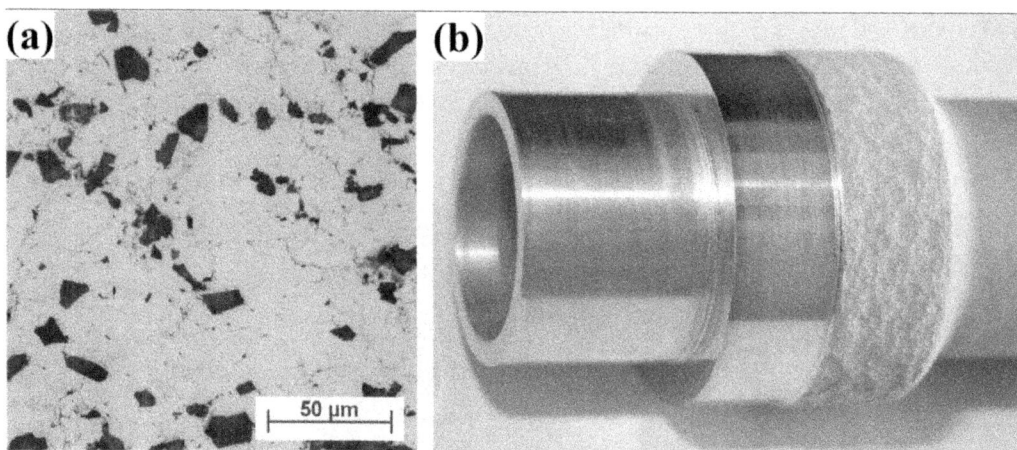

Figure 15 As polished coating microstructure (a) and a thick, machined layer of an Al-Al2O3 composite. The aluminum tube serving as substrate has a diameter of 40 mm. The intial thickness of the as sprayed coating attained 7 mm and was machined down to a thickness 6.5 mm.

be processed, which allow further maching and final finshes (figure 4b). The top side of the tube shows the kinematic turning point of the robot system. The sharp angle of about 80° to build up coatings demonstrates the high potential for rapid prototyping. On the other side of the sample, where the robot was driven into the overspray regime, the angle to build up coatings is roughly perpendicular to the substrate. Similar results are obtained for other metallic materials. As demonstrated in figure 15, also feedstock blends of metallic and ceramic powders can be used for building up dense coatings and thick structures. The detailed view into coating cross sections reveals only a minimum amount of porosity. Nevertheless, it must be noted that by using a powder blend containing about 50 vol. % of ceramics results in a coating hard phase content of only about 15 %. Nevertheless, that comparatively small hard phase contents reduced the abrasive wear by 30 to 50 %, as compared to bulk alumiunum alloys. By cold spraying, thick coatings of such aluminum – alumina composite coatings can be processed, which can be further machined by conventional techniques.

4. Summary

Cold spraying can produce dense coatings in thicknesses between 50 µm and several centimeters from a wide variety of ductile materials. Bonding in cold spraying can be attributed to the occurance of shear instabilities at particle surfaces and recently developed concepts allow predicting impact conditions for successfully building up coatings. The properties of cold sprayed coatings critically depend on particle impact conditions, i.e. particle velocity and temperature. Size effects in impact dynamics are considered by an optimised size distribution.

Acknowledgments

The Linde Gas AG (Höllriegelskreuth, Germany) and CGT Cold Gas Technology GmbH (Ampfing, Germany) are acknowledged for technical support. The authors thank Andrey Vishnevsky, Helmut-Schmidt-University for support in performing the MFT testing.

References

1. A.P. Alkhimov, A.N. Papyrin, V.F. Kosarev, N.I. Nesterovich, M.M. Shushpanov, *Gas- Dynamic Spray Method for Applying a Coating*, U.S. Patent No. 5302414, 1994.

2. A.P. Alkhimov, V.F. Kosarev, A.N. Papyrin: Sov. Phys. Dokl, Vol. 35, 1990, p. 1047 (Transl: American Inst. of Phys., 1991).

3. J. Karthikeyan: Spraytime, Vol. 12, 2005, p. 1.

4. T. Stoltenhoff, H. Kreye, H.J. Richter: J. Therm. Spray Technol, Vol. 11, 2002, p. 542.

5. F. Gärtner, T. Stoltenhoff, T. Schmidt, H. Kreye: in press by J. Thermal Spray Technology, 2006.

6. S.V. Klinkov, V.F. Kosarev, M. Rein: Aerospace Science and Technology, Vol. 9, 2005, p.582.

7.   H. Assadi, F. Gärtner, T. Stoltenhoff, H. Kreye, Acta Mater, Vol. 51, 2003, p. 4379.

8.   T. Schmidt, F. Gärtner, H. Assadi, H. Kreye: Acta Mater. Vol. 54, 2006, p. 729.

9.   H.J. Kim, C.H. Lee, S.Y. Hwang: Materials Science and Engineering *A* Vol. 391, 2005, p. 243.

10.   Schmidt, F. Gärtner, H. Kreye: in press for Proc. *International Thermal Spray Conference,* 2006.

Corrosion Resistance of Friction Stir Welded AA6061 Aluminium alloy: Electrochemical Polarization and Impedance Study

K.Ratnakumar* G. Madhusudhan Reddy[1] and K.Srinivasa Rao[2]

*Government Polytechnic, Visakhapatnam, INDIA
[1]Metal Joining Group; Defence Metallurgical Research Laboratory; Hyderabad, INDIA
[2]Andhra University College of Engineering Visakhapatnam-INDIA
Email: rkkancharla@yahoo.com

Abstract_Friction stir welding (FSW) a solid state welding process is currently considered to be most prospective welding process particularly for aluminum alloys. In this work, the microstructural and corrosion properties of friction stir welded AA6061 Al alloy were studied. The corrosion behaviour of the nugget zone(NZ), thermomechanically affected zone(TMAZ) and heat affected zone(HAZ) of friction stir welds of AA6061 alloy was studied. Dynamic polarisation and impedance testing were used to determine the pitting and general corrosion resistance of the welds respectively. Optical microscopy (OM) and scanning electron microscopy (SEM) studies were carried out to find the mechanism of corrosion. Vickers hardness profile was obtained across the weld. Friction stir welding of this alloy resulted in fine recrystallized grains in weld nugget which has been attributed to frictional heating and plastic flow. The process also produced a softened region in the weld nugget, which may be due to the dissolution and growth of possible precipitates. Corrosion resistance of nugget zone has been found to be better than that of TMAZ and base metal. Corrosion resistance of naturally aged (T4) alloy exhibit better corrosion resistance than that of artificial aged (T6) alloy._

Keywords: Pitting corrosion, General corrosion, Dynamic polarization, Impedance, Friction stir welding

1. Introduction

Friction stir welding is a solid state joining process developed by The Welding Institute (TWI), UK in the year 1991. The process goes on commercial in late nineties. The technique was primarily developed for joining aluminum alloys and other soft materials like magnesium alloys but consequently extended for steels[1]. The process was based on friction heating at the faying surfaces of two pieces to be joined, results in a joint created by interface deformation, heat and solid-state diffusion[2]. It is essentially a hot-working process where a large amount of deformation is induced into the work piece through pin and shoulder and the temperature never exceeds 0.8 T_m[3]. It results in a distinguished microstructure in precipitation hardenable aluminum alloys[4,5&6]. The first attempt of classifying microstructures was made by P.L.Threadgill. This was further revised and accepted by the friction stir welding licenses association. This system divides the weld zone into unaffected material or parent metal (PM), heat affected zone (HAZ), thermo-mechanically affected zone (TMAZ) and weld Nugget of Nugget Zone (NZ).

The resistance to corrosion of aluminium alloy welds is affected by the alloy being welded, the filler alloy and the welding process used. When localized corrosion does occur in aluminium welds, it may take the form of preferential attack of the weld bead, pitting, intergranular attack, or exfoliation may occur in the heat affected zone (HAZ) a short distance from weld bead. Welds in Al–Mg–Si alloys (AA6061) generally have good resistance to atmospheric corrosion, but in specifically corrosive environments, like seawater, localized corrosion may occur. The intermetallic phases like Mg_2Si in AA6061 Al alloys are harmful with respect to weldability and corrosion resistance. These intermetallic particles induce liquation in the partially melted zone (PMZ) of the weldment, leading to cracking and galvanic coupling of these phases with the surrounding matrix resulting in poor corrosion resistance[7].

The microstructure resulted by friction stir process in precipitation hardenable aluminum alloys are different from that of base metal microstructure or cast structure of fusion welds. These changes can bring a difference in corrosion properties of the weldment. Generally a passive oxide film can be readily formed on the surface of aluminum alloys, when exposed to air or water. However the corrosion rate could be very high due to the presence of chloride ions[8]. Further the corrosion behaviour of Al alloys largely depends on heterogeneity of their microstructures. As the friction stir process induces a dramatic change in microstructures, there is every need to understand the microstructure and corrosion behaviour of friction stir welds. In friction stir welding, the local frictional heating causes the metal to flow, forming an almost seamless weld with excellent strength. Although the microstructural heterogeneity of FSW is less than conventional fusion welds, there are still concerns about sucessptibility to corrosion[9]. Anodic attack of the weld can take place in the weld nugget, where mixing of the metal has taken place, or in the heat affected zone (HAZ) either side of the weld[10, 11, 12, 13 &14]. It is usually in the form of intergranullar attack., although some pitting has been observed[10,14].Hannour et al[15] reported that the Heat affected zone was the region susceptible to corrosion of friction stir welded alloys.

In the present work age-hardenable alloy AA6061is subjected to friction stir welding because this alloy is extensively used for aerospace and defence applications like body panels, engine components and wheels. The present work was aimed at studying the influence of prior thermal temper on the microstructure, pitting corrosion behaviour and AC impedance of Nugget zone (NZ), thermo mechanically affected zone (TMAZ) and heat affected zone(HAZ) of friction stir welded AA6061 alloy.

2. Experimental Details

Wrought AA6061 alloy plates of thickness 6mm in T6 (solution treatment at 540^0C for 1h and aged at 160^0C for 12 h) and T4 (solution treatment at 540^0C for 1 h and aged at room temperature for 30 days) tempers were used. Chemical composition of the base metal is given in the Table 1. The alloy is subjected to friction stir welding. The formation of joint during friction stir welding was shown in Fig. 1. The parameters used for the friction stir welding of the AA6061 alloy is given in the Table 2. The weld bead was made perpendicular to the sheet rolling direction. The top and bottom view of a typical joint of FSW AA6061 alloy was given in Fig. 2 . Prior to welding, the base metal coupons was brushed and thoroughly cleaned with acetone. The samples of base metal and welds were polished on Emery papers and disc cloth to remove the very fine scratches. Polished surfaces had been etched with Keller's reagent. Optical Microscopy (OM) was performed on the top surface of the weld. The microstructures were recorded with an image analyzer attached to the metallurgical microscope. Scanning electron microscopy (SEM) and transmission electron microscopy (TEM) with Energy dispersive X-ray spectroscopy studies were carried out to find the mechanism of formation of TMAZ and corrosion. Vickers hardness testing was carried out on the nugget zone(NZ), thermo mechanically affected zone(TMAZ) and heat affected zone (HAZ) of the samples with 5 kgf load and 15 minutes dwelling time by using Leco superficial Vickers hardness tester. The hardness profiles were drawn across the distance from the center of the weld. A software based PAR electrochemical weld tester system was used to make potentiodynamic polarization tests and AC impedance tests to study the pitting corrosion and general corrosion corrosion behavior of the base metal, weld metal, HAZ and PMZ regions. A saturated calomel electrode (SCE) and carbon electrode were used as reference and auxiliary electrodes respectively. All experiments were conducted in aerated 3.5% NaCl solutions with PH adjusted to 10 by adding potassium hydroxide. The potential scan was carried out at 0.166 mVs^{-1} with the initial potential 0f –0.25 V (OC) SCE to the final pitting potential. The exposure area for these experiments was 0.6 cm$^{2.}$ The potential at which current increased drastically was considered to be the critical pitting corrosion E_{pit}. Specimens exhibiting relatively more positive potential (or less negative potentials) were considered to have better pitting corrosion resistance. Optical microscopy on dynamically polarized samples was carried out to understand the mechanism of pitting.

Table 1 Chemical composition of the base metal AA6061

Element	Mg	Si	Mn	Fe	Cu	Al
Wt.%	0.689	0.531	0.331	0.230	0.305	Remaining

Table 2 Tool size and welding parameters used in experiments

Tool Size			Welding parameters	
Shoulder Diameter (mm)	Pin diameter (mm)	Pin length (mm)	Rotation speed (rev min⁻¹)	Travel speed (mm min⁻¹)
10	5	6.4	1000	15

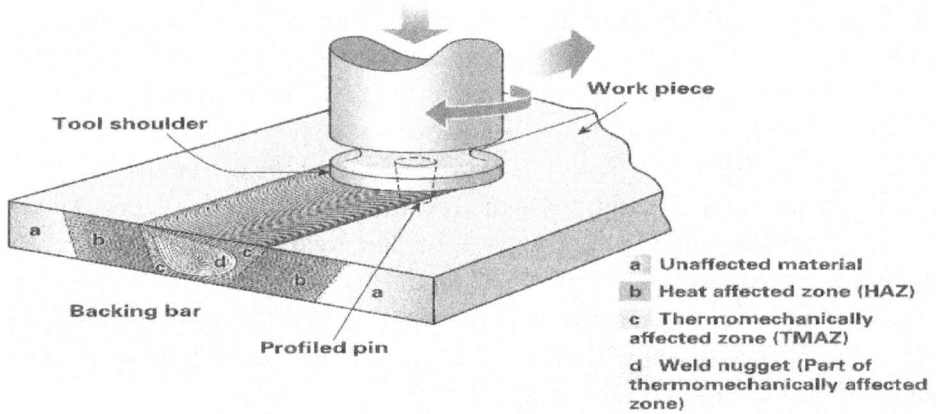

Fig. 1 Different zones of Friction stir weld joint.

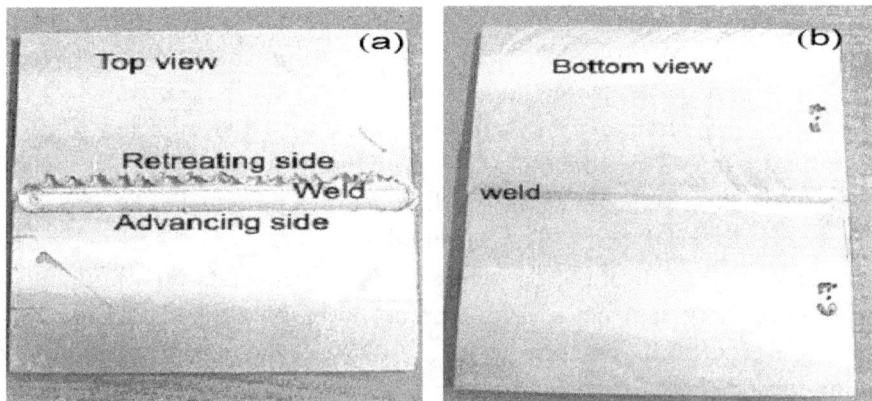

Fig.2 Top and bottom views of a typical joint of FSW AA6061 alloy
(a) top view (b) bottom view.

3. Results and Discussions

3.1 Base Metal

Optical micrographs of the base metals AA6061-T4 and AA6061-T6 are shown in fig.3a and 3b respectevely. The grain structure was well developed and grains of AA6061-T6 (133microns) are slightly coarser than that of AA6061-T4 (113microns) Coarse particles of Mg_2Si (dark) and Fe_3SiAl_{12} (grey) are clustered in stringers and aligned along the rolling direction (Fig. 4). Micrographs revealed that more number of Mg_2Si particles were present in artificially aging (T6) alloy when compared to that of naturally aged alloy (T4).

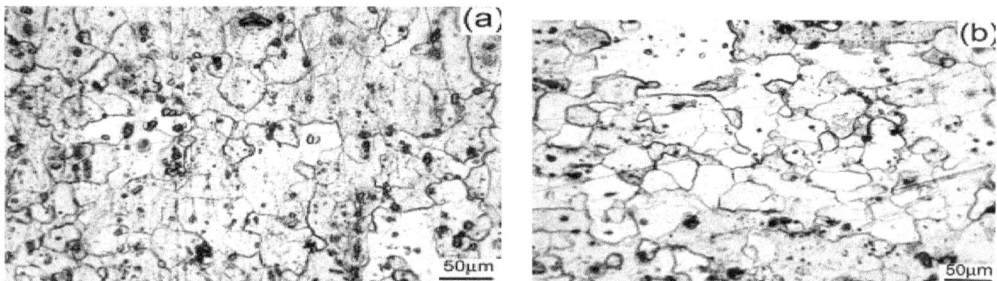

Fig. 3 Optical micrographs of AA6061 alloy (a) T4 Condition (b) T6 Condition.

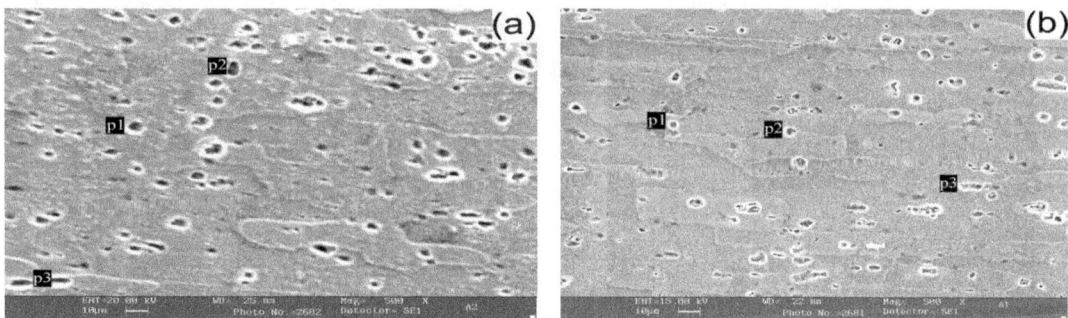

Fig. 4 Scanning electron micrograph of base metals: (a) AA6061-T6 (b) AA6061-T4.

3.2 Weld Nugget

Friction stir welding produces a far less heterogeneous microstructure than conventional fusion welding. It results in three microstructural regions: the nugget, where the material has undergone severe stirring and heating, the thermo-mechanically affected zone (TMAZ)which is subjected to both deformation and heating, and the heat affected zone (HAZ) where the material experiences only a thermal cycle.

337

Microscopically, these three regions are significantly different. Macrostructure which shows the formation of nugget zone and thermomechanically affected zone in friction stir welded AA6061 alloy was given in Fig. 5. Optical micrographs of various zones of FSW AA6061 in T6 condition is shown in the Fig. 6. The weld nugget has a recrystalized, fine equiaxed grain structure of the order of 4-5 microns as shown in the Fig. 7. This indicates that, during the process of FSW, the precipitates have dissolved into the solution and reprecipitated on subsequent cooling. The recrystalization of the weld nugget and the redistribution of the precipitates indicate that the temperature obtained during the process is above the solutionizing temperature but below the melting temperature of the alloy [8]. The finer precipitates are absent in weld nugget because the cooling rates are such that larger precipitates could nucleate and grow but not the finer ones. TEM of nugget zone was shown in Fig. 8. TEM bright field image showing the microstructure of the nugget zone consisting of fine recrystallized grains of alumjinium with Mg_2Si precipitates. Bright field image showing equilibrium Mg_2Si precipitates (arrow marked) at higher magnification. Micro diffraction pattern obtained from the Mg_2Si particles corresponds to <011> zone axis of FCC with an average lattice parameter of a = 0.64 nm.

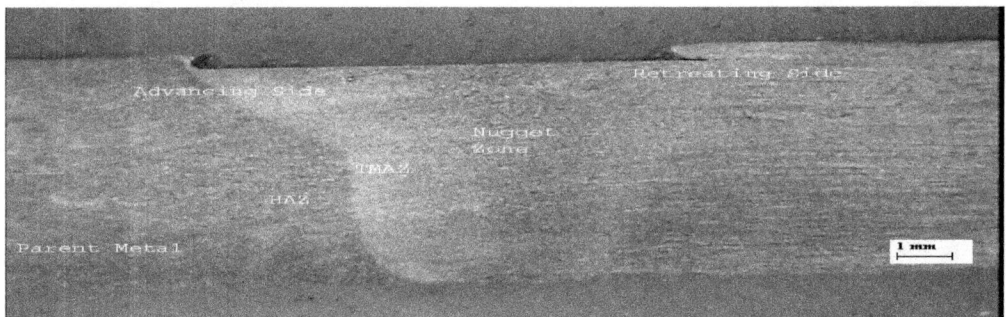

Fig.5 Typical microstructure of the as-welded sample cross section perpendicular to the welding direction

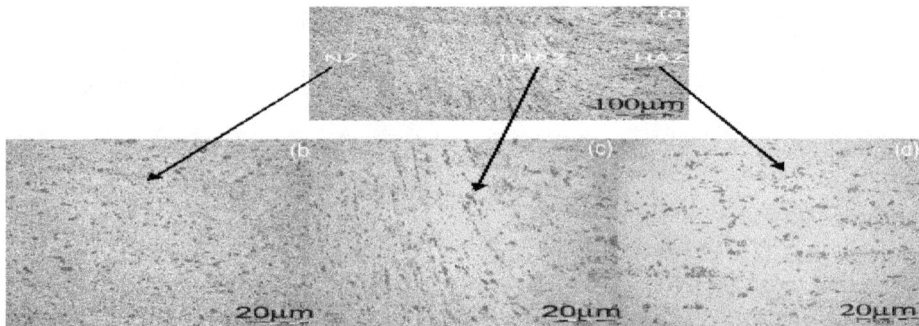

Fig. 6 Optical micrographs of various zones of FSW AA6061-T6 alloy

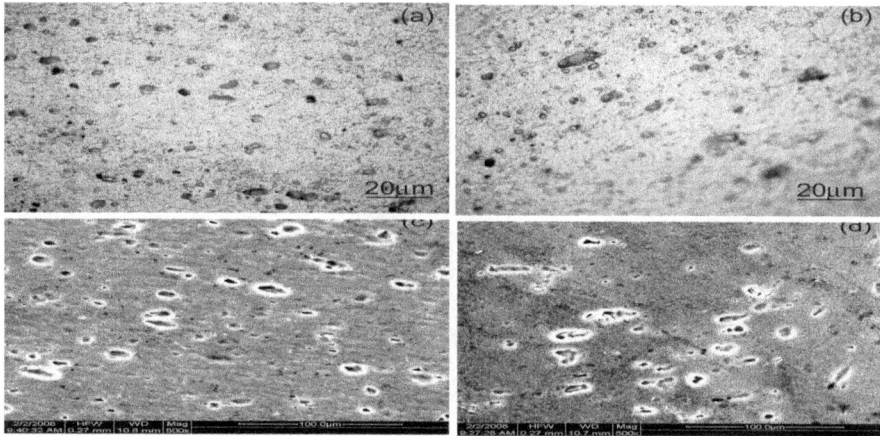

Fig. 7 Optical and SEM micrographs of nugget zone in AA6061-FS Welds

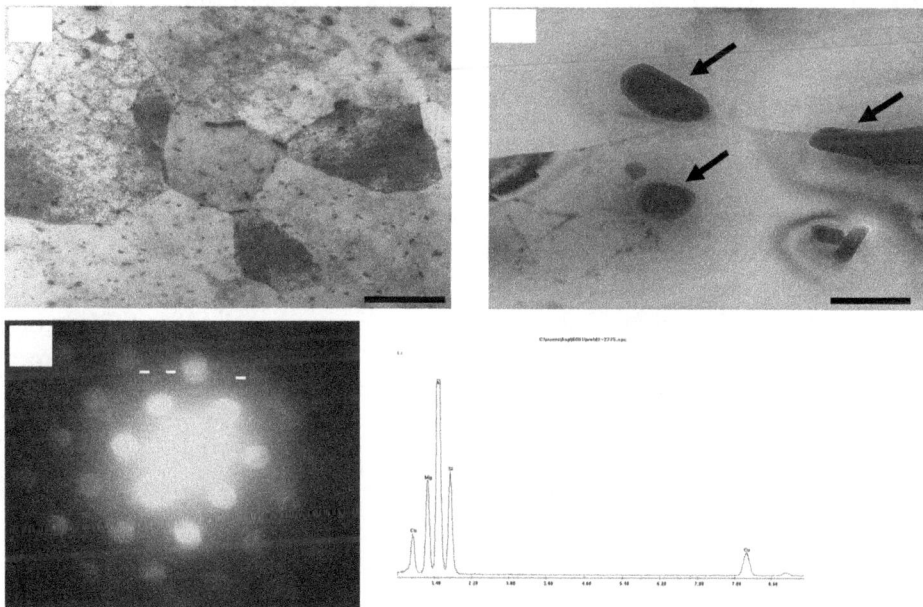

Fig.8 (a) TEM bright field image showing the microstructure of the weldnugget consisting of fine recrystallized grains of alumjinium with Mg2Si precipitates. (b) Bright field image showing equilibrium Mg2Si precipitates (arrow marked) at higher magnification. (c) Micro diffraction pattern obtained from the Mg2Si particles corresponds to <011> zone axis of FCC with an average lattice parameter of a = 0.64 nm. (d) Energy dispersive X-ray spectrum (EDXS) collected from the particles shown in (b) T4-retreating side

3.3 Thermo-mechanically affected Zone

This is the transition zone between the base metal and the weld nugget, characterized by a highly deformed structure as shown in fig. 9. Optical micrographs show a banded structure. There is no recrystalization in this region.. There is no significant change in the size and morphology of coarser precipitates, but their orientation along rolling direction like in parent metal is absent in TMAZ. The precipitates are quite random. The precipitates observed in parent metal are elongated and coarsened. It is characterized as having elongated and newly formed grain structure. Each weld zone exhibited a sharp [retreating side] and diffused [advancing side] macrostructural transition region between the TMAZ and NZ. These regions were formed as a result of the relation between rotation of the tool and the weld direction. The edge of the weld where the direction of the tool rotation is opposite to the travel [anti parallel] is referred to as the retreating side. The opposite case, where the direction of tool rotation is same as the travel, would be referred as the advancing side (Fig. 10). The thickness of TMAZ is small in T4 alloy when compared to that of T6 alloy. The difference in two regions was confirmed by slight differences in the optical microstructure and hardness tests. The microstructure of the advancing side of TMAZ is characterized by a sharp boundary between the nugget and the TMAZ. The retreating side of the FSW A356 alloy joint has a more complex microstructure, with no clear boundary between the nugget and the TMAZ. Although the nugget here has a smaller grain size than the TMAZ it also includes a number of large grains. Temperature in the thermo mechanically affected zone is higher in the advancing side compared to that of retreating side. TEM images of advancing side (Fig.11) clearly indicated coarsened precipitates and reduction in the dislocation density compared to that of retreating side

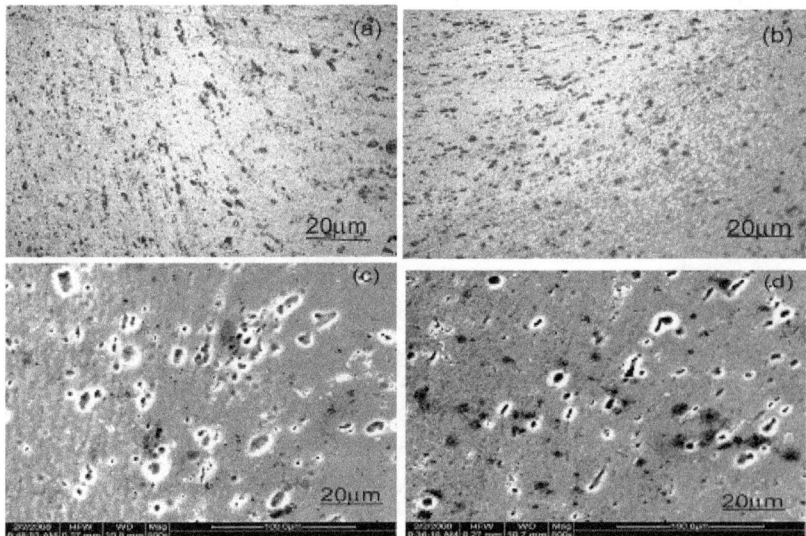

Fig. 9 Optical micrographs of TMAZ in F S welds of AA6061- alloy (a) optical of T6 alloy (b) optical of T4 alloy (c) SEM of T6 alloy (d) SEM of T4 alloy

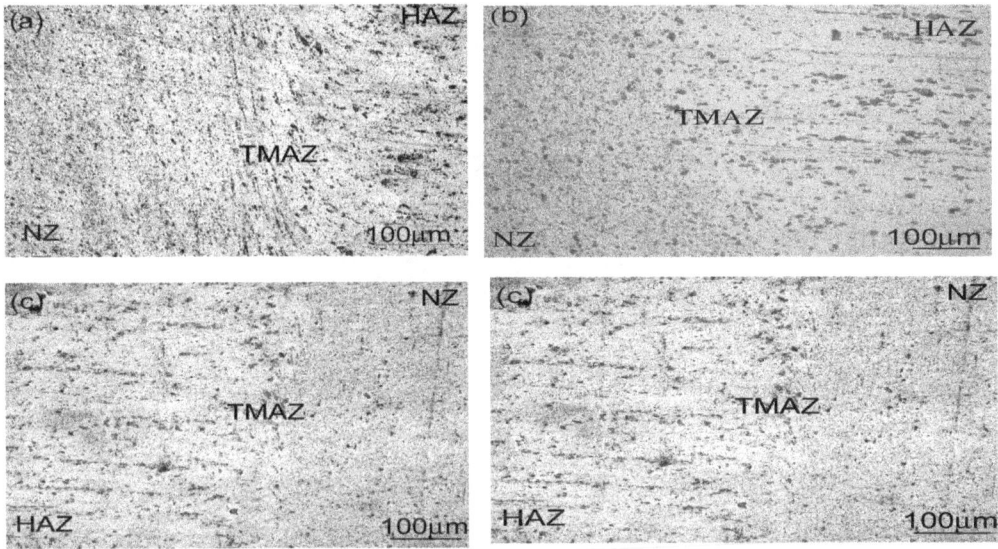

Fig. 10 Optical micrographs of NZ, TMAZ and HAZ in F S welds of AA6061- alloy
(a) T6-advancing side   (b) T6- retreating side (c) T4- advancing side (d)

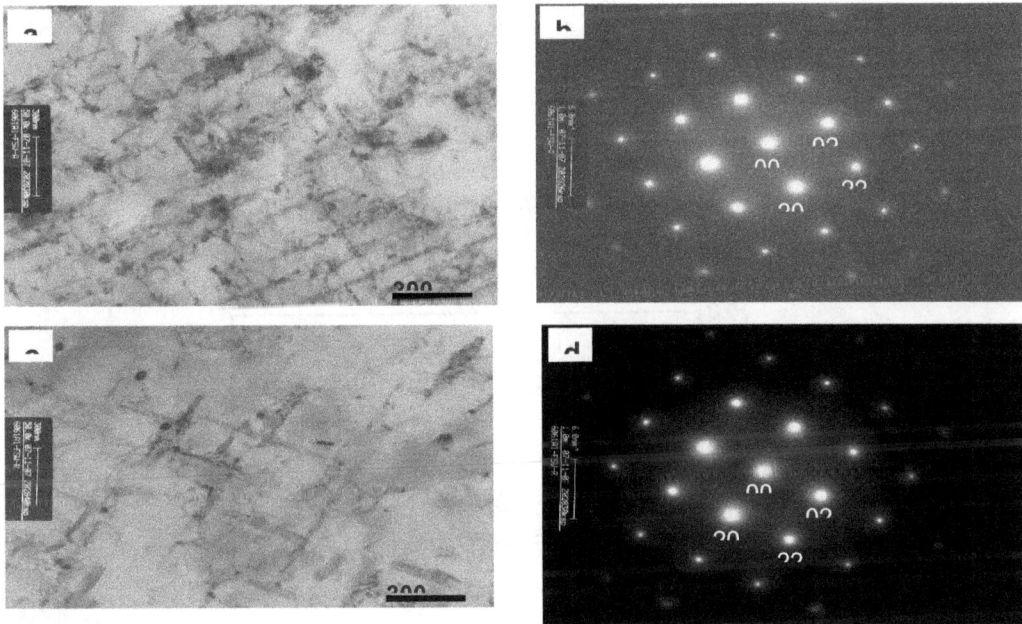

Fig.11 (a) & (c) TEM bright field image showing the microstructure at the retracting and advancing sides of the weld respectively consisting of metastable Mg2Si precipitates. in <001> direction. (b) & (d) Selected area electron diffraction obtained from (a) and (c) respectively consisting of diffraction spots belongs to Mg2Si precipitates.

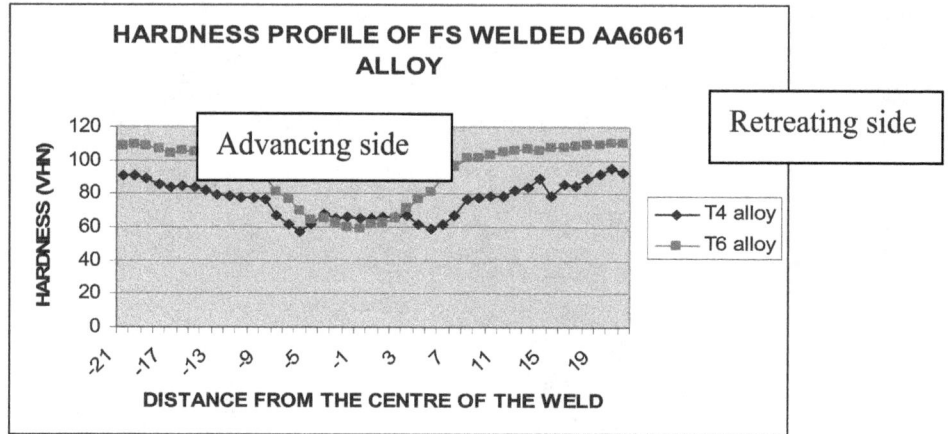

Fig. 12 Hardness profile of friction stir welded AA6061 alloys.

3.4 Hardness Profile

The hardness profile indicates a decrease in the weld hardness as shown in the Fig 12. This has been attributed to the dissolution of precipitates into solution and subsequently the weld cooling rates do not favour nucleation and growth of all the precipitates. The transition region indicates a reduction in hardness because of coarsening of precipitates. The hardness on one side of weld center is higher than the other side. This difference could be explained as follows. In the leading side for the rotating tool where the rotational velocity vector and the forward motion vector are in the same direction and due to this there is higher heating on one side of weld center and hence higher the hardness. Even though there is reduction in hardness of weld, it still meets classification societies requirements.

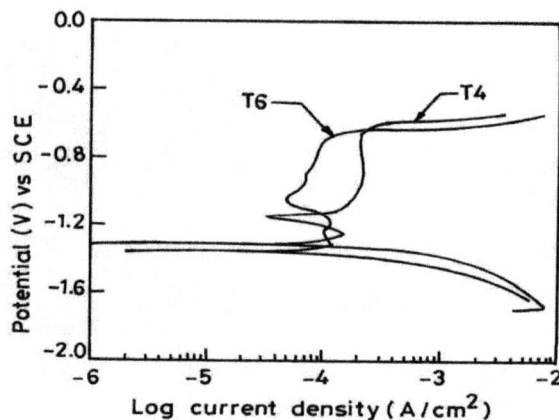

Fig. 13 Potentiodynamic polarization curvees of Base metal AA6061 alloy

3.5 Corrosion Studies

The potentiodynamic polarization curve for base metal in T4 and T6 conditions are given in fig.13 and Epit values are given in table 5. Base metal in T4 condition shows high pitting corrosion resistance than that of T6 condition. In T-6 condition, the corrosion could be attributed to the galvanic action between the Mg_2Si and the surrounding matrix. Coherent precipitates in age hardenable al-alloys were reported to be having non-galvanic as far as the electrochemical corrosion is concerned[16]. In view of the presence of G P Zones rather than Mg_2Si precipitates, specimens in T4 condition of AA6061 alloy would have exhibited better corrosion resistance. Polarization resistance of the alloy in T6 condition was lower than that in T4 condition. Hence the general corrosion resistance of the alloy was better in naturally aged condition (T4). The potentiodynamic polarization curves for WN, TMAZ and HAZ are given in Fig. 14. The critical pitting potentials of test specimens (Table 7), clearly indicates a greater pitting corrosion resistance of weld nugget than base metal. This is because the precipitates present in the alloy promote matrix dissolution through selective dissolution of aluminum from the particle. These precipitate deposits are highly cathodic compared to the metallic matrix, which initiates pitting at the surrounding matrix and also enhances pit growth. As discussed earlier, during FSW only the coarser precipitates could nucleate and grow but not the finer ones. This aids in formation of passive film, which remained more intact on surface of the sample. Aging produces a microstructure of uniform distribution of precipitates in aluminum matrix. This condition creates inhomogeneity on a microscopic scale. The precipitates are noble and promote anodic dissolution of the matrix. The TMAZ shows a poor pitting corrosion resistance and general corrosion resistance (Figs 14 &15 and Table 6) as compared to base metal. During FSW process this region heats up to a temperature just below the solutionizing temperature of the alloy. This results in coarsening of precipitates, as evident by the decrease in the hardness. The slightly more negative critical pitting potential of TMAZ than base metal can be attributed to the residual stresses induced during the process of FSW. The microstructures of pitted surfaces weld joint is shown in Fig. 16. It reveals clearly that pit density of weld region is much less than that of base metal and TMAZ. Hence it can be concluded that pitting resistance of weld region is better than that of base metal and TMAZ. Optical micrographs of pitted surfaces of NZ, TMAZ and HAZ of AA6061 alloy in T4 and T6 conditions are given in Figs. 17-19. These shows that the pitting density is low in T4 alloy than T6 alloy. Hence it can be concluded that naturally aged alloy (T4) friction stir welds have better corrosion resistance than artificially aged (T6) friction stir welds.

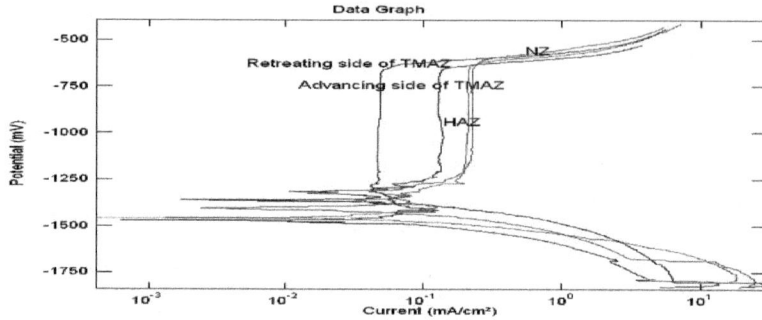

Fig. 14 Potentiodynamic polarization curves of NZ,TMAZ and HAZ in T6 condition.

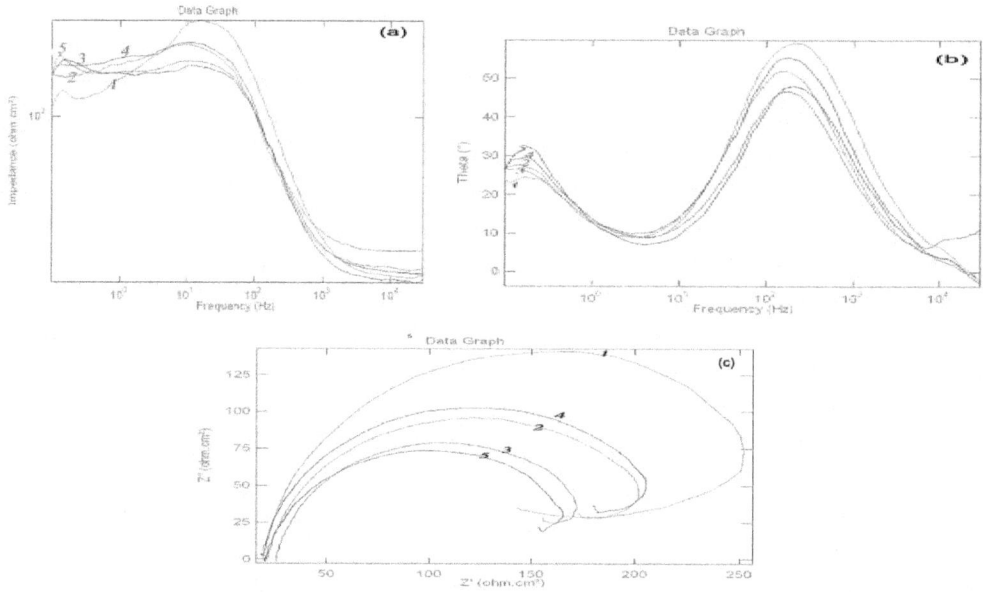

Fig. 15 Bode and Niquist plots of NZ,TMAZ,HAZ and BM in FSW AA6061 T6 alloy
(a) Bode plots   (b) Theta plots (c) Niquist plots and (1) Nugget zone
(2) Advancing side of TMAZ (3) HAZ   (4) Base metal (5) Retreating side of TMAZ.

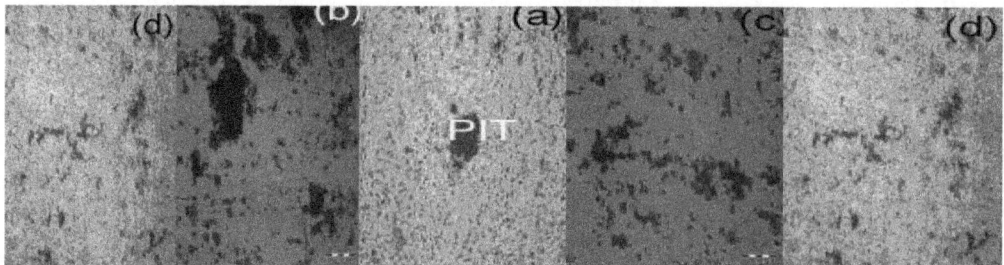

Fig. 16 Post corrosion Optical micrographs of cross section of FSW AA6061 T4 alloy at
(a) Nugget Zone (b) Retreating side of TMAZ (c) Advancing side of TMAZ (d) HAZ.

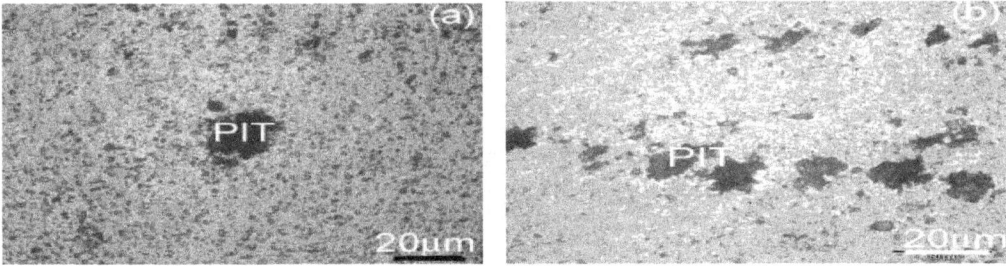

Fig. 17 Optical microphotographs of NZ in FSW AA6061 alloy after corrosion
(a) T4 condition (b) T6 condition.

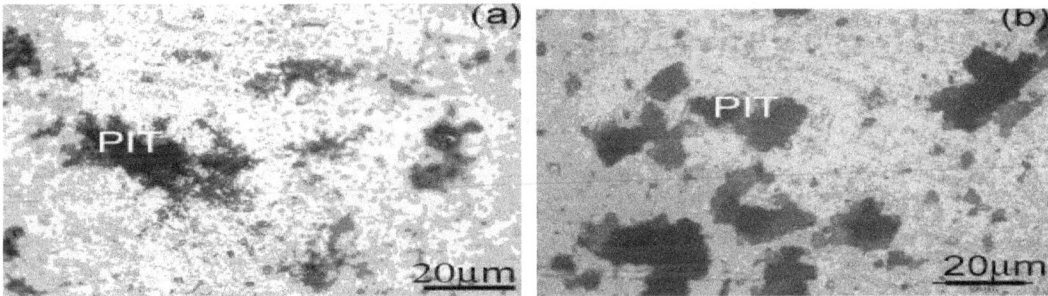

Fig. 18 Optical micrographs of TMAZ in FSW AA6061 alloy after corrosion
(a) T4 condition (b) T6 condition.

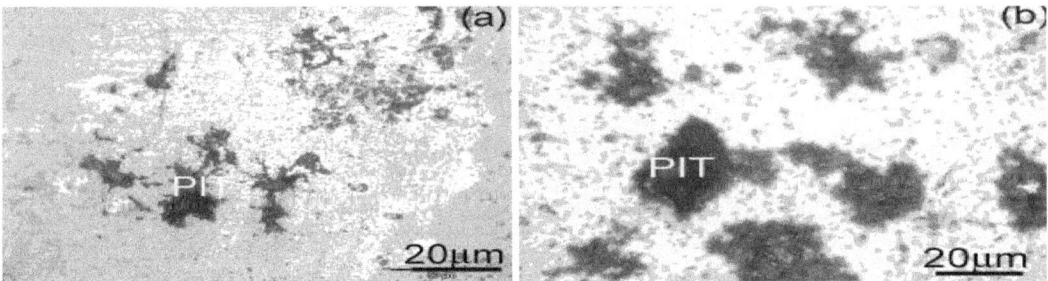

Fig. 19 Optical microphotographs of HAZ in FSW AA6061 alloy after corrosion
(a) T4 condition (b) T6 condition

Table 3 Composition (Wt%) of particles in base metal (SEM-EDX)
(a) AA6061-T6 Base Metal

Position	Mg	Si	Fe	Mn	Cu
P1	5.30	28.51	0.28	0.02	3.38
P2	0.26	49.81	0.28	----	3.38
P3	0.33	10.61	0.22	0.33	1.88

345

(b) AA6061-T4 Base Metal

Position	Mg	Si	Fe	Mn	Cu
P1	0.1738	12.98	0.01	0.20	0.83
P2	22.8	30.34	0.23	0.15	0.39
P3	0.14	1.31	0.18	0.11	0.55

Table 4 Composition (in Wt%) of particles in TMAZs (SEM-EDX) in AA6061 welds
a) AA6061-T6-FSW

Position	Mg	Cu	Fe	Mn	Si	Al
P1(Precipitate)	2.05	6.65	1.24	2.55	0.36	86.99
P2 (Matrix)	2.03	0.26	0.41	0.51	0.20	95.71

b) AA6061-T4-FSW

Position	Mg	Cu	Fe	Mn	Si	Al
P1(Precipitate)	2.01	2.24	0.32	0.52	0.87	93.88
P2 (Matrix)	1.73	0.29	0.16	0.39	0.64	95.37

Table 5 Epit and impedance values of base metal AA6061 alloy

Prior thermal temper values(Ohm.cm^2)	Epit values(mV)	Z
T4	-613	198.0
T6	-635	169.5

Table 6 Epit values of NZ, TMAZ, and HAZ of FSW AA6061 alloy

Zone	T4 (mV)	T6 (mV)
Nugget Zone	-581	-595
Advancing side of TMAZ	-625	-642
Retreating side of TMAZ	-637	-688
Heat Affected Zone	-630	-658

Table 7 Impedance values of NZ, TMAZ, and HAZ of FSW AA6061 alloy

Zone	T4 (ohm.cm^2)	T6 (ohm.cm^2)
Nugget Zone	358.4	248.0
Advancing side of TMAZ	182.7	156.4
Retreating side of TMAZ	153.6	130.4
Heat Affected Zone	169.7	143.7

4. Conclusions

1. Microstructure in nugget zone has a recrystallized, fine equi-axed grain structure of the order of 4-5 microns.

2. TMAZ is the transition zone between base metal and the nugget zone, characterized by a highly deformed structure. There is no recrystallization in this region. The orientation is along thickness direction which is similar to rolling direction in parent metal

3. Microstructures shown that precipitation of eutectics in TMAZ of friction stir welded AA6061 alloy are less in T4 temper than T6 temper.

4. Nugget zone exhibits better corrosion resistance than base metal. The higher pitting corrosion resistance of the nugget zone can be attributed to the partial elimination of inhomogeneity of microstructure in this region.

5. The TMAZ showed poor pitting corrosion resistance when compared to base metal. This may be due to coarsening of precipitates during FS welding as this region heats up to a temperature just below the solutionizing temperature of the alloy.

6. It is observed that advancing side of TMAZ exhibited more pitting corrosion resistance than that of retreating side. This is attributed to higher temperatures generated in advancing side of TMAZ than retreating side.

7. Corrosion resistance of friction stir welded T4 alloy is more than that of T6.

References

1. Thomas W H, Threadgill P L and Nicholas E D: Science and Technology of Welding and Joining, Vol. 4, 1999, p. 365 .

2. Rhodes C G, Mahoney M W, Bingel W H, SpurlingRA and Bampton C C,Scripta Materialia, Vol. 36, 1997, p. 69.

3. Dawes C J and Thomas W M, Welding Journal, Vol. 75, 1996, p. 41.

4. Flores, Olga VK, Christine MLE, Brown D P, Brook M, and McClure J C, *Scripta* Materialia, Vol. 38, 1998, p. 703.

5. MurrL E, Li Y, Elizabeth A, Flores R D and McClure J C, J. of Materials Processing and Manufacturing Science, Vol. 7, 1998, p. 145.

6. Liu G, Murr L E, Niou C S, McClure J C, Vegaf R, Scripta Materialia, Vol. 37, 1997, p. 355.

7. K. Srinivasa Rao and K. Prasad Rao, Materials Science and Technology, Vol. 21, No. 10, 2005, p. 1199.

8.   C. Huang and S. Kou, Weld. Journal, Vol. 79, 2000, p. 113.

9.   A.J.Davenport, R.Ambat, M.Jariyaboon, P.C.Morgan, D.Price, A.Wescott and S.Williams, J. of Corrosion Science and Engineering, Vol. 6, p. 023.

10.   F.Hannour, A.J Davenport and M.Strangwood, Proc. on the II Int. Symposium on Friction Stir Welding, 26-28 June 2000, Gothenburg, Sweden,.

11.   G.Billias, R.Braun, C.Dalle Donne, G.Staniek and W.A.Kaysser, I Int. Symposium on Friction stir Welding, 1999, TWI, UK, PP1.

12.   J.B.Lumsben,M.W.Mahoney, G.Pollock and C.G.Rhodes, Corrosion, Vol. 55, 1999, p. 127.

13.   G.S.Frankel and Z.Xia, Corrosion, Vol. 55, 1999, p. 139.

14.   Ram Ambat, M. Jariaboon, A.J.Davenport, S.W.Williams, D.A.Price and A.Wescott, Proc. of the15[th] Int. Corrosion Congress, Granada, Spain, September 22-29,2002.

15.   Szklarska,Z and Smialowska, *Pitting corrosion of aluminium*, Corrosion Science, Vol. 41, 1992, p. 1743.

Processing and Characterization of Nanostructured Ni-Al Coating on Superalloy for High Temperature applications

Atikur Rahman[a], R.Ambardar[a], R.Jayaganthan[b,*], S.prakash[b], V. Chawla[b], R. Chandra[c]

[a] Department of Metallurgical Engineering, National Institute of Technology Hazratbal, Srinagar, India
[b] Department of Metallurgical and Materials Engineering, Indian Institute of Technology Roorkee, Roorkee
[c] Institute Instrumentation Centre, Indian Institute of Technology Roorkee, Roorkee
*Corresponding author: rjayafmt@iitr.ernet.in, atikurrhmn@gmail.com

Abstract *The high temperature oxidation behavior of magnetron sputtered Ni-Al coating on the superalloy substrate has been studied in the present work. The micro structural feature of the coating was characterized by FE-SEM and AFM. XRD was used to identify the formation of different phases in Ni-Al coating. Thermo gravimetric technique was used to investigate the oxidation behaviour of the coating, in air at 900^oC. The growth kinetics of oxide layers was predicted from the weight changes of the coating samples measured during oxidation. It was found that the corrosion rate of nanostructured Ni-Al coated superalloy was lower than that of the uncoated superalloy due to the formation of continuous, dense, adherent and protective oxide scale over the surface of the coating. The morphological features and phases of the corroded coating were characterized by FE-SEM and XRD, respectively to elucidate the mechanism of high temperature oxidation. A continuous thin layer of protective oxide films such as NiO and Al_2O_3 has formed over the Ni-Al coating exposed to air at high temperature, 900^oC.*

Keywords: *Ni-Al nanostructured coating, Microstructural Characterization, RF- magnetron Sputtering, High Temperature Oxidation*

1. Introduction

Superalloys find extensive use in gas turbines, especially for fabricating hot section components to provide superior strength and creep resistance in high temperature environment. During operation, blades and vanes of gas turbines are subjected to severe thermal cycling and mechanical loads. In addition, they are also attacked chemically by both high temperature oxidation and hot corrosion. It is impossible to impart both high temperature strength and high temperature oxidation resistance to the superalloys simultaneously along with ease of manufacturing. It is possible only with the composite materials to meet such requirements, where the base material provides the necessary mechanical strength and the coating provides the protection against wear, erosion- corrosion, and oxidation [1-3]. Therefore, protective superalloy coating is currently used on superalloys in energy conversion and gas turbines to protect their surfaces from oxidation and hot corrosion. However, to realize the enhanced efficiency, which is unattainable with the conventional coatings, of gas turbine engines, nanostructured coatings with superior performance in terms of high temperature oxidation resistance and thermo-mechanical fatigue life are required. It is well known in the literature that the nanostructured coatings provide a better corrosion resistance at high temperature due to the formation of continuous protective layer assisted by the enhanced diffusivity of atoms in the coating [4-9].

Nanostructured Ni-Al coating is one of the potential coating could be exploited for providing high temperature oxidation protection to the superalloys. The B2 intermetallic compound (β-NiAl), possesses many attractive properties such as low density (approx. 5.9gm/cm^3) as compared to MCrAlY base coating, high melting point (approx. 1995 K), good oxidation resistance and metal like electrical and thermal conductivity [10-12]. Hence, thin film materials based on β-NiAl have been used for a wide variety of engineering applications such as under layers in magnetic recording media and high temperature protective coatings [13-14]. For instance, NiAl is the basis of a family of oxidation and corrosion-resistant coatings that have been used on Ni-based and Co-based superalloys. However, the high temperature oxidation behavior of nanostructured Ni-Al coating on Ni-based super alloy substrate is scarce in the literature. Therefore, the present work has been focused to study the oxidation behaviour of magnetron sputtered Ni-Al coating on the Superni 718 superalloy. The Ni-Al coating was deposited by DC/RF magnetron sputtering. The micro structural features of the coating were characterized by FE-SEM/EDS and AFM. XRD was used to identify the different phases in the Ni-Al coating. Thermo gravimetric technique was used to study the kinetics of high temperature oxidation behaviour of the Ni-Al coating as well as bare superalloy substrate in air at 900°C. The morphology of corroded products of coating and bare superalloy substrate was characterized by FE-SEM. The improved corrosion resistance of nanostructured Ni-Al coating exposed to high temperature environment is discussed in the present work.

2. Experimental

2.1. Deposition and characterizations of Ni-Al coating on Superni-718

Ni-based super alloy namely Superni- 718 has been chosen to deposit Ni-Al coating in the present work. The super alloy was procured from Mishra Dhatu Nigam Limited, Hyderabad, India in annealed and cold rolled sheet form and its chemical composition is shown in Table 1. Each specimen measuring approximately: 18mm (length) X15mm (width) X3mm (thickness) were cut from the rolled sheet and polished by using 220, 320, 1/0,2/0,3/0,4/0 grid SiC emery papers. Commercially available Ni Targets (99.99% pure) with its dimension 2" diameter, 1.0 mm thickness was used. Similarly, commercially available Al Target (99.99% pure) with 2" dia and 5.0 mm thickness was used. Before coating all samples were cleaned in acetone, ethanol and deionized water. The process parameters used in RF magnetron sputtering are shown in Table 2.

Table 1 Chemical composition of the super alloy used in the study

Midhani grade	Chemical composition (wt %)												
	Fe	Ni	Cr	Ti	Al	Mo	Mn	Si	Co	Nb	P	C	S
Superni-718	19.8	Bal	17.6	0.96	0.53	3.23	0.02	0.03	0.01	4.91	0.005	0.02	0.007

Table 2

Sputtering Condition	Parameters
Target	Ni (99.99% pure, 2" dia & 1mm thickness), Al (99.99% pure, 2" dia & 5mm thickness)
Base pressure	3×10^{-6} Torr
Sputtering gas pressure	10 mTorr (Argon)
Sputtering power	75 W / 85 W (Ni/Al target)
Substrate	Superni-718
Substrate temperature	350 ^{0} C
Deposition time	1 hr

The Ni-Al coating deposited on Superni 718 were subjected to XRD, FE-SEM, and AFM analysis.

2.2. High temperature oxidation studies of Superni-718 substrate and Ni-Al coating

The bare and coated superalloy samples were kept in the alumina boats and then inserted inside SiC tube furnace. The bare substrate specimens were mirror-polished before oxidation. Cyclic oxidation studies were performed in air with each cycle consisting of 1 hr of heating at 900^0C followed by 30 minute of cooling at room temperature for up to 50 cycles. The purpose of imposed cyclic loading was to create the aggressive conditions, similar to the actual conditions, for corrosion testing. The weight changes were monitored, using an electronic balance Model CB-120(Contech, Mumbai, India) with a sensitivity of 1 mg during each cycle for both bare and coated samples and the data were used to calculate the corrosion rate. As spalled scale if any was also included in weight change measurements. The kinetics of oxidation was determined from the weight change measurements. After oxidation studies, the corroded samples were analyzed by using XRD and FE-SEM/ EDAX techniques.

3.Results and Discussions

3.1 Visual observation

The bare alloy upon exposure to high temperature oxidation showed a brownish grey color after the completion of 5th cycles and subsequently it transformed into dark grey color after the completion of 20th cycles. The coated sample showed a blue color and brown color after 5th cycles and 20th cycles respectively. The appearance of different color over the samples indicates the formation of different oxide scale on the materials.

3.2 Microstructural characterization of Ni-Al coating

Field emission scanning electron microscopy graph with EDS at different points on the surface of Ni-Al coating at 350°C were taken and they are shown in Fig. 1. A very dense coating with the equiaxed grain structure is evident from the micrograph.

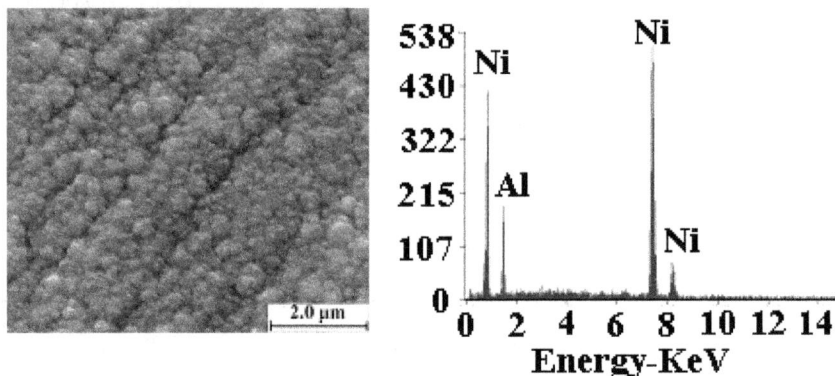

Fig. 1 FE-SEM micrograph with EDS of a deposited Ni-Al coating on superni-718 substrate.

The formation of high dense coating is due to the influence of the deposition temperature and power used during sputtering.

3.3. XRD and AFM Analysis of the coating

The XRD pattern of the as deposited coating on Superni 718 is shown in Fig. 2 (a). The formation of phases such as Ni_3Al, NiAl, and Al_4Ni_3 are observed from this figure. The XRD results for the corroded products of bare and coated samples are shown in Fig. 2 (b). The oxide scales such as Cr_2O_3, NiO, Fe_2O_3 are formed on the bare superalloy subjected to high temperature air oxidation at 900°C. The corroded products of Ni-Al coating showed the presence of Cr_2O_3, NiO, Al_2O_3, , Fe_2O_3, $NiCr_2O_4$ as observed from the Fig. 2(b).

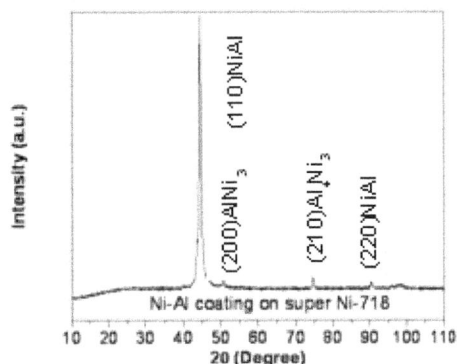

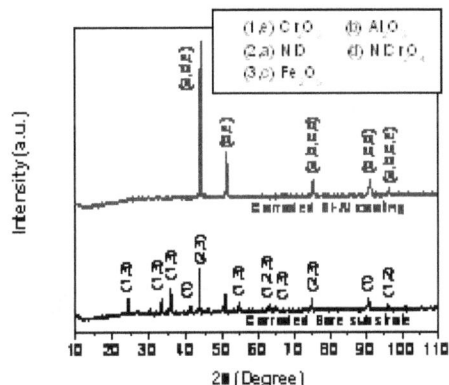

Fig.2. (a) XRD pattern of Ni-Al coating on Bare Superni 718

Fig.2. (b) XRD pattern of corroded and Ni-Al coating on Superni 718.

The 2D and 3D AFM images of the Ni-Al coating are shown in Fig. 3a and 3b. The Ni-Al coating exhibits equiaxed and elongated grain morphology as evident from this figure. The average particle size and roughness of the coating were found to be 10.44 nm and 15.19 nm respectively. The smooth surface of the coating is essential for the formation of adherent and protective oxide scales when it is exposed to high temperature environments.

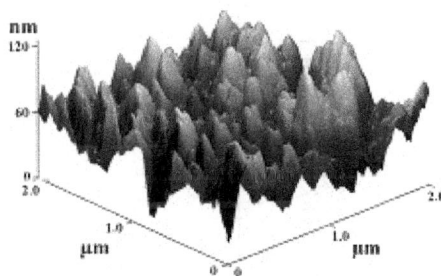

(a)   (b)

Fig.3 (a) AFM images of Ni-Al coating (2D), (b) AFM of Ni-Al coating (3D)

3.4 Oxidation Studies of Bare alloy and Ni-Al coating in Air

The weight gain per unit area versus number of cycle's plots for the bare substrate and Ni-Al coating on Superni-718 subjected to high temperature air oxidation studies is shown in Fig. 4a. It is shown that under cyclic conditions, the bare substrate is more prone to corrosion/oxidation attack. The weight changes data as function of number of cycles are analyzed to obtain the nature of its kinetics. The parabolic rate constant is calculated from the slope of the linear regression fitted line from (weight changes/area) ^{2} versus number of cycles shown in Fig. 4b. There is a visible deviation from the parabolic rate law in case of bare Superni-718 substrate up to 50 cycles. The weight gain data for the bare samples shows that it is prone for continuous oxidation.

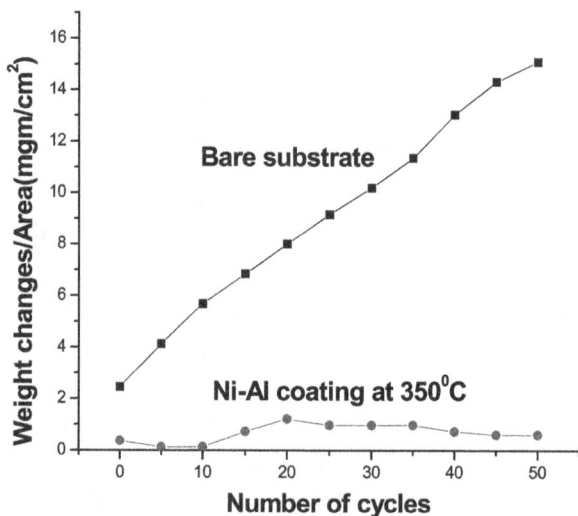

Fig.4 (a) Weight changes/area (mgm/cm^2) versus number of cycles for bare and Ni-Al coating Oxidized in air at 900°C.

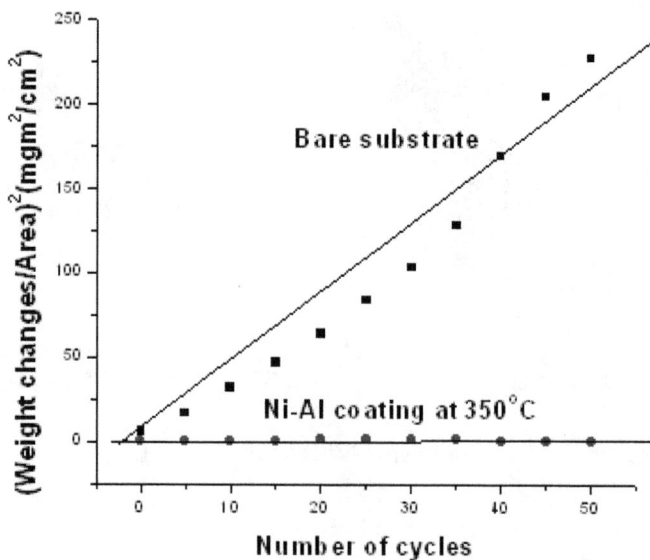

Fig 4 (b) (Weight changes/area) ^{2} mgm^2/cm^4 versus number of cycles for the bare and Ni-Al coating Oxidized in air at 900°C.

The higher weight gain of bare alloy after 30 cycles is due to spallation of initial oxide scales and the continuous exposure of fresh surface of the materials during oxidation. The evaporation of volatile impurities in the coating, during high temperature exposure, result in the formation of pores, which might also contribute to the enhanced reaction between the bare sample and oxygen. Ni-Al coated Superni-718 has shown a minimum weight gain as compared to bare sample. The calculated K_P for the Ni-Al coated superalloy is found to be 0.0222, which is very less as compared to the K_P value of 12.51 obtained for the bare super alloy. Thus, Ni-Al coating has provided a better protection to the Superni-718 during high temperature oxidation at 900^0 C in air environment.

3.5 Surface Scale analysis

SEM micrographs (LEO 534) at the selected points of interest of corroded bare substrate Superni-718 are shown in Figs 5 (a-d). The scale formed on bare Superni-718 has spalled and the cracks were formed as shown in Fig. 5(c) and (d). The large numbers of small pores on the surface of the specimen are observed from Fig 5(b).

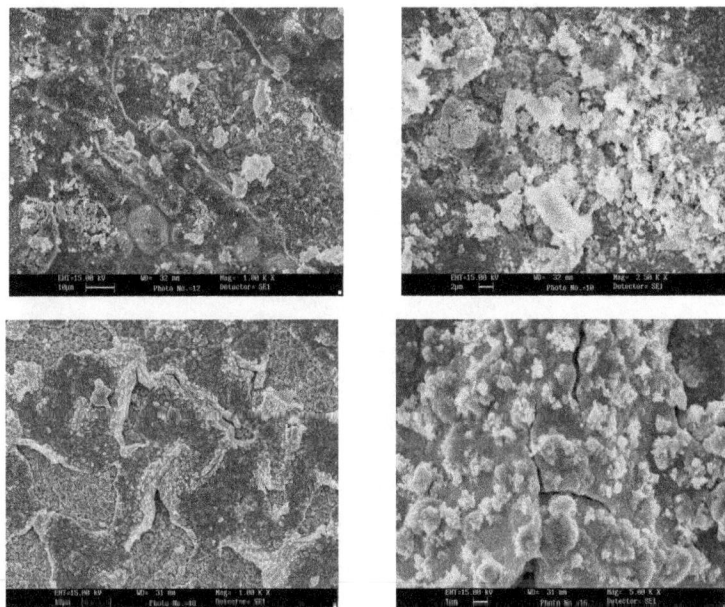

Fig. 5 (a-d): SEM micrographs of corroded bare Superni 718.

The FE-SEM micrograph of corroded coating at the selected point of interest is shown in Figs 6. along with its corresponding EDS results. The oxide scale such as NiO, Al_2O_3 Cr_2O_3 and Fe_2O_3 have formed on the surface of the coating as evident from the Fig. 6 at the selected spot on the coating. The formation of Cr_2O_3 and Fe_2O_3 indicate the interdiffusion of the substrate elements, Fe and Cr into the coating. The scale formed on the surface of the coating is adherent and its spallation has not occurred, indicating the protective nature of the coating obeying parabolic rate of oxide growth. The nanosized grains found in the Ni-Al coating helps in the enhanced reactivity of Al and Ni with oxygen for the formation of continuous layer of their corresponding oxide, which is essential for providing an adequate protection to the Superni 718 in the high temperature environment.

Element	Wt %	At %
OK	49.70	52.47
AlK	49.40	41.88
CrK	04.33	01.90
FeK	00.60	00.25
NiK	08.98	03.50

Element	Wt %	At %
Al_2O_3	80.24	78.12
Cr_2O_3	06.50	04.25
FeO	00.30	00.41
NiO	12.96	17.22

Fig. 6 FE-SEM/EDX Micrographs of corroded Ni-Al coating on Superni 718 .

4. Conclusions

The oxidation behaviour of magnetron sputtered Ni-Al coating on the super alloy substrate exposed to air environment at 900°C under cyclic conditions has been investigated in the present work. The oxidation kinetics of Ni-Al coating was compared with that of the bare Superni 718 and found that the parabolic rate constant was very less for the former, which is due to the formation of adherent and protective oxide scale over the surface of the coating. The continous mass gain of bare superni 718 is due to the spallation of initial oxide scales, renewal of fresh surface of the materials, and the pores resulting from the evaporation of volatile impurities in the alloy at high temperature. The XRD analysis of the corroded products of the Ni-Al coating indicated the formation of protective oxide scale such as Cr_2O_3, NiO, Al_2O_3, Fe_2O_3, $NiCr_2O_4$. The morphological features of corroded coating and bare samples were characterized by FE-SEM and it was shown that the scale formed on the surface of coatings were adherent and no spallation has occurred unlike in bare samples.

Acknowledgments

One of the authors, Dr. R. Jayaganthan would like to thank CSIR, NEW DELHI, INDIA for their financial support to this work.

References

1. B.S. Sidhu, D. Puri, S. Prakash, Mater. Sci. Technol., Vol. A, No. 368, 2004, p. 149.

2   M.H. Li, X.F. Sun, J.G. Li, Z.Y. Zhang, T. Jin, H.R. Guan, Z.Q. Hu, Oxid. Met. 59 (5/6),   2003, p. 591.

3. P.S. Sidky, M.G. Hocking, Br. Corr. J., Vol. 34, No. 3, 1999, p. 171.

4. S.Geng, F. Wang, S. Zhang, Surf. Coat.Technol., No. 167, 2003, p. 212.

5. T.J.Nijdam, L.P.H.Jeurgens, J.H.chen, W.G.sloaf, oxidation of Metals, Vol.64, No., 516 Dec.2005.

6. X.Ren, F.Wang, Surf.Coat.Technol.No. 201, 2006, p. 30.

7. G.Chen, H.lov, Corr.Sci. No. 42, 2000, p. 1185.

8. X.Peng, F. Wang, Corr. Sci. No. 45, 2003, p. 2293.

9. L.Ajdelsztajn, J.A.Picas, George E.Kim, F.L.Bastian, J. Schoenung, V.Provenzano, Mater. Sci. Eng. Vol. A, No. 338, 2002, p. 33.

10. R. Darolia, JOM Vol. 43, 1991, p. 44.

11. D.B. Miracle, Acta Metall. Mater. Vol. 41, 1993, p. 649.

12. R.D. Noebe, R.R. Bowman, M.V. Nathal, Int. Mater. Rev. No. 38, 1993, p. 193.

13. T.J.S. Anand, H.P. Ng, A.H.W. Ngan, X.K. Meng, Thin Solid Films No. 441, 2003, p. 298.

14. B. Ning, M.L. Weaver, Surf. Coat. Technol. Vol. 11, 2004, p. 177.

Effect of Thermal Oxidation on Wear Behaviour of Titanium Alloys used for Automobile and other Structural Applications

G. V. S. Nageswara Rao* and K. Chandramouli
Department of Metallurgical and Materials Engineering
National Institute of Technology, Warangal, Andhrapradesh, India
* Email: nr_gvs@yahoo.com

Abstract: *Thermal oxidation (TO) is a recent surface modification process developed for improving wear resistance of titanium and its alloys. This TO technique produces a thin, hard, wear resistant coating on the surface of the titanium alloy, thus significantly improving the tribological properties of the titanium alloys. The objective of the present work is to study the wear behaviour of titanium based alloys in untreated and TO treated conditions and compare them. In the present study, TO treatments were conducted on the polished samples of commercial pure titanium, Ti-6Al-4V and T-10V-2Fe-3Al using air at 6500C for 6, 12 and 24 hours in a fluidized bed furnace. After each cycle time, the oxidized samples were observed for weight gain and the changes in surface roughness values (Ra values) using a MITUTOYO surface roughness tester. X-ray diffraction analysis was carried out on untreated and TO treated specimens using Philips X-ray diffractometer. Microsturctural examinations were conducted using an optical microscope and a Leo 440i scanning electron microscope. Hardness tests were conducted using a Shimadzu Micro Hardness Tester and dry sliding wear tests (as per ASTM G99 standard) were conducted using a DUCOM pin-on-disc wear testing machine. After each wear test, the wear tracks developed on the untreated and oxidized pins as well as on rotating disc were observed by light optical microscope. The results indicate that TO treatment produces a rutilebased surface layer which significantly enhances the surface hardness and wear resistance of these alloys. Based on the experimental results, in conjunction with systematic analyses, the wear-reduction mechanisms involved in the TO-treated material are discussed. It was found that the significantly reduced tendency to adhesive wear has contributed to the enhanced wear resistance of the TO-treated titanium alloys.*

1. Introduction

Titanium and its alloys have been used for many years due to their excellent strength to weight ratio and corrosion resistance [1, 2]. Unfortunately titanium is also known for its poor tribological characteristics, particularly in sliding-contact situations [3, 4]. Titanium components fail due to their inherent tendency to seize and gall when exposed to sliding-contact situations [5]. Thus the use of titanium components within the systems has been limited, as there are components that involve relative sliding, such as nuts, valves, and pipe connectors. Routine disassembly and reassembly of these components can result in failure of the titanium parts. Hence, it is very important that the tribological response of titanium be significantly improved for its future exploitation. Once this tribological barrier has been successfully overcome, titanium alloys can realize their full potential as an engineering material and no longer be restricted to structural applications only.

One important factor attributed to titanium's poor tribological performance is its electron characteristic property. It has been found the lower the value or the percentage of d-bond character that a metal possesses; the more active is its surface [6, 7]. The more active the metal the higher is the coefficient of friction. Titanium has a very low value of d-bond character; hence it is very active.

The frictional characteristic properties of titanium are also due to its crystal structure. Alpha titanium is hexagonal. Most hexagonal metals exhibits good frictional properties, but titanium has a c/a axial ratio of 1.588, which is less than ideal for closest packing. This situation tends to make the basal plane less favorable for slip and allows prismatic slip to occur. This increases the number of operational slip systems and promotes junction growth in a contact situation, which requires slip on different planes [8, 9]. Junction growth is the growth of asperity junctions through plastic deformations, formed by 'real' surfaces (each with asperities) come in direct sliding contact. Junction growth of contacting metal systems is promoted by high ductility values [10–12].

Another important factor of titanium's frictional properties is its poor response to lubrication. An oil or oil-type films on titanium is easily broken down because of poor adherence of the oil to the metal surface [13, 14]. Additionally, it was found that the only lubricants to give any decrease in friction in bare titanium were halogenated compounds, especially those containing fluorine. This was thought to be due to chemical attack of the surface by the lubricant, which then held in itself in place. Unfortunately, the halogenated lubricants only provided a thin carpet of molecules on the surface; thus, the lubricant was easily penetrated.

All of these aforementioned factors play a great role in the overall properties of titanium. Together, these factors produce a poor tribological situation, which is well documented [3-5]. As a result, the tribological situation can be improved effectively by modifying the nature of the contact surfaces. Hence, the discipline of surface engineering must evolve techniques that allow titanium to fulfill its engineering potential. Also to withstand the high stresses encountered in engineering components such as bearings and gears, deep case hardening is necessary. There is no hardening reaction in titanium alloys comparable to the martensitic hardening transformations in steels; so, to develop a cost effective surface engineering technique for deep case hardening of titanium alloys is a major challenge to the surface engineers for both the scientific and technological view point. Many techniques have been applied and developed for titanium, such as laser–surface alloying treatments, thermal spray or PVD coating processes [15, 16]. Unfortunately, problems such as distortion or poor adhesion have prevented these processes from unlocking the full potential of titanium.

Recently, a novel surface engineering process called **T**hermal **O**xidation (**TO**) was successfully developed for the Ti-6Al-4V alloy [17, 18]. The **TO** process is a thermo chemical treatment carried out in an atmosphere containing oxygen and nitrogen within the temperature range of 600oC to 650oC. **TO** process can simultaneously produce a thin (≈ 2 ⌠m), hard, rutile oxide layer and develop an oxygen rich case hardened zone (≈ 20 ⌠m). **TO** treatment not only improves surface characteristics of titanium and its alloys but also induces the formation of a thicker oxide layer with increase in temperature and which is accompanied with the dissolution of oxygen beneath it. The development of the **TO** wear-resistant layer will be the first step that aims to develop a titanium gear for the motor sport industry.

In the present work, it is proposed to make an attempt to adapt and develop **TO** process for use on titanium based alloys. The objective of the present work is to study wear behaviour of titanium based alloys in untreated and surface modified conditions and compare them. This work also involves characterization studies of modified surface layers by means of microscopic examination (optical and SEM) and X-ray diffraction analysis. Mechanical performance of the thermally oxidized titanium alloys was studied by micro hardness measurements and sliding wear tests.

2. Materials used

The materials selected are commercial pure titanium (CP-Ti), Ti–6Al–4V (Ti64) and Ti-10V- 2Fe-3Al alloys. These materials were selected based on the consideration that Ti-6Al-4V and Ti-10V- 2Fe-3Al are versatile titanium alloys with a combination of good properties and ease of production, and that there is ever increasing interest in improving their tribological properties to meet varying requirements arising from diverse application conditions. Ti-6Al-4V was TITAN 31, which confirms to international specifications of ASTM B 265/B 348 GRADE 5. This alloy was purchased from MIDHANI, HYDERABAD in the form of hot rolled bars of diameter 12mm. It has high strength to weight ratio and high specific strength at elevated temperatures. It as excellent resistance at temperatures up to 520°C along with high fatigue strength and toughness. Ti-6Al-4V is also known for its excellent corrosion resistance, but possesses relatively poor wear resistance. Ti-10V-2Fe-3Al alloy was supplied by Titanium Alloy Group of DMRL, HYDERABAD in hot rolled (at 7600C) and solution treated (7000C and 8500C) condition in cylindrical bars of 8mm diameter. Ti-10V-2Fe-3Al is meta-stable beta titanium (MSBT) alloy, developed to provide weight savings over high-strength steels in forged aircraft undercarriage components, like landing gears of aero-planes, helicopter rotor applications and automobile applications such as springs in gear boxes etc. In spite of its superior mechanical properties, like Ti-6Al-4V, this alloy also suffers from poor resistance to wear. The chemical compositions and initial microstructures of these alloys are presented in Table 1 and figure 1 respectively.

Table 1 Chemical composition of selected Ti-Alloys

Ti-alloy / Element (%wt)	Al	V	Fe	Ti
CP-Ti	-	-	-	99.5
Ti-6Al-4V (Ti64)	6	4	-	Balance
Ti-10V-2Fe-3Al[*]	3	10	2	Balance

[*] Ti-10V-2Fe-3Al alloy was taken in two heat treated conditions
   a) hot rolled at 760^0C and solution treated at 700^0C/WQ – abbreviated as MSBT1
   b) hot rolled at 760^0C and solution treated at 850^0C/WQ – abbreviated as MSBT2

3. Experimental work

The specimens for micro hardness (8mm dia and 10 mm height) and wear test (cylindrical samples, called pins, of 8mm dia and 30mm height) were prepared and polished on 400# SiC paper, washed and degreased in acetone. Thermal oxidation treatments were then conducted on the polished samples using air at 650^0C for 6, 12 and 24 hours in a fluidized bed furnace. Air was supplied continuously both for fluidisation as well as oxidation purpose. Prior to the thermal oxidation, the initial weight and initial surface roughness of the samples were measured to observe the changes in these parameters due to thermal oxidation. Before the thermal oxidation treatment, the furnace is thoroughly cleaned in order to remove the presence of any foreign matter or moisture. Temperature recorder and controller were set to the pre-set temperature. After the temperature was attained to the set point, the samples were loaded into the furnace. Samples were tied with binding wire (mild steel) to facilitate hanging in the furnace.

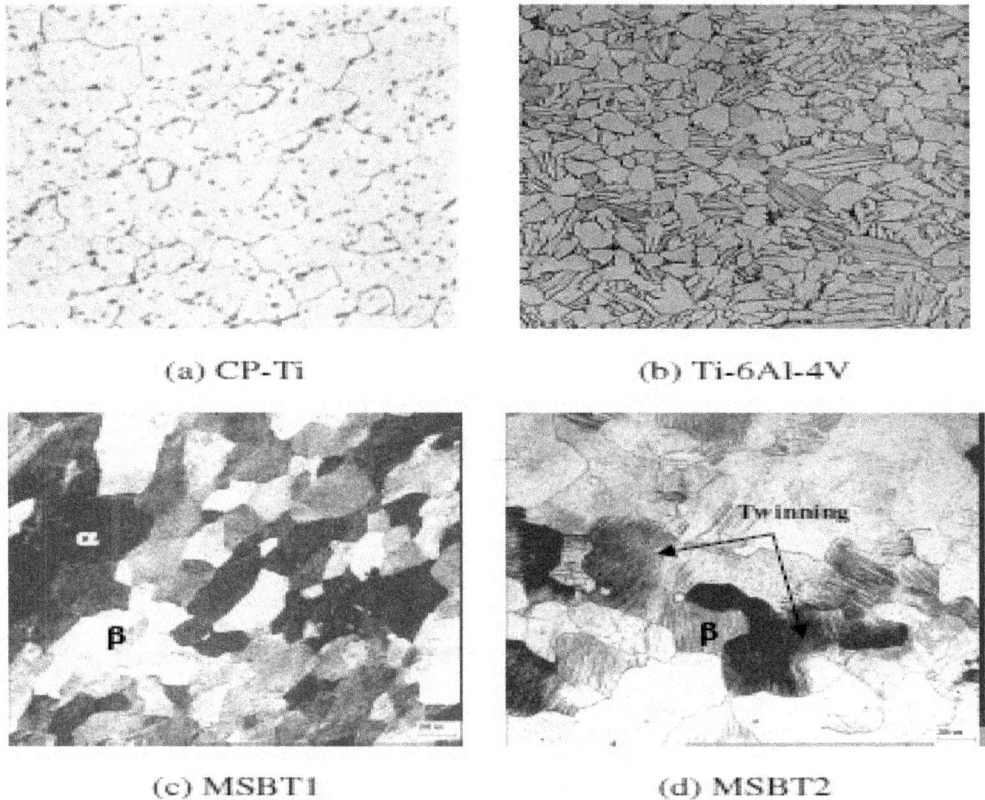

(a) CP-Ti

(b) Ti-6Al-4V

(c) MSBT1

(d) MSBT2

Fig. 1 Initial microstructure of selected alloys (Etching agent: Kroll's reagent).

4. Characterization and Evaluation studies

After each cycle time, the oxidized samples were cooled to room temperature in air and cleaned with acetone. The samples were observed for weight gain using a precision digital balance up to 4 decimal accuracy and the changes in surface roughness values (Ra values) were measured using a MITUTOYO surface roughness tester (Surf Test-SJ 301). X-ray diffraction analysis was carried out on untreated and thermally oxidized specimens using Philips X-ray diffractometer (Cu-kα radiation of wavelength 1.54 A). The rotating crystal method was used for surface phase determination of oxidized samples. The incident beam angle was 200 and the diffraction angle was between 20^0 and 100^0 with a count time of 1 sec. The cross sectioned samples were hot mounted, polished and etched with a solution of 10%HF+5%HNO3 in distilled water for about 15 seconds. Microsturctural examinations were conducted using an optical microscope and a Leo 440i scanning electron microscope. Hardness tests were conducted using a Shimadzu micro hardness tester (Model: HMV- 2000). Hardness measurements were performed on the surfaces of the samples before and after thermal oxidation using 50gf

362

load for 15 seconds. For each specimen at least 5 successive measurements were made and average was taken. Dry sliding wear tests (as per ASTM G99 standard) were conducted using a DUCOM pin-on-disc wear tester (Model: TR-20LE). The disc was made of EN31 steel, hardened to 65 Rc. The pin was fixed against a rotating disc and tightly clamped while ensuring the level of loading arm. Wear, frictional force and coefficient of friction were acquired on line using WINDUCOM-2006 software. The test conditions used in the present study are presented in Table 2. During the tests, wear and friction force were continuously measured by means of a linear variable differential displacement transducer (LVDT) and the data were recorded as a function of normal load and time. The wear characteristics were evaluated by the weight loss of the specimens. Prior to and after each test both pin and disc were cleaned with acetone separately and the pins were weighed before and after each test and weight loss of the samples were measured. After each wear test, the wear tracks developed on the untreated and oxidized pins as well as on steel disc were observed by light optical microscope.

Table 2 Sliding wear test conditions

Parameter	Details
Pin material	Ti64, MSBT1 and MSBT2
Pin diameter	8 mm
Disc material	EN31 steel
Disc diameter	180mm
Disc hardness	65 Rc
Disc speed	200 rpm
Total sliding distance	2 km
Test temperature	Room Temperature
Test condition	Dry condition
Normal load	2.5kgf, 5kgf and 7.5kgf

5.0 Results and Discussions

5.1 Weight gain measurements

The weight gain of the samples measured after thermal oxidation for different times is shown in figure 2. From these results it can be observed that there is a significant and progressive increase in weight of the samples after thermal oxidation. Further, it can be concluded that higher the oxidation time higher is the weight gain. This is because of the availability of fresh oxygen that is supplied continuously during the entire period of oxidation. This oxygen diffuses into the surface of the samples and responsible for the formation of an oxide coating on the surface.

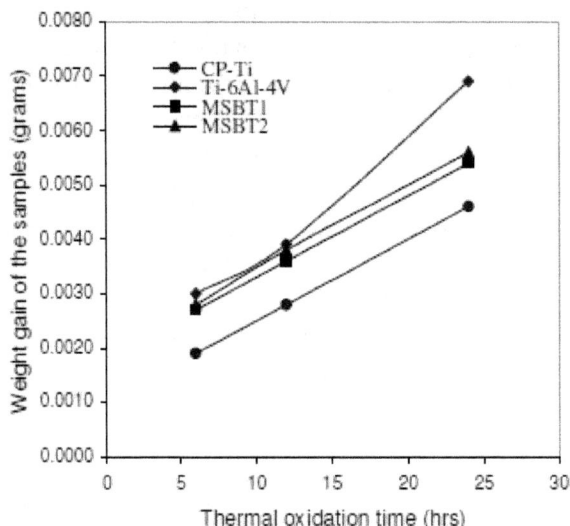

Fig. 2 Thermal oxidation time Vs. Weight gain of the samples.

5.2 Surface roughness measurements

After TO, the surfaces of the Ti alloys were covered with a dark grey colored oxide layer. The surface roughness (Ra values) of the samples, recorded before and after thermal oxidation is presented in Table 3.

Table 3 The surface roughness values of the samples before and after thermal oxidation

Surface roughness (Ra) values							
Cp-Ti		Ti64		MSBT1		MSBT2	
Before TO	After TO	Before TO	After TO	Before TO	After TO	Before TO	After TO
0.29	0.30†	0.22	0.29†	0.20	0.30†	0.22	0.28†
0.27	0.31*	0.24	0.32*	0.22	0.28*	0.22	0.29*
0.31	0.34#	0.20	0.32#	0.24	0.31#	0.24	0.29#
† 6 hrs, * 12 hrs, # 24 hrs							

From these values, it can be observed that there is a considerable increase in surface roughness after thermal oxidation. This can be attributed to the formation of oxide

364

layer, which will form peaks and valleys on the surface. Further, the increase in oxidation time does not influence the surface roughness much. This is because of the rubbing or polishing action of the fluidizing media (alumina) used in the fluidized bed furnace. It also can be observed that there is almost no variation in the surface roughness of the selected materials after thermal oxidation in fluidized bed furnace.

5.3 XRD Analysis

The XRD patterns obtained on Cp-Ti and Ti-6Al-4V samples subjected to thermal oxidation in fluidized bed furnace at 6500C are shown in figure 3. These XRD patterns clearly indicated that the oxidized surface layer is principally comprised of rutile (TiO2). Similar patterns have also been obtained on the samples of MSBT1 and MSBT2 after thermal oxidation in fluidized bed furnace.

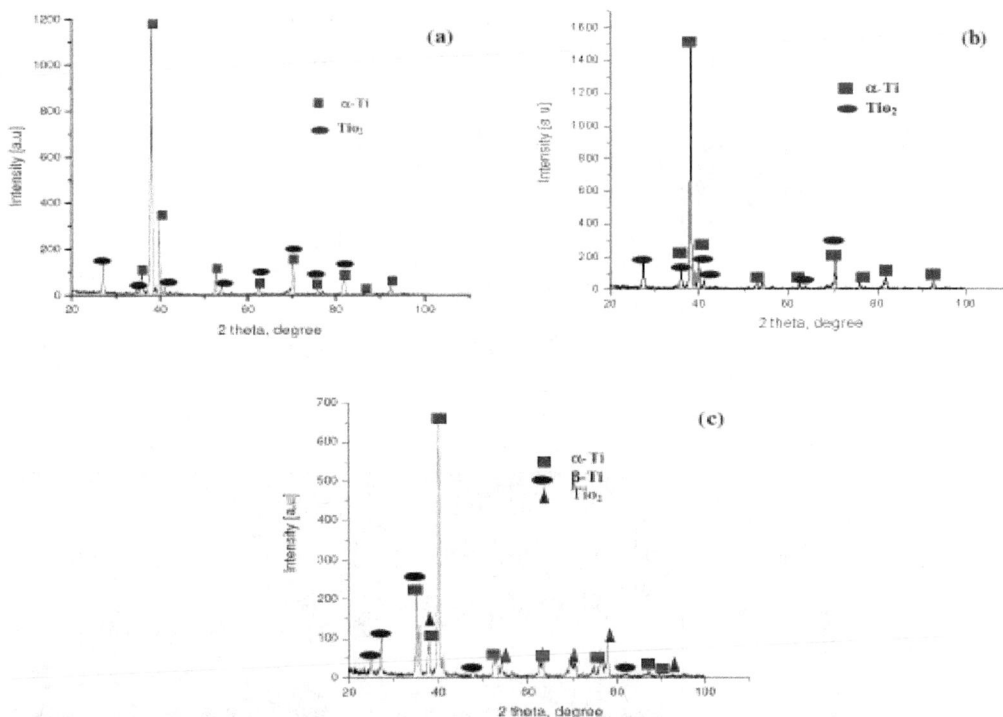

Fig. 3 XRD patterns of titanium samples subjected to thermal oxidation at 650 °C (a) Cp-Ti (6 hrs). (b) Cp-Ti (12 hrs) and (c) Ti-6A1-4V (12 hrs).

5.4 Microstructural analysis

The initial microstructures of the selected titanium alloys are shown in figure 1. The crosssectional optical micrographs of CP-Ti samples after thermal oxidation for 6 hrs, 12 hrs and 24 hrs are given in top and the corresponding SEM images are shown at the bottom of figure 4. The presence of TiO2 layer is clearly visible in these images. However, the oxide layer formed is very thin ($\approx$1-2$\lceil$m thick) and almost uniform in thickness in the samples treated for 6 hrs and hardly seen under optical microscope even at 400X. The corresponding SEM image is shown at the bottom. This higher magnification image reveals the presence of very thin oxide layer ($\approx$1-2$\lceil$m thick). Increasing oxidation time promoted the formation of thick oxide layers. The oxide layers are clearly seen in the samples treated for higher cycle times i.e. 12 and 24 hrs. This oxide layer exhibits good adherence with the substrate material. The energy dispersive x-ray analysis (EDAX) of CP-Ti samples before and after TO is shown in figure 5 and these confirms the presence of oxygen in the surface layers.

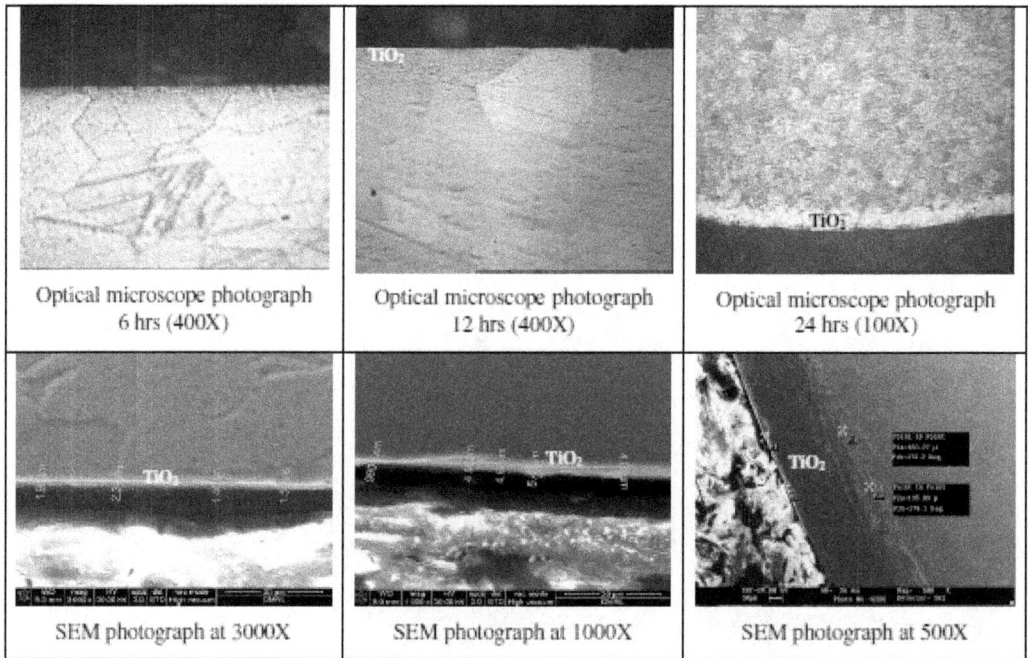

| Optical microscope photograph 6 hrs (400X) | Optical microscope photograph 12 hrs (400X) | Optical microscope photograph 24 hrs (100X) |
| SEM photograph at 3000X | SEM photograph at 1000X | SEM photograph at 500X |

Fig. 4 Microstructures of CP-Ti samples after TO at 6500C at different cycle times.

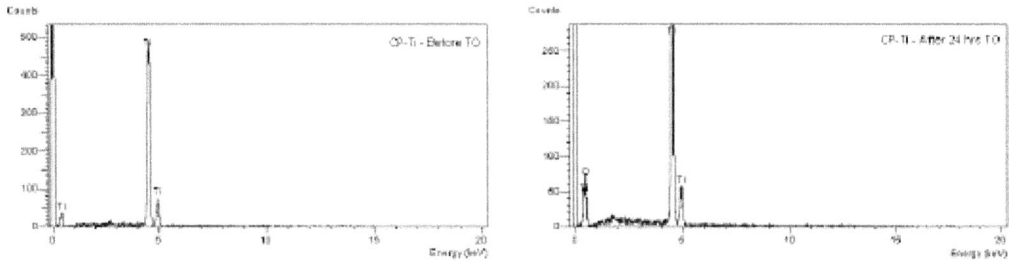

Fig. 5 Energy dispersive x-ray analysis of CP-Ti samples before and after TO.

The microstructures of Ti-6Al-4V specimens after thermal oxidation for 6 hrs and 24 hrs are shown in figure 6 and the corresponding SEM images are presented in figure 7. From these figures it is evident that the surface layers in TO-treated Ti-6Al-4V samples comprise a thin rutile oxide layer (~2-10 ⌠m) and an oxygen diffusion zone (~200 ⌠m) beneath it. The thickness of the oxide layer and oxygen diffusion zones are time-dependent. Increasing oxidation time the thicknesses of these two zones increases. The increase is rapid at higher treatment times. This is in agreement with earlier researches [17, 18].

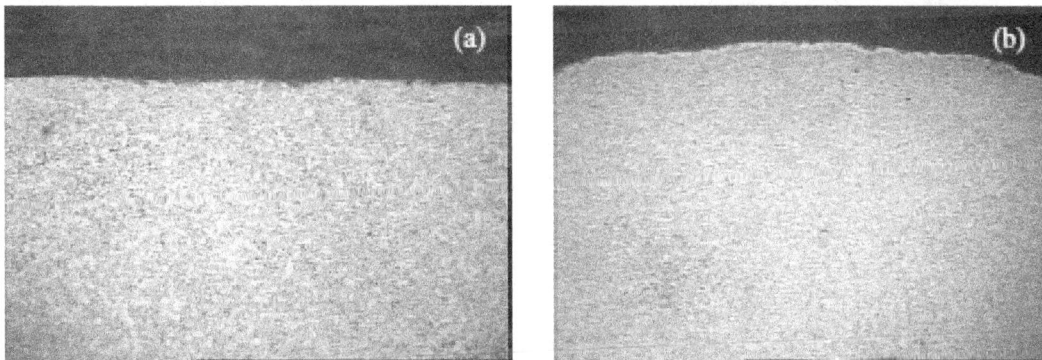

Fig. 6 Optical microscope images of Ti-6A1-4V samples after thermal oxidation at 650 ºC (a) 6 hrs treatment (at 100X), and (b) 24 hrs treatment (at 100X)

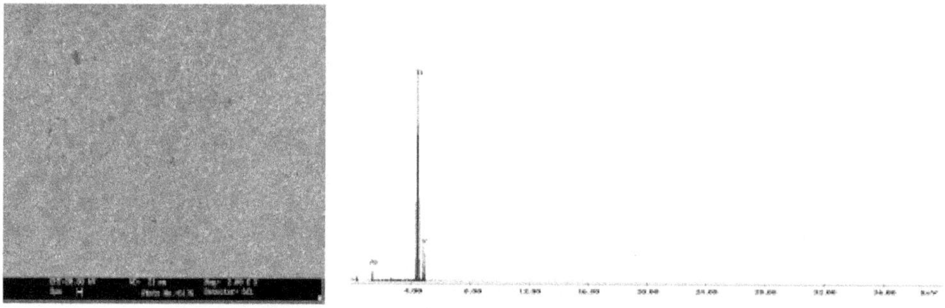

Before thermal oxidation

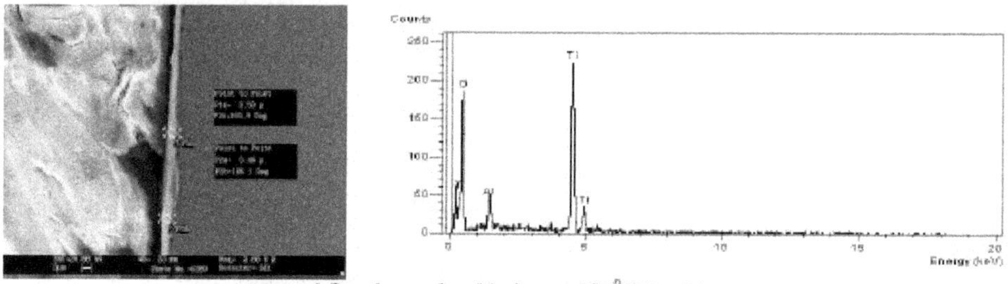

After thermal oxidation at 650⁰C for 6 hrs

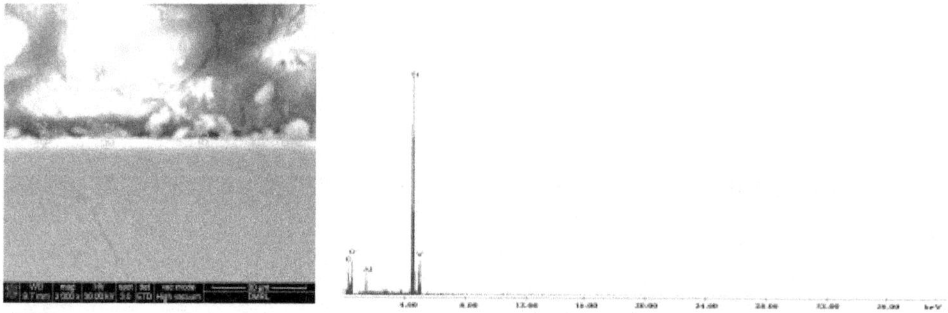

After thermal oxidation at 650⁰C for 12 hrs

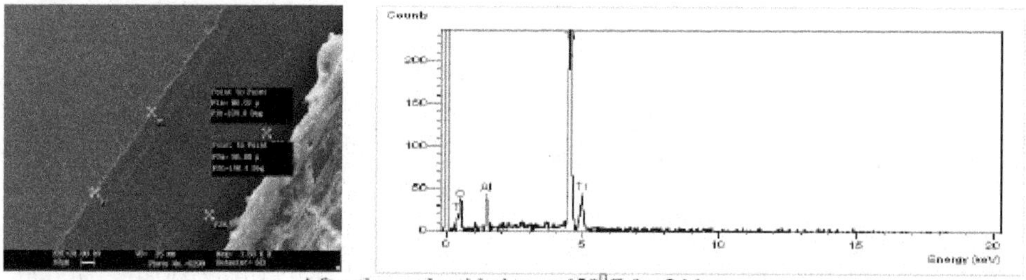

After thermal oxidation at 650⁰C for 24 hrs

SEM   photographs of Ti-6Al-4V   Corresponding elemental analysis (EDAX) of Ti-6Al-4V samples
samples

Fig. 7 SEM images of Ti-6Al-4V samples before and after thermal oxidation.

The micro hardness values measured on the surfaces of Cp-Ti, Ti-6Al-4V, MSBT1 and MSBT2 alloys before and after thermal oxidation are presented Table 4. The hardness depth profiles recorded on Cp-Ti, Ti-6Al-4V samples from surface to interior, oxidized for 6 hrs, 12 hrs and 24 hrs are shown in figure 8. Similar trend can be expected in the case of MSBT1 and MSBT2 alloys also.

Table 4 Surface hardness before and after thermal oxidation

Material	Surface hardness before thermal oxidation	Surface hardness after thermal oxidation		
		6 hrs	12 hrs	24 hrs
CP-Ti	186	612	745	811
Ti-6Al-4V	380	748	762	774
MSBT1	390	764	782	796
MSBT2	420	772	792	812

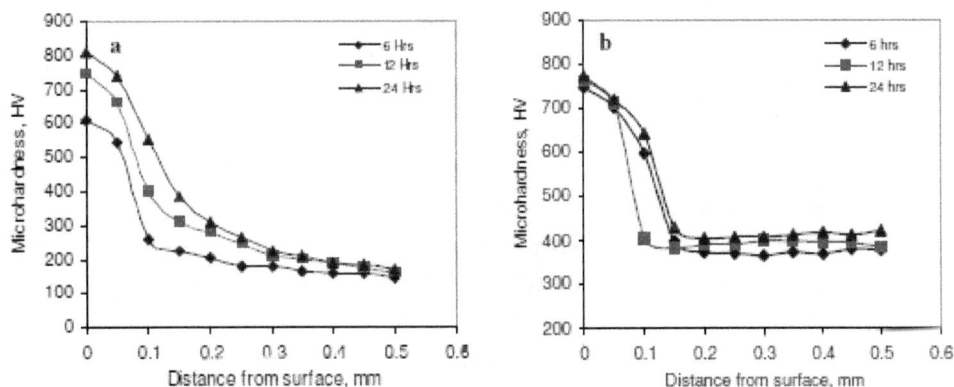

Fig. 8 Variation of micro hardness as a function of distance from surface after thermal oxidation (a) Cp-Ti and (b) Ti-6Al-4V.

The surface hardness was increased by oxidation treatment. There was 2-4 fold increase in surface hardness due to TO treatment depending on the alloy and treatment time. The increase is larger for CP-Ti samples. Further, the increase in hardness was increased with oxidation time. The samples treated for higher oxidation time (24 hrs) yielded more hardness. This can be attributed to the presence of strong and adherent dense oxide layer on the surface. There also exists a gradually decreasing hardened region on the cross-section due to interstitial solutes. The hardness-depth profiles demonstrated that that significant solid solution hardening had taken place beneath the

TiO2 layers. These oxygen diffusion zones extend about 100-150 ⎰m below the surface. It can therefore be concluded that the great increase in surface hardness is due to the presence of dense, hard TiO2 layer on the surface, which significantly improves the wear resistance.

5.6 Wear Studies

The results of sliding wear tests conducted on untreated pins of CP-Ti, Ti-6Al-4V, MSBT1 and MSBT2 alloys are presented in figure 9. From these figures it is clear that the wear of untreated pins of these materials is increased with increasing the applied normal load. All these materials exhibit very poor friction behaviour, characterized by high and unstable friction coefficients, which fluctuated greatly during the sliding cycles. This is typical of poor tribological materials, which experience "stick-slip" during dry sliding. The wear of untreated pins of CP-Ti at higher load (7.5 kgf) is very much different and experiences seizure compared to other materials. This situation is caused by adhesive wear in which microscopic asperities on the metal surfaces come into contact as a result of relative sliding and they tend to weld together forming a bond at the junction which can have rupture strength greater than the strength of the underlying metal. Fracture then takes place at one of the asperities causing metal to be transferred from one surface to the other. The debris so formed gives rise to the accelerated wear that occurs with titanium.

The effect of applied load on the wear of the untreated pins of Ti64, MSBT1 and MSBT2 materials is also shown in figure 9. It is clearly apparent that the wear resistance of the titanium alloys also varied with the loading conditions in untreated condition. Under low load (2.5 kgf) and intermediate load (5 kgf) conditions, wear of all three selected materials were low. However, when a higher load of 7.5 kgf was applied, a much higher wear has been observed for all the materials.

Further, untreated Ti64 and MSBT2 specimen did not show any scuffing at all selected loads (figure 9). However, the pins of MSBT1 shows very high scuffing at very low load (2.5kgf), moderate scuffing at medium load (5 kgf) and no scuffing at high load. This can be ascribed to their initial microstructural features. Figure 10 shows the comparison of wear of untreated pins of Ti-6Al-4V, MSBT1 and MSBT2 alloys at different loads. The wear behaviour appears as dependent on load. There is a progressive increase in wear of these materials with load.

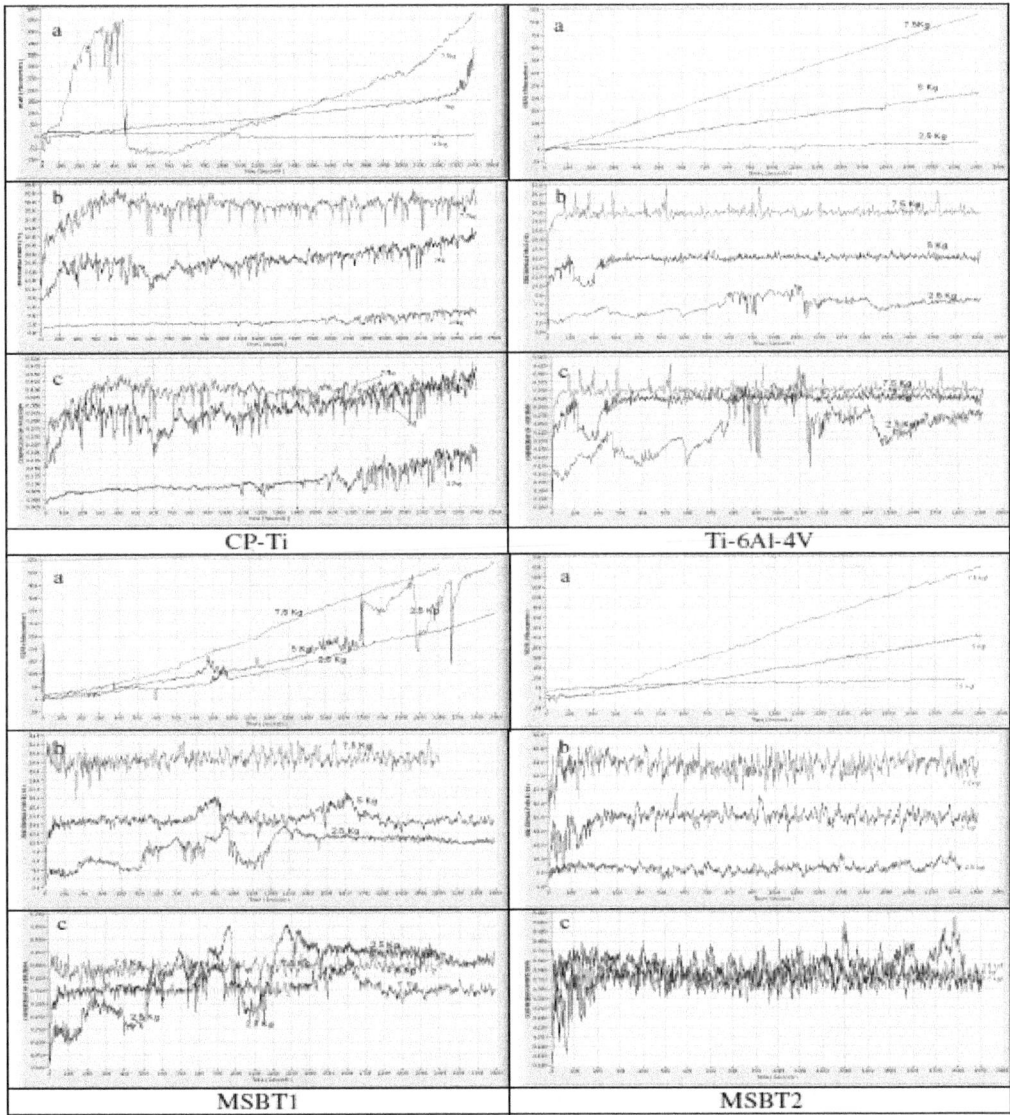

Fig. 9 Pin-on-disc wear test results of selected materials before TO at different loads (a) wear, (b) frictional force, (c) coefficient of friction.

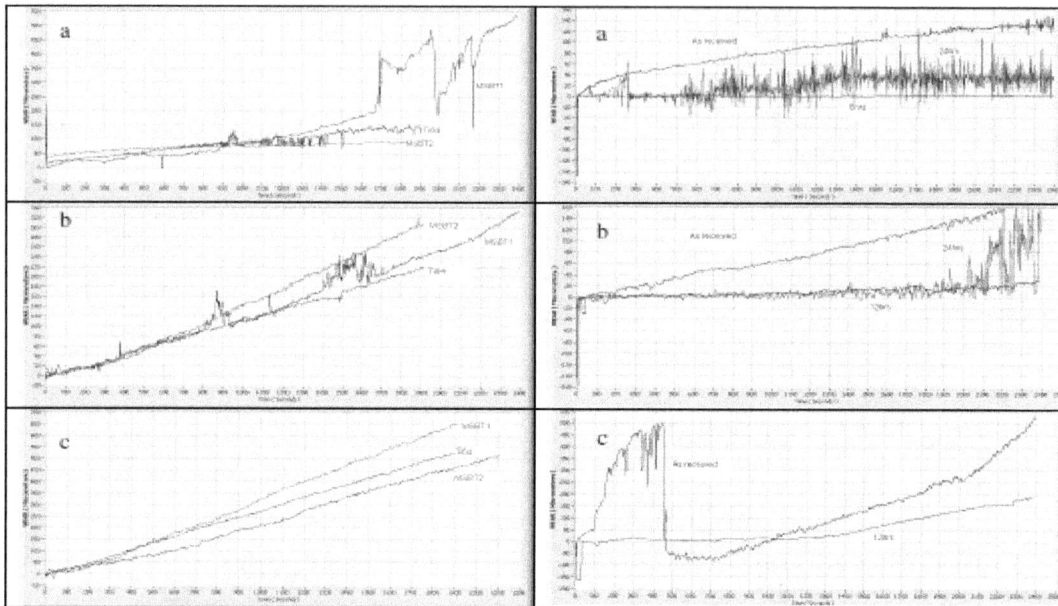

Fig. 10 Comparison of wear of untreated Ti64, MSBTI and MSBT2 samples at different normal loads. (a) 2.5 kgf, (b) 5.0 kfg, (c) 7.5 kgf, (c) 7.5 kgf.

Fig. 11 Comparison of wear behaviour of CP-Ti samples after TO at different loads. (a) 2.5 kgf, (b) 5.0 kgf, (c) 7.5 kgf

The wear behaviour of CP-Ti samples before and after thermal oxidation treatment are shown in figure 11 and the wear resistance (WR, km/gm) of the untreated and oxidized samples of CP-Ti are presented in figure 12.

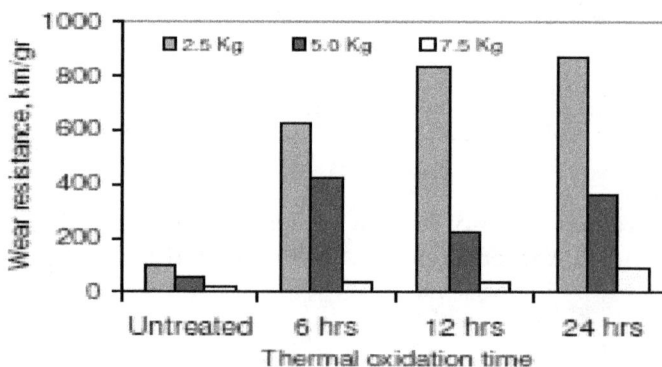

Fig. 12 Wear resistance of CP-Ti samples at different normal loads before and after TO at different cycle times.

Thermal oxidation of the CP-Ti samples improved the frictional characteristics significantly. It is observed that the overall friction coefficient was reduced during the test for lower treatment times, but increased for higher treatment times. The improvement in tribological behaviour by thermal oxidation is further demonstrated by the mild wear mechanism involved in the thermally oxdised CP Ti (Fig.11) and much increased wear resistance (Fig. 12). This is mainly due to the increase in surface micro hardness. Examination of the wear tracks revealed that the unoxidised CP-Ti was worn severely by the adhesive wear mechanism, such that the wear tracks were wide and deep with a very rough surface appearance (Fig. 13(a)). On the contrary, the oxide layer experienced only mild wear. The resultant wear track is smooth and polished appearance (Fig. 13(b)). The superior wear resistance of the rutile layer is derived from its good adhesion with the substrate, high hardness and much improved frictional behaviour.

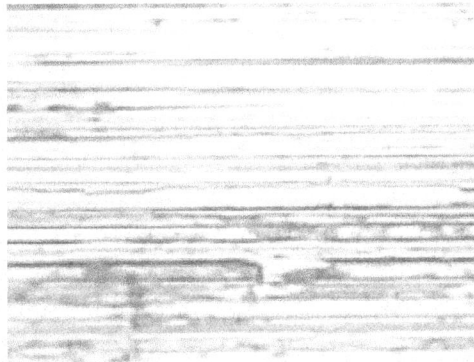

(a) Before TO (100X)   (b) After TO (100X)

Fig. 13 Morphology of wear tracks obtained with CP-Ti samples.

Comparison of wear of Ti64 alloy before and after thermal oxidation for different cycle times is presented in figure 14. From this figure it can be observed that thermal oxidation treatment has greatly improved the wear resistance as compared to the untreated samples. The calculated wear rates for the selected materials at different applied loads are presented in figure 15.

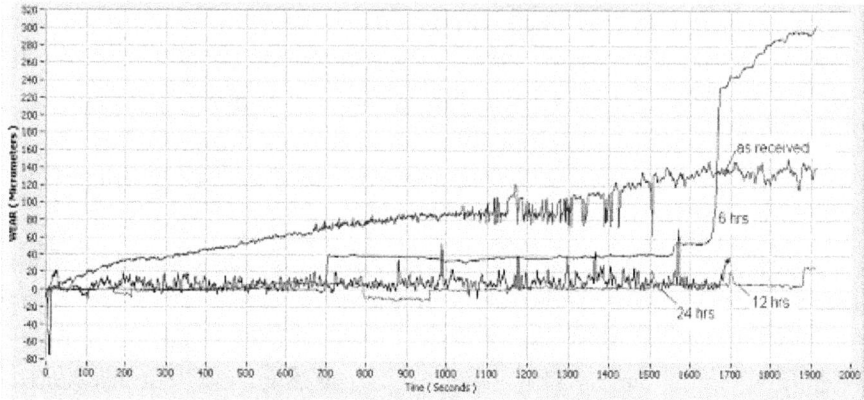

Fig.14 Comparison of wear of Ti64 before and after thermal oxidation for different cycle times.

The untreated Ti64 material was characterised by a very high wear rate (Fig.15.), which may be associated with the preferential transfer of Ti64 onto the steel counterpart and the strong abrasive action of the transferred material. After wear testing, the worn surfaces were examined using light optical microscopy and they were found to be very rough, with typical adhesive wear features evidenced by numerous adhesive craters and deep ploughing grooves (Fig.16). The wear rate of the TO-treated specimen was dramatically reduced by more than 4-5 orders of magnitude over the untreated material (Fig.15). This benefit can be ascribed to a lubricous rutile oxide layer and hardened oxygen diffusion zone. This can be understood by the fact that the intermetallic contact (in case of Ti against steel) was replaced by a steel / ceramic tribological pair. It is the surface oxide layer that significantly reduces the friction of the TO-treated material.

Scuffing is a form of severe sliding wear that is associated with local solid-state welding between the rubbing surfaces operating under heavy loading conditions. It is generally acknowledged that scuffing occurs when the boundary lubricant film breaks down due to a high surface temperature generated by friction, resulting in exposure of a critical fraction of the contacting surface and thus leading to local solid-state welding. The excellent anti-scuffing capacity of the TO-treated materials [19] could be ascribed to the following mechanisms. Firstly, the surface oxide helps to establish a tenacious boundary lubricant film, presumably due to the fact that the oxidised titanium surface has a higher wettability [20]. Secondly, the surface oxide formed in the TO treatment confers a low friction and thus a low friction energy dissipation, which is of crucial importance for maintaining a stable boundary lubricant film in the contacting area. Thirdly, the TO treatment has changed the nature of the sliding surface, in that the friction pair oxide/steel has low mutual solubility and the rutile (ionic material) differs in bonding type with steel (metallic material); thus, there is a low metallurgical compatibility and hence less propensity for adhesion at asperity contacts. Even if the oil film were eventually broken down under progressively higher loads, the surface oxide layer could

374

effectively prevent metal-to-metal contact and thus solid-state welding, provided that the surface oxide layer could maintain its integrity.It is a tenacious oxide layer coupled with a strong support beneath that contributes to the best anti-scuffing performance of the TO-treated material. This can be understood on the basis of the bonding strength between the surface oxide layer and the underlying material, the hardness and the depth of the hardened diffusion zone.

Careful examination of the worn surfaces of the untreated pins against En31 steel counterpart, revealed essential information regarding the wear mechanism involved. Untreated Ti64 alloy is characterized by a very high wear rate, which is in good agreement with the morphology of the worn surfaces (Fig.16). Adhesive wear has obviously occurred with these materials, and the worn surfaces have very rough with deep ploughing grooves and many craters. Similar pictures have been observed with MSBT1 and MSBT2 alloys also. The observed high wear is thought to be the result of the preferential transfer of these materials to the steel counterpart. This can be understood by the fact that titanium and its alloys are chemically active and have a high ductility, which gives rise to the strong tendency to adhesion [18]. Therefore, the adhesive strength of the junctions formed is usually much higher than the strengths of these materials, and such junctions will rupture within the weaker titanium asperities, which accounts for the many craters on the worn surface of untreated material.

Moreover, the transferred titanium becomes work-hardened after the multiple contacts in the wear couple, which in turn results in severe abrasive wear damage to the titanium surface, as is demonstrated by the deep abrasive grooves (Fig.16). The TO treated materials have shown significant improvement in wear resistance due to the strong, tough and adherent rutile oxide film formed on the TO treated surface. This rutile oxide layer can effectively eliminate adhesive action and enhance boundary lubrication, thus giving rise to low wear rate [21]. This is evident from the worn surfaces of the TO treated Ti64 samples in figure 16.

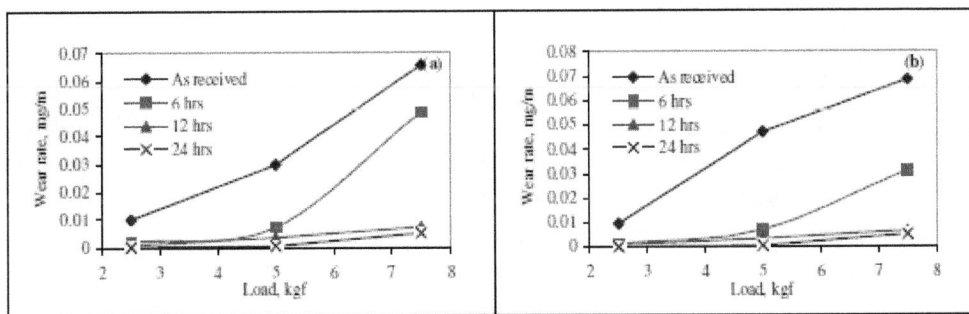

Fig. 15 Wear rate Vs applied load of (a) Ti64 samples, (b) MSBTI samples before and after TO at different cycle times.

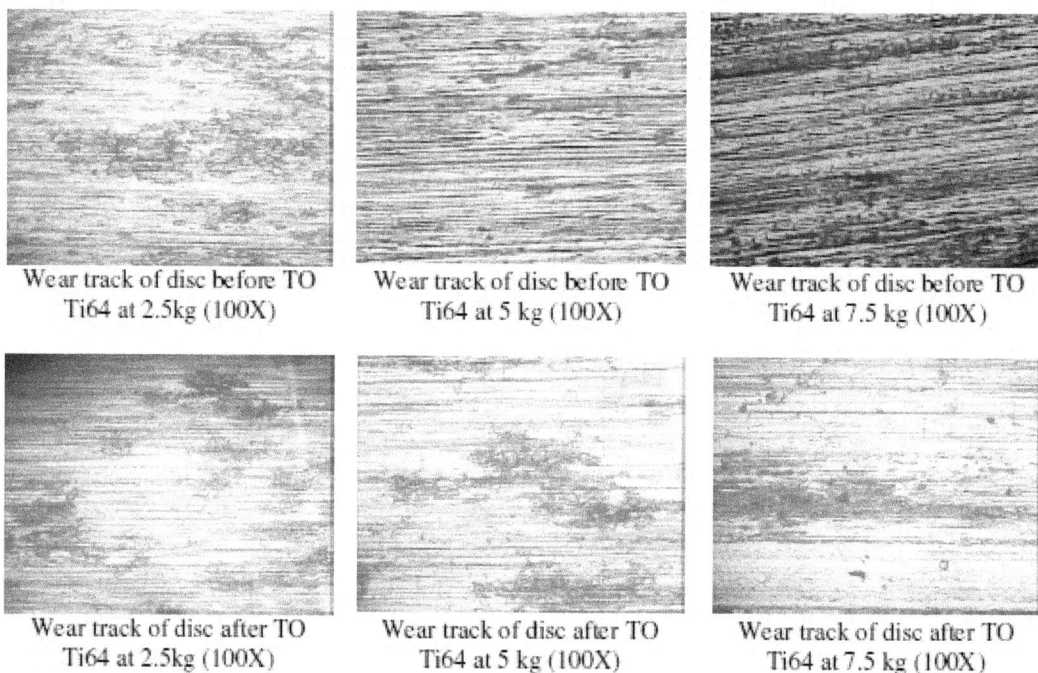

| Wear track of disc before TO Ti64 at 2.5kg (100X) | Wear track of disc before TO Ti64 at 5 kg (100X) | Wear track of disc before TO Ti64 at 7.5 kg (100X) |
| Wear track of disc after TO Ti64 at 2.5kg (100X) | Wear track of disc after TO Ti64 at 5 kg (100X) | Wear track of disc after TO Ti64 at 7.5 kg (100X) |

Fig. 16 Morphology of wear tracks formed on the disc.

6. Conclusions

Untreated CP-Ti, Ti-6Al-4V and Ti-10V-2Fe-3Al cannot be used under dry sliding conditions because of their strong tendency to severe adhesive wear. The effect of thermal oxidation (carried out in fluidized bed furnace) on the wear behaviour of these alloys have been investigated. Oxidation at 650 °C for 6 hrs, 12 hrs and 24 hrs resulted in the formation of outer $TiO2$ layer and oxygen diffusion zone beneath it. The thickness, surface roughness and surface hardness of oxidized alloys increased with increasing oxidation time. The TO treatment has shown a significant improvement in wear resistance of these alloys in sliding contact with a steel surface under dry conditions. A 10-12 order of reduction in wear rate of CP-Ti has been observed compared with untreated material and 4-5 orders in the other materials. The improved sliding wear resistance of the TO-treated titanium alloys can be attributed to the tough, adherent rutile oxide film formed on the treated surface during the TO treatment. This rutile oxide layer can effectively eliminate adhesive action and enhance boundary lubrication, thus giving rise to low wear rate.

Acknowledgements

The authors would like to acknowledge the financial support of Ministry of Human Resource Development, Government of India. The authors also would like to thank DMRL, Hyderabad for their help and support.

References

1. R. Boyer, G. Welsch, and E.W. Collings, Materials properties Hand book-Titanium alloys, ASM INTERNATIONAL Materials Park, OH, 1994.

2. H. B. Bomberg, F. H. Forces, P. M. Morton, *Titanium Technology: Present trends and Future Trends*, F. H. Froes, D. Eylon, N. B. Bomberger, (eds.), The Titanium Development Association, Dayton.

3. E. Rabinowicz, Met. Progr., 1954.

4. G. W. Rowe, Br.J.Appl. Phys.,1956.

5. K. G. Bdinski, Wear, 1991.

6. K. Miyoshi and D. H. Buckly, ASLE Trans., 1982.

7. N.Ohmae, T. Okuyama, and T. Tsukizoe, Tribology Int., 1980.

8. D. H. Buckley, T. J. Kuczkowski, and R. L. Johnson, NASA Technical Reports, 1965.

9. F. D. Rosi, C. A. Dube, and B. H. Alexader, J. met., 1952.

10. J. A. Greeenwood and J. B. P. Williamson, Proc. Roy. Soc. London, 1996.

11. D. Tabor, Proc. R. Soc. London, 1959.

12. D. Hull, D. J. Bacon, *Introduction to dislocations*, U.K., 1984.

13. P. D. Miller and K. W. Holladay, Wear, 1958.

14. E. Rabinowica and E. P. Kingsbury, Met. Progr., 1955, p. 243.

15. F. M. Kutas and M. S. Mishra, ASM Handbooom, vol. 18, *Friction, lubrication and wear Technology*, ASM INTERNATIONAL, Materials park, OH, 1992, p.778.

16. S. Yerramareddy and S. Bahadur, Wear, 1992, vol. 157, p. 243.

17. H. Dong, A. Bloyce, P. H. Morton and T. Bell, 8th World Conference on Titanium, Birmingham, UK, 1995, pp.1999.

18. H. Dong and T. Bell: Wear, Vol. 238, 2000, p. 131.

19. Ph. D. Thesis, H. Dong, The University of Birmingham, 1997.

20. R.M. Streicher, H. Weber, R. Schon and M. Semlitsch, Biomaterials Vol. 12, 1991, p. 125.

21. H. Dong, A. Bloyce, P. H. Morton and T. Bell, Surface Engineering, *Vol.* 13, 1997, p. 402.

WELDING TECHNOLOGY

Microstructural Characterization and Mechanical Properties of Spot Welded Dissimilar Advanced High Strength Steels

S.D. Bhole, D.L. Chen, and M.S. Khan
Department of Mechanical and Industrial Engineering, Ryerson University, Canada
E. Biro, G. Boudreau, and J. Van Deventer
ArcelorMittal Dofasco Inc., Box 2460, Hamilton, Ontario, Canada

Abstract *Resistance spot welding experiments were conducted on dissimilar materials combination of HSLA350/DP600 steels. The welds were characterized using optical and scanning electron microscopy. The fusion zone of the dissimilar materials spot weld was predominantly martensitic with some bainite. The DP600 weld properties played a dominating role on the microstructure and mechanical properties of the dissimilar materials spot welds. But, the fatigue performance of the dissimilar welds was similar to that of the HSLA welds. The dissimilar materials combination showed a wide welding current range for weldability.*

Keywords: Resistance spot welding; Dissimilar materials; Microstructure; Mechanical properties; Fatigue; Fracture mode, Nugget diameter and Hardness.

1. Introduction

Advanced high strength steels (AHSS) have been used in the automotive industry for weight reduction, safety improvements and cost saving. The higher initial work hardening rate with excellent uniform and total elongation combine to give dual phase (DP) steels a much higher ultimate tensile strength and a lower ratio of yield strength to tensile strength than conventional low carbon steels. These characteristics have made DP steels attractive for automotive applications [1]. A significant body of work has been completed to document the resistance spot weldability of AHSS however most of the work has focused on welding a specific grade of steel to itself. Khan *et al.* [2] compared the microstructure and mechanical properties of zinc coated DP600 spot welds produced by resistance spot welding (RSW) and friction stir spot welding. They found that the microstructure of the HAZ and hardness were similar in the fusion and stir zones for both types of welds. The weld efficiency of both processes was similar for DP600 sheet steel when compared on the basis of fracture load versus energy or bonded area.

In automotive applications, a dissimilar combination of material type and thickness are specified in order to tailor material properties for local requirements [3]. The dissimilar materials combinations are very common in automotive construction and adoption of dissimilar metal combinations provides possibilities for the flexible design of the product by using each material efficiently [4]. Kacar *et al.* [5] investigated failure behaviour of dissimilar resistance spot welds between low carbon steel and austenitic stainless steel. They concluded that for spot welds made at low welding currents, low fusion zone hardness and small fusion zone size led to interfacial mode of failure during shear–tensile test. For spot welds made at high welding currents, higher hardness of fusion zone due to martensite formation and larger fusion zone led to pullout failure mode during tensile–shear test. Hasanbasoglu *et al.* [6] determined the weldability of austenitic and interstitial free steels. They found that the nugget size on interstitial free steel was larger than the austenitic steel side and the failure occurred by tearing of interstitial free steel side of the spot welded materials under all conditions. The hardness of the weld nugget was higher in interstitial free steel side.

The present investigation was carried out to examine the microstructure, welding behaviour and mechanical properties of dissimilar welds between galvannealed DP600 and HSLA350 steels. In RSW of DP600 steel to HSLA350 steel, due to their different chemical compositions, the final fusion zone composition will be a combination of the microstructure of the two materials. The final weld metal composition will have a different hardness than the respective fusion zones in the spot welds of the DP/DP and HSLA/HSLA combinations. Thus, the weld strength and the performance of the dissimilar materials weld are expected to be different than the welds in the individual materials.

2. Experimental Procedure

The materials selected in this study were hot-dip galvannealed (HDGA) HSLA350 steel and HDGA dual phase 600 (DP600). The chemical composition and tensile properties of the test materials are given in Tables 1 and 2. A single-phase 50 kVA resistance spot welder with an attached pneumatic cylinder was used. The materials were welded according to AWS D8.1M [7] in different combinations, viz, DP/DP, HSLA/HSLA and HSLA/DP, at different welding currents. The HSLA/DP combination HSLA350 was placed on top of DP600. The welding parameters are given in Table 3. Electrodes with 7.0 mm diameter flat tip ends made of CuCr alloy of Resistance Welder Manufacturers Association (RWMA) class II were used. Test coupons for tensile shear and cross tension were prepared according to AWS D8.1M [7] as shown in Figures 1 and 2. Tensile shear and cross tension tests were carried out according to AWS D8.1M [7] at a crosshead speed of 10 mm/min. For the metallographic examination spot welded specimens were carefully sectioned from the centre, mounted, ground, and polished with sand paper and diamond paste. The polished samples were etched with a 2% Nital reagent and examined via both optical microscope and scanning electron microscope (SEM). The microhardness measurements were made using the Vickers scale at an applied load of 500 g. An Instron universal testing machine was used to conduct load controlled tension–tension fatigue tests. Two groups of samples having nugget diameters of 7.5 mm and 5.5 mm were subjected to various sinusoidal loads with load-ratios of R = 0.1 and R=0.5 at a frequency of 50 Hz at room temperature. The maximum and minimum dynamic loads used for both groups were 60% and 10% respectively of the maximum load obtained in the tensile shear test (18.9 kN).

Table 1 Chemical composition of the test materials (in wt %)

HDGA HSLA350 steel

C	Mn	Si	Ni	Cr	V	M o	Ti	P	S	Cu	Nb
0.05	0.6	0.05	0.01	0.04	0.003	0.004	0.001	0,03	0.004	0.043	0.01

HDGA DP600 steel

C	Mn	Si	Ni	Cr	V	M o	Ti	P	S	Cu	Sn
0.10	1.5	0.19	0.01	0.18	0.005	0.24	0.02	0.009	0.002	0.02	0.007

Table 2 Tensile properties of the test materials

HDGA HSLA350 steel

Zinc coating weight (g/m^2)	Sheet thickness (mm)	Yield strength (MPa)	Ultimate tensile strength (MPa)	Total elongation (%)
57.2	1.57	370	442	36.3

HDGA DP600 steel

Zinc coating weight (g/m^2)	Sheet thickness (mm)	Yield strength (MPa)	Ultimate tensile strength (MPa)	Total elongation (%)
45.8	1.57	399	641	25.4

Table 3 Welding parameters for the dissimilar combination HDGA HSLA350/HDGA DP600 steels

Squeeze time (cycles)	Weld force (kN)	Welding current (kA)	Weld time (cycles)	Hold time (cycles)
35	4.2	7.5	18	10

* Note: One cycle is equivalent to 0.01667 second, calculated on the basis of a current frequency of 60 Hz.

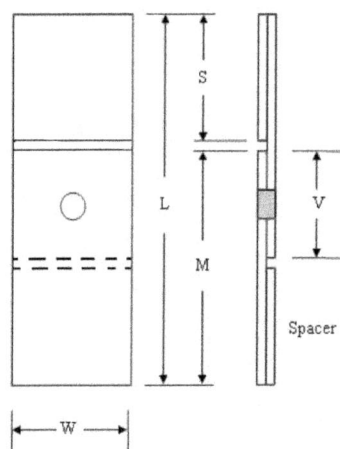

M = 127 mm, W = 45.2 mm, L = 203.2 mm, V = 50.8 mm, S≥40 mm.

Fig. 1 Schematic illustration of shear tensile test specimen.

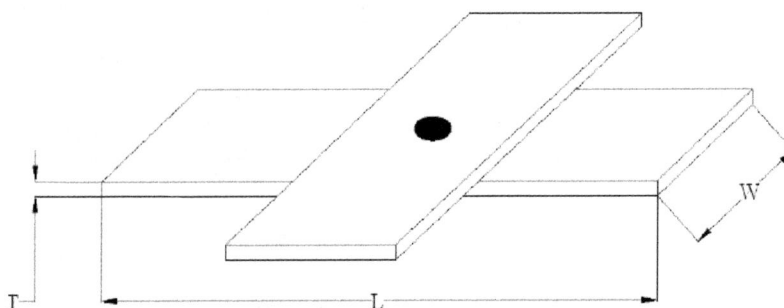

W = 49.02 mm, L = 152.4 mm and T=1.55mm

Figure 2 Schematic illustration of cross tension test specimen

3. Results and Discussions

3.1 Microstructure

The micrographs shown in Figures 3 and 4 represent variations in the microstructure of a dissimilar spot weld nugget from the base metal (BM), across the heat-affected zone (HAZ) to the fusion zone (FZ). The microstructure of the BM of DP600 steel is composed of ferrite, bainite and martensite [8, 9] and that of the HSLA350 consists of polygonal ferrite, acicular ferrite, lath martensite and fine copper and niobium-carbonitride precipitates [10]. The volume fraction of martensite in the HAZ near BM and HAZ near FZ on the DP side was higher as compared to the HAZ on HSLA side as shown in Figures 3 (c) and 4 (c). This is because the DP steel has higher amount of C as compared to the HSLA steel (0.104% and 0.058% respectively).

The FZ is predominantly martensitic with some bainite, as shown in Figures 3 (e) and 4 (e). The transition region between the FZ and HAZ is shown in Figures 3 (d) and 4 (d) . The microconstituents with martensite, bainite and ferrite in the HAZ were finer than in the FZ. This is because the austenitizing was incomplete in the HAZ, and even when austenite grains formed, grain growth was restricted by the formation of martensite and bainite, and because of the thermal cycles [8, 11, and 12].

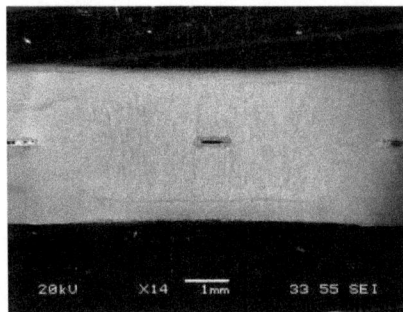

(a)

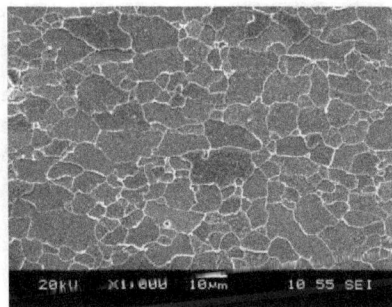

(b)

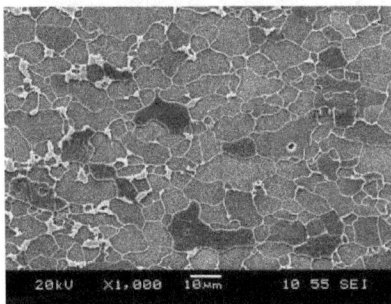

(c)

(d)

385

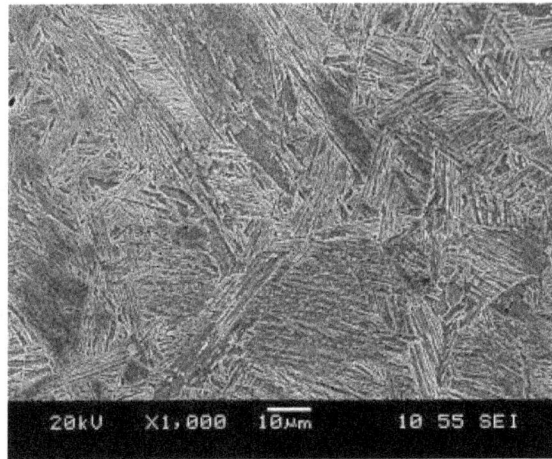

(e)

Figure 3 SEM micrographs of spot weld regions on HSLA side: (a) cross-section of nugget, (b) base metal HSLA 350 steel, (c) HAZ and base metal boundary in HSLA 350 steel, (d) HAZ near fusion zone and (e) fusion zone in centre.

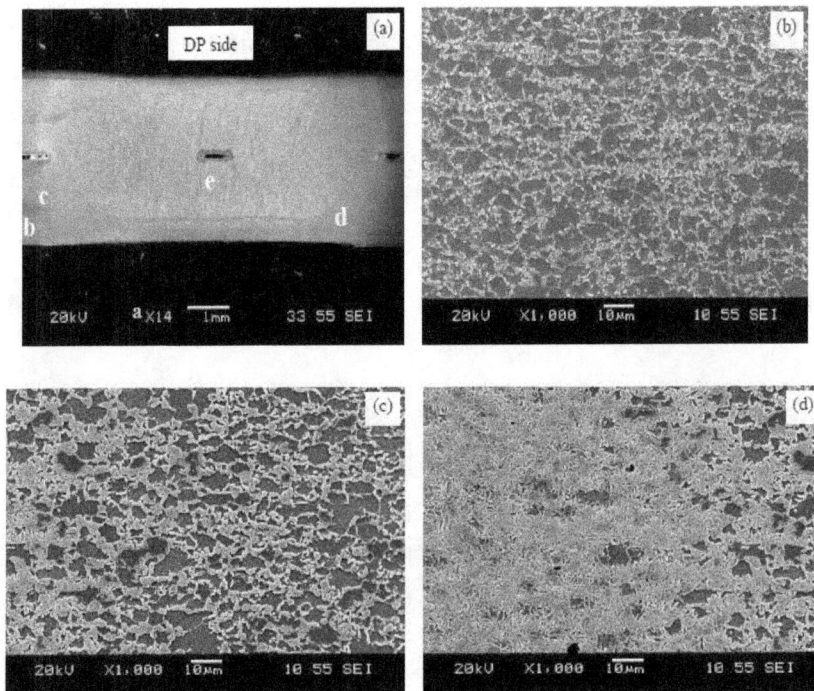

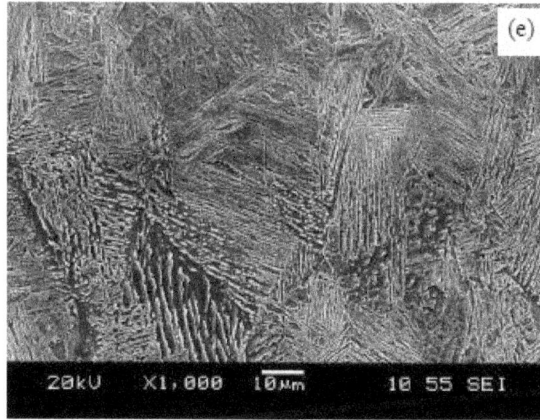

Figure 4 SEM micrographs of spot weld regions on DP side: (a) cross-section of nugget, (b) base metal DP 600 steel, (c) HAZ and base metal boundary in DP 600 steel, (d) HAZ near fusion zone and (e) fusion zone in centre.

In RSW, the cooling rates are very high, leaving insufficient time for carbon diffusion. Hence, the lath martensite is believed to form containing very thin regions of retained austenite between the laths, or pockets of laths, and possibly some lower bainite [12, 13]. Moreover, the higher carbon and manganese contents in the BM result in a higher hardenability. Therefore, the high cooling rate during RSW, coupled with the higher carbon and manganese contents, leads to the formation of martensite and bainite in the FZ and the HAZ.

3.2 Microhardness profile

Figure 5 shows the microhardness profiles of the DP/DP, the HSLA/HSLA and the HSLA/DP resistance spot welds at 7.5 mm nugget diameter. The hardness of the HSLA/DP combination weld nugget is in-between the hardness of the similar material combination nuggets. According to the following carbon equivalent (CE) formula [14]

$$CE = C + \frac{Si}{30} + \frac{Mn + Cu + Cr}{20} + \frac{Ni}{60} + \frac{Mo}{15} + \frac{V}{10} + B \qquad (1)$$

the CE's of the DP steel and the HSLA steel are 0.208 and 0.141 respectively. The higher CE of the DP steel makes the microstructure more martensitic and harder as compared to the HSLA steel. It is believed that the effect of HSLA steel was to dilute the DP steel and reduce its hardenability as can be seen from the hardness profile of the HSLA/DP nugget. Thus, the final composition of the HSLA/DP nugget will be a combination of the two steels HSLA and DP. The difference in the peak HAZ hardness of the HSLA/DP nugget is an indication of the difference in hardenability

387

between the base metals. Similar results have been reported by Hernandez *et al.* [8] for the Trip/HSLA combination. The higher hardness in the weld nuggets is attributed to the formation of martensite, as shown in Figure 3. The 5.5 mm nugget exhibited higher hardness than the 7.5 mm nugget because a smaller nugget diameter experiences a higher cooling rate and thus a more martensitic microstructure and higher hardness [15].

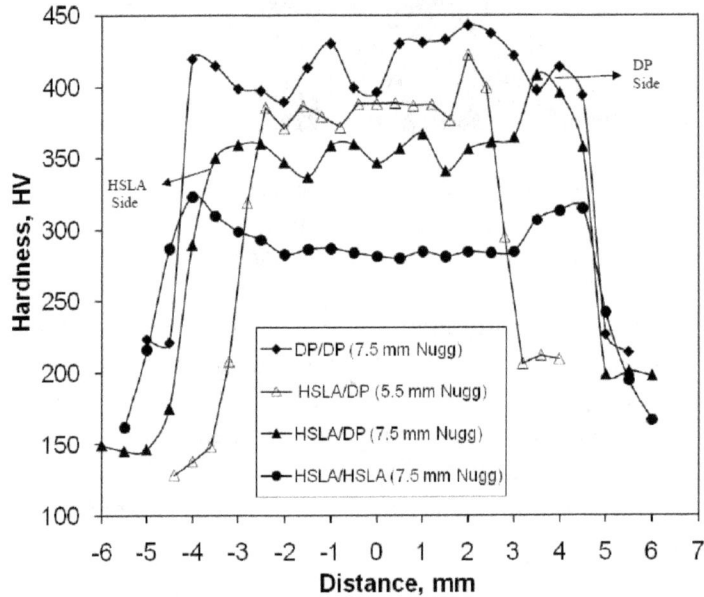

Figure 5 Microhardness profiles of DP/DP, HSLA/DP and HSLA/HSLA spot welds.

3.3 Welding behaviour

Figure 6 shows the minimum button formation and expulsion points as a function of welding current for the similar and dissimilar materials combinations The HSLA/DP combination exhibited a different welding behaviour than the similar materials combinations. To characterize the welding behaviour of the steel, the determination of the useful welding current range is important as it provides a range of the welding current over which the welds with buttons of acceptable size can be produced. The welding current range is defined as [16]

$$I_{Range} = I_{max} - I_{min} \qquad (2)$$

388

where I $_{max}$ is the current at the onset of expulsion, and I$_{min}$ is 4 t as per AWS D8.1M [7] where t is thickness of the sheet. The differences in the welding range for the different material combinations is because of the difference in bulk electrical resistivity of each combination. This difference in the bulk electrical resistivity is mainly due to differences in the chemistry of the materials, especially in Si which has the highest elemental electrical resistivity [17]. The DP steel has a higher Si content as compared to the HSLA steel (0.19 % and 0.052 % respectively), and hence a higher bulk electrical resistivity than the HSLA steel.

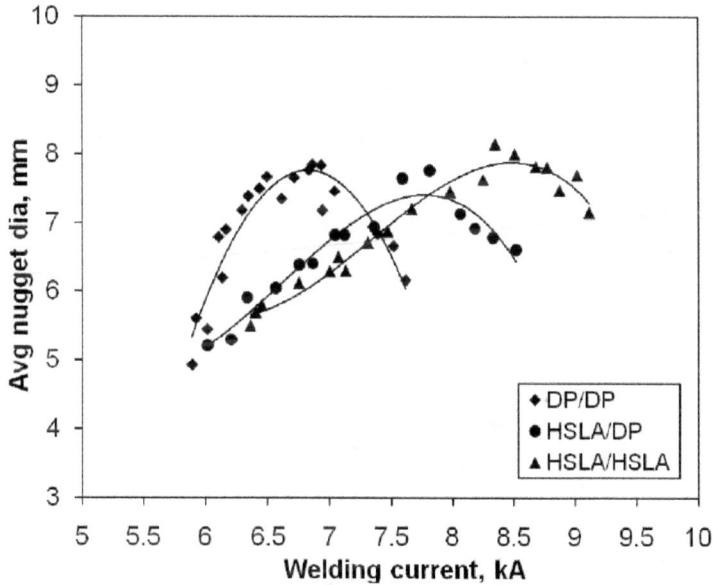

Figure 6 Welding current vs. nugget diameters of similar and dissimilar materials combinations.

3.4 Tensile shear and cross-tension properties

Figure 7 shows the tensile shear and cross-tension failure load vs. welding current for the similar and dissimilar materials combinations. It can be seen that as the welding current increases the failure load also increases. Similar results have been reported by Aslanlar *et al.* [18]. This enhancement in tensile shear and cross-tension load-bearing capacity of the weldment is primarily attributed to the enlargement of the nugget with increase in current [19]. The difference in failure loads for the different combinations at the same nugget size and cross-head speed is due to the differences in the microstructures, which influenced the microhardness as discussed in section 3.2, which in steels is proportional to the ultimate tensile strength [20] and the yield strength [21]. In case of the HSLA/HSLA and HSLA/DP spot welds, interfacial fractures were observed at the lower welding currents in tensile shear geometry. The mode of failure changed from interfacial fracture to button pull-out failure as the welding current was increased. This is likely because the smaller nugget has less energy absorption capacity than the larger nugget [22].

However the DP/DP combination exhibited interfacial fracture at all current levels. In cross-tension geometry for HSLA/DP spot welds two different modes of failure were observed, viz, interfacial button-pullout and button-pullout. At lower welding current corresponding to smaller nugget diameter, the mode of failure was interfacial button-pullout, and the nugget stayed in the HSLA sheet. At higher welding current the mode of failure was button pullout and the nugget stayed in the DP sheet. This is because of the difference in the microstructure and hardness. It has been reported that as the hardness increases the fracture toughness decreases [23]. The observed tensile shear strength was higher than the cross tension strength.

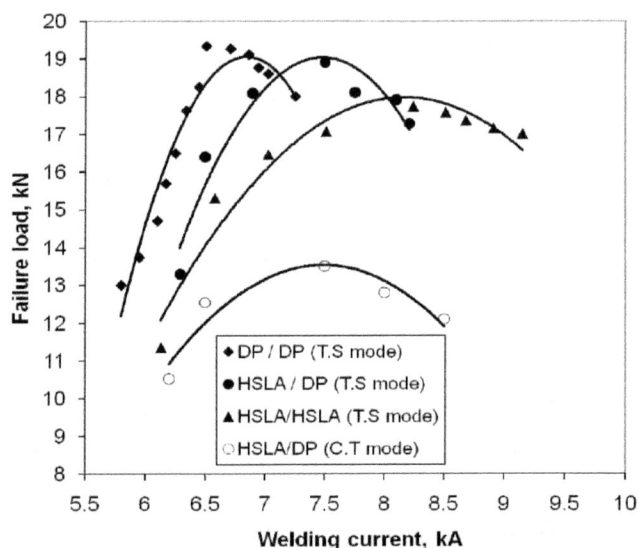

Figure 7 Welding current vs. tensile shear (T.S.) and cross-tension (C.T.) failure loads of similar and dissimilar materials combinations.

The fracture surfaces shown in Figure 8 are in agreement with this result. The fracture surface of the DP/DP spot weld represented by Figure 8 (a) shows a shear as well as ductile mode of failure because of high hardness of the nugget. The fracture surfaces of the HSLA/DP and HSLA/HSLA spot welds show ductile mode of failure as shown in Figures 8 (b) and (c) which indicates low hardness of the nuggets.

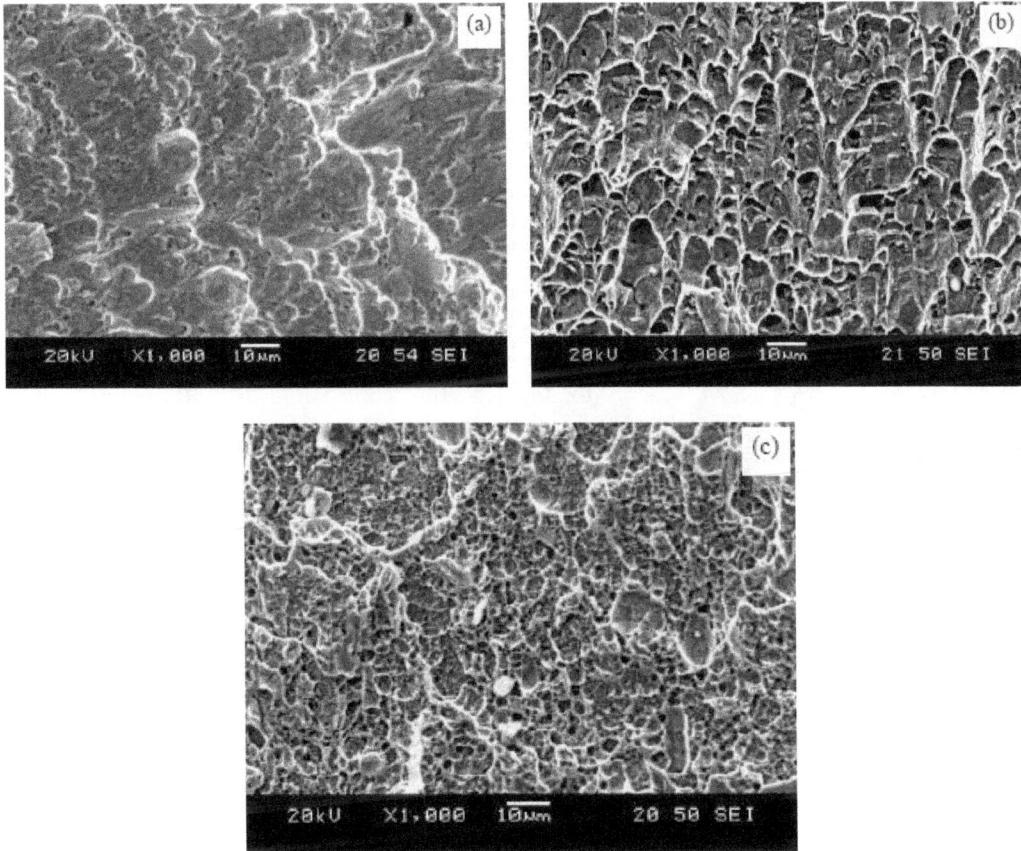

Figure 8 Interfacial fractures of tensile shear samples of different combinations DP/DP, (b) HSLA/DP and (c) HSLA/HSLA.

3.5 Fatigue properties

One of the important aspects of the present study was to investigate the effect of nugget diameter on the fatigue strength of resistance spot welds in HSLA/DP combination. In tensile shear geometry the shear stresses act along the cross-section of the spot-weld nugget. Thus, the state of stress in a spot-welded joint is given by the Equation (3) [24]

$$\tau = 4F \Big/ \pi D^2 \qquad\qquad (3)$$

where, D is diameter of the nugget (mm) and F is the applied load (N). The load amplitude vs. cycles to failure data was converted to stress amplitude vs. number of reversals to failure by calculating the shear stress using Equation (3). The following Basquin type equation was used for curve fitting of fatigue data

$$\Delta \tau/2 = \tau'_f \left(2N_f\right)^b \qquad (4)$$

where $\Delta \tau/2$ is shear stress amplitude, τ'_f is shear fatigue strength coefficient, b is the shear fatigue strength exponent and $2N_f$ is number of reversals to failure. The values of b and τ'_f for 7.5 mm and 5.5 mm nugget sizes with different R values are given in Table 4. Figure 9 shows the shear stress-amplitude vs. $2N_f$ curves for the 7.5 mm and 5.5 mm diameter nuggets of the dissimilar spot welds in HSLA/DP. It can be seen that as the shear stress amplitude decreases the number of cycles to failure increases for both groups. The 5.5 mm diameter nugget exhibited higher fatigue strength (stress) than the 7.5 mm diameter nugget, because, as the nugget diameter decreases the hardness of the nugget increases (as discussed in section 3.2).

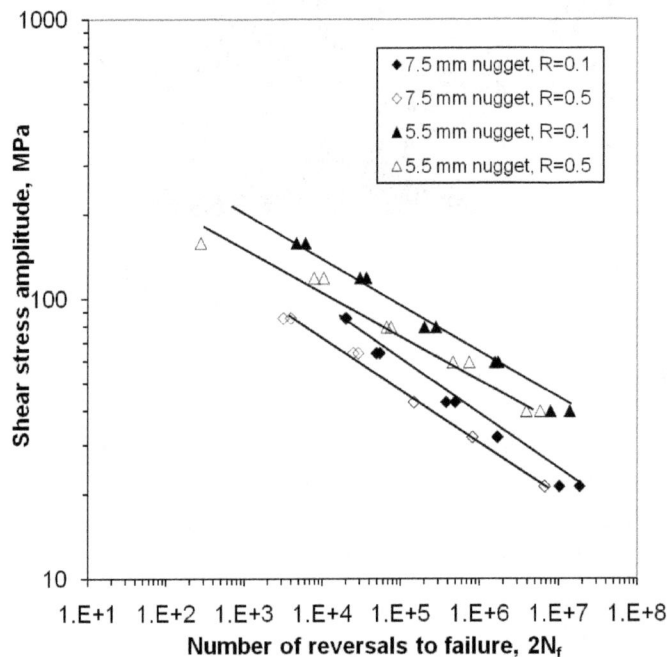

Figure 9 Comparison of fatigue strength of 7.5 mm. and 5.5 mm. diameter nuggets at different fatigue loads with R=0.1 and 0.5.

Table 4 Basquin equation constants for 7.5 mm and 5.5 mm nugget sizes with different R values

Nugget size (mm)	Shear stress ratio (R)	Shear fatigue strength exponent b	Shear fatigue strength coefficient τ'_f (MPa)
7.5	0.1	-0.21	639
7.5	0.5	-0.19	407
5.5	0.1	-0.15	523
5.5	0.5	-0.15	423

3.5.1 Fatigue crack initiation

Cracks normally initiate at the edge of the nugget or in the HAZ, as shown in Figure 10. The opening between the two sheets in the nugget area is seen as an initial crack due to the stress concentration [25], as shown in Figure 10 (a). The boundary of the weld nugget near the opening was also weakly bonded and relatively brittle likely from the molten zinc coating, as shown in Figure 10 (b). SEM observations showed that cracks initiated at the boundary of the nugget and at the HAZ, and then the cracks enetrated through the thickness and propagated to the outer surface.

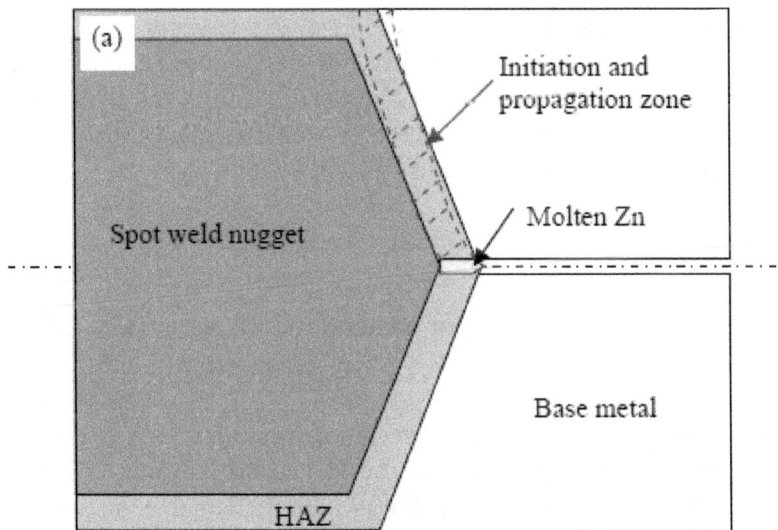

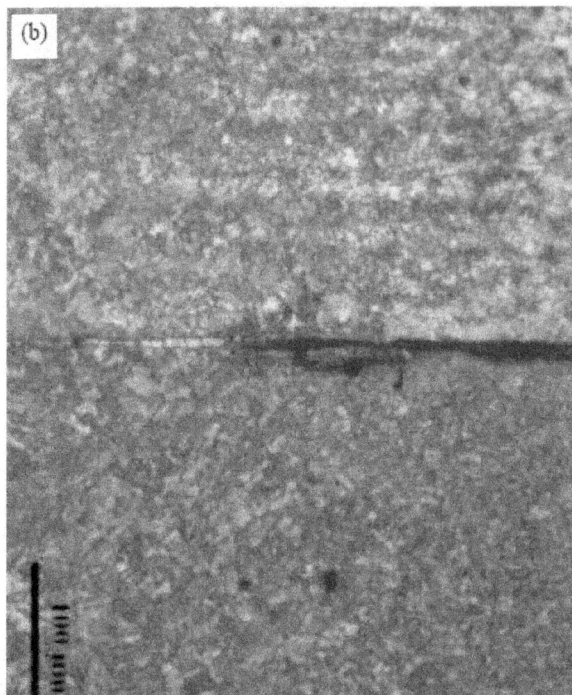

Figure 10 Images of a fatigue crack: (a) schematic illustration and (b) optical image of the cross section near the opening of the weld.

3.5.2 Effect of stress ratio on fatigue properties

Figure 9 also shows the effect of stress ratio on the fatigue strength of different nugget sizes. It can be seen that at any given stress level the samples with R = 0.1 exhibited higher number of cycles to failure (longer fatigue life) as compared to samples with R = 0.5. Similar results have been reported by Ning *et al.* [26]. As the stress ratio R increases from 0.1 to 0.5, the mean stress also increases, and the number of cycles to initiate a crack decreases with the increase of mean stress [27, 28].

4. Conclusions

In the present study of resistance spot welding of dissimilar HSLA350/DP600 steel, the following conclusions may be made.

1. Fusion zone and heat-affected zone microstructures were martensitic, with varying amounts of martensite in the HAZ and lath martensite and bainite in the FZ. The hardness in the FZ was higher than that in the BM due to the formation of lath martensite, and the nugget exhibited different HAZ hardness.

2. A wide welding current range was established, indicating good weldability.

3. The tensile-shear load bearing capacity of dissimilar material weld increased with increasing peak weld current. The load bearing capacity was higher than that of similar material combination HSLA/HSLA but less than that of DP/DP combination, the mode of failure changed from interfacial fracture to button pull-out as the welding current was increased.

4. The cross-tension load bearing capacity of the dissimilar material weld nugget was higher than that of similar material combination HSLA/HSLA but slightly less than that of DP/DP combination. The mode of failure changed from partial interfacial button-pullout failure to button-pullout failure as the welding current was increased.

5. At a given stress amplitude, the number of cycles to failure were higher for 5.5 mm diameter nugget than the 7.5 mm diameter nugget. At very high stress levels, the fracture was either interfacial failure, or initiated in the HAZ near the nugget and failed with some plastic deformation. At intermediate stress levels, the cracks initiated at the nugget boundary and propagated along the circumference and finally into the base metal. At very low stress levels, the cracks basically initiated at the nugget boundary, penetrated through the thickness and propagated along a straight line in the base metal in a direction normal to the loading axis.

6. Fatigue performance of the dissimilar materials HSLA/DP spot welds was similar to that of the HSLA/HSLA spot welds.

Acknowledgements

The authors would like to thank the Natural Sciences and Engineering Research Council (NSERC) of Canada for providing financial support. One of the authors (D.L. Chen) is also grateful for the financial support by the Premier's Research Excellence Award (PREA), Canada Foundation for Innovation (CFI), and Ryerson Research Chair (PRC) program. The authors would also like to thank R. Churaman, J. Amankrah, Q. Li, and A. Machin for easy access to the facilities in the laboratories of Ryerson University. They a re also grateful to Karol Balaz and ArcelorMittal Dofasco Inc. for the provision of materials and welding facilities.

References

1. Advanced High Strength steel (AHSS) Application Guideline, Version. 3, 2006, International Iron & Steel Institute Committee on Automotive Applications.

2. M. I. Khan, M. L. Kuntz, P. Su, A. Gerlich, T. North and Y. Zhou, Science and Technology of Welding and Joining, Vol. 12, 2007, p. 175.

3. V.H.B Hernandez, L.K. Michael, Y.Z. Norman and R.K. Chan, N. Scotchmer, Sheet Metal Welding Conf. XIII, Detroit, 2008, p. 1.

4. Z.Sun and R.Karppi, Materials Processing Technology, Vol. 59, 1996, p. 257.

5. R. Kacar and A. Hasanbasoglu, Materials and Design, Vol. 28, 2007, p. 1794-.

6. A. Hasanbasoglu and R. Kacar, J. of Materials Science & Technology, Vol. 22, 2006, p. 375.

7. AWS D8.1M: 2007, *Specification for Automotive Weld Quality Resistance Spot Welding of Steel*, an American National Standard.

8. M. Marya and X.Q. Gayden, Welding Journal, Vol. 84, 2005, p. 172-s.

9. Technical Transfer Dispatch #6 - Body Structure Materials, ULSAB-AVC Consortium, 2001.

10. M. Shome and O.N. Mohanty, Metallurgical and Materials Transactions A, Vol. 37 A, 2006, p. 159.

11. K. Easterling, 1992, Introduction to the Physical Metallurgy of Welding, 2nd Edition, Butterworth Heinemann, p. 76-89 and p. 191.

12. J.E. Gould, S.P. Khurana, and T. Li, Welding Journal, Vol.85, 2006, p. 111-s.

13. M.D. Tumuluru, Welding Journal, Vol. 85, 2006, p. 31.

14. John F. Lancaster, Handbook of Structural Welding, Woodhead Publishing, ISBN 1855733439, 9781855733435, 1997, p. 205.

15. S.H. Avner, 1997. Introduction to Physical Metallurgy, 2nd edition, p. 256.

16. M. Vural, and A. Akkus, Materials Processing Technology, Vol. 153 &154, 2004, p. 1.

17. U. Bohnenkamp and R. Sandstrom, Steel Research, Vol. 71, 2000, p. 410.

18. S. Aslanlar, A. Ogur, Ozsarac, U. Ilhan, E. Demir, Z. S. Aslanlar, A. Ogur b, U. Ozsarac a, E. Ilhan a, Z. Demir c, Material and Design, Vol. 28, 2007, p. 2.

19. P.K. Ghosh, P.C. Gupta, R. Avtar and B.K. Jha, ISIJ International, Vol. 30, 1990, p. 233.

20. W.D. Callister Jr., Materials Science and Engineering - An Introduction, 7th edition, JohnWiley and Sons, New York, 2007, p. 155.

21. J.R. Cahoon, W. H. Broughton, A. R. Kutzak, Metallurgical Transactions, Vol. 21, 1971, p. 1979.

22. X. Sun, E.V. Stephens, and M.A. Khaleel, Engineering Failure Analysis, Vol. 15, 2008, p. 356.

23. Ibrahim Sevim, Materials and Design, Vol. 27, 2006, p. 21.

24. V.M. Gonçalves and P. A. F. Martins, Materials and Manufacturing Processes, Vol. 21, 2006,

    a. p. 774.

25. C. Ma, D.L. Chen, S.D. Bhole, G. Boudreau, A. Lee, and E. Biro, Materials Science and Engineering A, Vol. 485, 2008, p. 334.

26. N. Pan, S.D. Sheppard, and J.M. Widmann, Fatigue and Fracture Mechanics, Vol. 29, 1999, p. 802.

27. A.R. Jack and A. T. Price, Intern. J. Fract. Mechs, Vol. 6, 1970, p. 401-.

28. S.J. Maddox, International J.of Fracture, Vol. I1, 1975, p. 389.

Welding Aspects of Advanced Materials

G. Madhusudhan Reddy

Defence Metallurgical Research Laboratory, Hyderabad

Abstract: *Many applications of conventional and modern engineering materials are limited by their ability to be joined. As new materials emerge and higher levels of performance are placed on these materials, a clear understanding of process-microstructure-property relations in these welded/joined materials is necessary to assure adequate performance in service. Many of the advanced materials being developed today cannot be used, and perhaps will never be used, in great quantities because insufficient attention has been given to their manufacturing properties. Indeed, very few materials, if any, are used in an isolated and elementary form. They must be integrated into the product or structure. For this reason, welding is widely used both in the manufacturing of the smallest components and for the building of the largest. Advances in welding science must therefore keep pace with progress in materials science.*

This article deals with scientific principles that were unearthed during the course of scientific investigations in the area of welding of technologically important materials such as super alloys, stainless steels, metal matrix composites, aluminium, titanium alloys and ultra high strength steels in similar and dissimilar metal combinations used in defence applications. The problems in the joining of many of these materials are related to metallurgical characteristics influenced by welding processes and filler materials. In dissimilar metal combinations metallurgical compatibility is the main issue. In this article the areas covered include the selection of filler materials for welding of high strength aluminium alloys and ultrahigh strength steels. The dependence of the strength and ductility properties of dissimilar metal joints on the extent and nature of inhomogeneities is shown. The optimum combinations for welding various metal combinations are presented. Joining aspects of incompatible materials such as alluminium alloys to stainless steels are also discussed.

Weld zone grain refinement with inoculants and advanced welding techniques to improve weld zone properties also forms a part of this paper. Effect of heat input and welding processes on ballistic behaviour of armour steels and improvement in ballistic properties of soft welds with hard-facing are outlined. Effect of electron beam oscillation patterns on the structure and properties of titanium, Inconel 718 and stainless steel are also discussed.

1. Introduction

The progress in material science and engineering over the past two decades has not been matched by improvement in science and technology of welding. Welding is essential for almost any engineering material application, and is often a critical issue for new and advanced materials. The ability to make economic and reliable joints in these materials will be an advantage in their commercial uptake, and without this capacity these materials will be restricted to fewer applications. Very little work is being done within the country in the area of advanced materials joining. Welding being a major route to the fabrication of engineering components, the generation of an exhaustive database on the welding aspects of these materials optimises their applications for best results. This paper deals with the some of the problems encountered in welding/joining of Al-Zn-Cu-Mg (AA 7010) alloy, Al-Mg-Si (AA 6061) alloy, metal matrix composites, Al-Li alloys, Super alloy 718, titanium alloys, ferritic stainless steels, maraging steels, titanium aluminides high strength low alloy steels. And dissimilar metal combination of compatible and incompatible materials

2. Improving Mechanical Properties of Aluminium Alloy Gas Tungsten Arc Welds Through Modification of Fusion Zone Chemistry

2.1 Al-Zn-Mg-Cu Base Alloy (AA 7010) [1-2]

Commercial 7000 series aluminium alloys are based on medium strength Al-Zn-Mg and Al-Zn-Mg-Cu systems. The medium strength alloys are weldable, whilst the Cu containing alloys are non- weldable. This is because; Cu additions in excess of 0.3 wt% give rise to hot cracking during the solidification of welds, where, joining of components by welding is an essential step. Investigation were carried out at DMRL on the influence of scandium bearing filler materials on the weldbility of an Al-Zn-Mg-Cu base 7010 alloy using conventional gas tungsten arc welding process and known *Varestraint* and *Houldcroft* tests for weldability. It is observed that the hot cracking tendency is pronounced for the welds deposited without scandium containing Al-Mg filler (AA 5356). On the other hand, the hot cracking tendency is greatly minimised for the welds deposited with scandium containing filler. The beneficial role of the scandium containing filler is found to be associated with the significant micro-structural refinement of the welds in terms of both the grain size as well as the morphology of the second phase particles present as solidification products

Role of trace additions of Ag (-0.1 at%) in increasing the nucleation frequency and stability of a variety of phase precipitates in several Al alloy systems is well known. However it is not known whether or in what ways such small additions of Ag could affect the weldbility of Al alloys. The presence of Ag in the filler alloy has the noticeable effect of equiaxing and refining the grain size of the fusion zone. Studies were carried out at DMRL on solidification cracking behaviour of 7010+Ag using varestarint test. Results indicated that solidification resistance is comparable to the commercially welded 7020 Al alloys.

2.1 AA 2219 [3-5]

The main reason for the low strength of the fusion zone of AA 2219, was the dissolution of all the strengthening precipitates. Other factors contributing to this were coarse columnar grains, connected network of eutectic and depletion of copper in the grain interiors. Rao et al [3-5] approached the problem by way of adding scandium to the fusion zone through filler. Addition of about 0.4 wt% Sc, resulted in substantial improvement in proof strength, ultimate tensile strength and percent elongation. Magnesium (0.8 wt%) was added to weld along with scandium.

3. Weld Microstructure - Property Correlation

3.1 Aluminium-lithium alloys [6-10]

Aluminium-lithium alloys offer weight savings in aerospace structural applications due to their lower density and higher elastic modulus as compared to the conventional aluminium alloys. Further weight savings are possible if mechanical fastening is replaced by welding as the joining method. The weldability of Al-Li alloys has received relatively limited attention so far because traditionally most aircraft parts are mechanically fastened. If weldable Al-Li alloys were available commercially, they could be considered for many other structural applications, such as marine hardware, light pressure vessels etc. Major weldability issues are (a) porosity (b) solidification cracking (c) low mechanical properties Currently, there is a great interest in the aerospace community to establish reliable welding techniques for joining Al - and Al-Li alloys. Extensive studies were carried out on the effect of filler metal composition and welding parameters of conventional as well as novel welding techniques, on the weld soundness, microstructure and mechanical properties. This work established for the first time that hot-crack free autogenous welds of aluminium-lithium alloys could be made applying the current pulsing (CP) and magnetic arc oscillation (MAO) techniques using alternating current gas tungsten arc welding (AC-GTAW) process. An altogether new technique involving a combination of pulsed current and arc oscillation was developed for the first time for refining the microstructure and improving the weld properties and applied to the welding of Al-Li alloy. The study clearly established that pulsed current arc oscillation techniques refine the fusion zone grain structure of (Al-Li) alloys very significantly and result in significant improvement in strength and ductility values in the as-welded condition it self . The study showed for the first time that combination of pulsed current and arc oscillation resulted in welds with much more uniform and fine microstructure than that what was possible with these techniques individually. The work showed that these techniques significantly reduce the micro-segregation of the major solute elements in the solidification structure leading to substantial improvement in their corrosion resistance. The fatigue crack growth resistance in CP and MAO welds was found to be the best. The effects of various filler metal compositions (including new compositions) on macro/microstructure, hot cracking, mechanical properties and corrosion behaviour of the welds have been investigated. A mechanism has been proposed for the absence of epitaxial growth in Al-Li alloy welds.

3.2 AA 6061 (Al-Mg-Si) [11]

AA 6061 aluminium alloy is extensively employed in aeronautical applications due to their low density, high specific strength and excellent corrosion resistance. Many of the aluminium alloys also exhibit excellent weldability a prime requirement for any engineering structure, where welding is a predominant fabrication route. Heat treatable wrought aluminium-magnesium-silicon alloys conforming to AA 6061 (Al-Mg-Si) are of moderate strength and possess excellent welding characteristics over the high strength aluminium alloys. Hence, alloys of this class are extensively employed in aircraft and missile applications. Although Al-Mg-Si alloys are readily weldable, they suffer from severe softening in the heat affected zone (HAZ) because of reversion (dissolution) of $\beta''(Mg_2Si)$ precipitates during weld thermal cycle. This type of mechanical impairment presents a major problem in engineering design. The kinetics of precipitate dissolution in the heat affected zone (HAZ) is primarily a function of weld thermal cycle, prior thermo-mechanical history of the parent metal and its chemistry. It will be more appropriate to overcome or minimize the HAZ softening to improve mechanical properties of weldments. To address heat affected zone softening phenomena in the weds of AA 6061 alloy a study has been taken up at DMRL to improve the properties in as-welded condition by resorting pulsed current and magnetic arc oscillation techniques. As these techniques are reported an improved properties as a result of refined microstructure as a consequence of effective weld pool stirring This study assumes significance as the influence of innovative techniques such as pulsed current and arc oscillation in the welds with filler addition has been attempted for the first time. The study pertains to welding of AA 6061 with AA 4043 filler using conventional current, pulsed current and magnetic arc oscillation techniques of gas tungsten arc welding process. Response of these welds to post weld solution treatment + ageing (STA) has also been investigated. In the as-welded condition conventional welds exhibited widest soft heat affected (HAZ) zone while pulsed current (PC) and magnetic arc oscillation (MAO) techniques considerably reduced HAZ softening. Fine microstructural features in PC and MAO welds resulting from effective convective currents in the weld pool exhibited better mechanical properties. PC and MAO welds also responded better to post weld STA. Arc oscillation welds exhibited better properties even in as-welded condition. This study established a route to obtain acceptable mechanical properties in the as-welded condition it-self.

401

4. Improving High Temperature Properties of Inconel 718 By The Application of Special Weld Techniques [12-17]

Inconel 718, the nickel base superalloy has an excellent reputation for weldability and resistance to post weld stress relief cracking. However, it has a major limitation in the form of niobium segregation during solidification in the interdendritic regions leading to the formation of Laves phase responsible for degradation of its mechanical properties. Very high temperature solutionising temperatures may dissolve Laves. However, it is not a practical solution to welds owing to oxidation and distortion problems. Hence, any attempt to control the niobium segregation at the welding stage itself is highly desirable. Low weld heat inputs such as used in electron beam and laser welding processes and techniques (such as electron beam oscillation) which increase weld pool agitation are expected to reduce niobium segregation. Weld pool agitation by pulsed current technique, arc oscillation as in gas tungsten arc welding are also expected to be beneficial.

Extensive work has carried out on the application of laser beam, electron beam with beam oscillation techniques and gas tungsten arc with pulsed current mode to study their effect on niobium segregation. The response of fusion zone of 718 to ageing was studied with and with out solutionising treatment. Stress-rupture tests were conducted to evaluate the high temperature behaviour.

Reddy et al [12] work has conclusively proved that electron beam oscillation technique and pulsed laser technique and the pulsed arc techniques in GTA welding reduce niobium very effectively. He proved that elliptical electron beam oscillation and pulsed laser weld techniques are the best from the point of view of reducing the Nb segregation, Laves content and the continuity of laves phase. The elliptically oscillated electron beam welds and pulsed laser beam welds exhibited the best response to post weld-ageing treatments. Through these studies, indicated that application of electron beam oscillation and pulsed laser beam are a powerful techniques in reducing the Laves phase at the welding stage itself and in improving the response to post weld ageing treatment and thereby improving the mechanical properties.

5. Weldability Studies of Steels [18-23]

During Fusion Welding Materials Are Prone To Solidification Cracking due to the presence of low melting point phases. In order to address this issue there is a need tailor the compositions of the materials during development itself. The susceptibility of materials to solidification cracking is assessed by tests like varestraint test in which the material is subjected to bending immediately at the completion of welding during the onset of solidification. The radius of bending is a measure of strain the material is subjected to. The cracking tendency is measured in terms of total length of crack, maximum crack length and number of cracks. The critical strain below which the cracking tendency is least is the measure of resistance of the material to solidification cracking. Our studies revealed that steels like AISI 4140 are resistant to cracking. Among the ultrahigh strength steels maraging steels are observed to be more resistant. In the welding of AISI 4140 it has been observed that pulse current welding results in more cracking than conventional welding.

Some of the materials are prone to cracking after welding after a delay time, at temperatures, below $150^{0}C$. This is known as cold cracking as it occurs at low temperatures or delayed cracking due to the occurrence of cracking after time lapse. Since this is aided by hydrogen in weldment it is also known as hydrogen assisted cracking. In order to overcome this problem the hydrogen levels in the parent materials as well as filler metals has to be kept low. The susceptibility of the materials to cold cracking is assessed by self-restraint tests as well as constant external load tests. Self-restraint tests are qualitative and are 'Go' or 'No Go' type. External load tests like implant tests aid in determining critical stress to which the weld can be subjected to with out cracking over a period of time. These tests are helpful in selecting suitable filler metals resistant to delayed cracking. Our studies on self-restraint tests revealed that maraging fillers are resistant to cracking compared low alloy steel filler in the welding of an ultra-high strength steel.

In the welding of armour steels E 312 (30Cr-10Ni) austenitic stainless steel is recommended to be employed and is being in use since 1940. The philosophy behind such use is that stainless steel has greater solubility for hydrogen and therefore, resistant to cold cracking. In addition, stainless being highly ductile can withstand restraint stresses by it self, undergoing deformation. However, experience shows that delayed cracking continues to occur in spite of use of, such filler. Our studies employing implant test has been able to identify fillers resistant to cracking (Fig. 6). Detailed analysis indicated that E 312 is more prone to cracking due to its tendency to form brittle carbides due to an affinity between carbon in the armour steel and Cr in the filler. The cracking was observed to be, not related to hydrogen.

6. Studies to Improve Ballistic Capability of Armor Steel Welds and Their Heat Affected Zones [20-24].

Armor steels belong to quenched and tempered class of low alloy high strength steels. These are employed in high hardness condition for better bullet penetration resistance. In the assembly of these armors welding is extensively employed. During welding the areas adjacent to the fusion zone are exposed to temperatures approaching inter critical region or AC1. Therefore, these regions get softened due to the decomposition of meta stable martensite. The extent of this softening is dependent on the weld thermal cycle to which these regions are exposed. The softening leads to reduction in ballistic capability of these regions. The thermal cycle is dictated by the welding process as well as heat in put in a particular welding process. For example, low heat input processes that lead to faster cooling rates exhibit least softening and therefore These weld HAZs possess better ballistic properties. Control of heat in put in SMAW can further improve the ballistic properties. Tensile residual stresses are beneficial for ballistic resistance. Acicular martensitic microstructures exhibit better ballistic properties than lath martensitic microstructures.

Since it is a practice to weld armor steels with austenitic stainless steel fillers the welds are soft and therefore exhibit poor ballistic properties. Through our detailed investigations we have been able to address this problem. The solution that is found is to incorporate hard material between the soft austenitic weld layers as a sandwich. This can be done, either, by depositing hard facing material as weld deposit or by in situ hard layer formation. For in-situ hard layer formation aluminium filler is deposited over stainless steel weld facilitating formation Fe_3Al or Ni_3Al and such other hard intermetallics.

7. Fusion Zone Grain Refinement

Weld fusion zones typically exhibit coarse columnar grains because of the prevailing thermal conditions during weld metal solidification. This often results inferior weld mechanical properties and poor resistance to cracking. While it is thus highly desirable to control solidification structure in welds, such control is often difficult because of the high temperatures and highly thermal gradients in welds in relation to castings and the epitaxial nature of the growth process Nevertheless, several methods for refining weld fusion zone have been tried with some success in the past: inoculation with heterogeneous nucleants, micro-cooler addition, surface nucleation induced by gas impingement and introduction of physical disturbance through techniques such as torch vibrations. The use of inoculants for refining the fusion zone was, as a matter of fact, not as successful as in castings because of the extremely high temperatures involved in welding and also due to the undesirable affects of the inoculating elements on weld mechanical properties at the levels required for producing grain refinement. Other techniques like surface nucleation and micro-cooler additions were also turned down because of the complicated welding set-ups and procedures associated with their use. In the process, two relatively new techniques- current pulsing and magnetic arc oscillation have gained wide popularity because of their striking promise with minor modifications to the existing welding equipment. The use of current pulsing (CP) and magnetic arc oscillation (MAO) for producing grain refinement in weld fusion zones has been the subject of several investigations in the recent past. The application of these techniques in reducing grain size and improving mechanical properties of some of the alloys used in defence application are discussed in some details in this paper.

7.1. Titanium alloys [25-27]

Weld fusion zones in α-β Ti alloys are known to exhibit poor ductility compared to the base material. This is usually attributed to an acicular, at least partially martensitic intragranualr microstructure and large prior-beta grain size. The as-welded ductility can not be improved by altering merely the weld energy input because a high heat input promotes coarsening of the prior β grains, while a low heat input accelerates cooling and increases the aspects ratio of the matrix microstructure. In an attempt to refine the weld solidification structure without adversely influencing the intragranular microstructure in α-β Ti alloys, MAO and CP have been used. Various oscillation frequencies from 1-20 Hz were studied at constant amplitude of 0.6 mm. Maximum grain refinement was observed at a frequency of 2Hz. In CP studies, both DC and AC were employed. Pulse frequencies were varied in such a way that the total heat input is same in all cases. Maximum grain refinement was observed at a frequency of 6Hz.

Poor impact toughness is the main problem in the conventional electron beam welds of Ti-6Al-4V. Beam oscillation welding technique has been investigated with a view to alleviate this problem. Microstructure and mechanical properties of the welds were evaluated in the as- welded and post weld heat-treated conditions. Beam oscillation in the electron beam welds exhibited improved impact toughness and ductility as compared to conventional welds. Post weld heat treatment in the sub-transus as well as super-transus regions led to improvement in impact toughness of conventional as well as beam oscillated welds. The observed impact toughness trends are well correlated to crack path deflection and deviation in that, increased crack path deviation results in enhanced toughness. In the as-welded condition residual stress was found to be compressive and is attributed to the predominantly martensitic microstructure. Post weld heat treatment that led to the decomposition of martensite to equilibrium α + β exhibited tensile residual stress as a result of accompanying volume expansion during α′ to α + β transformation.

7.2. Aluminium alloys

One of the major problems in the welding of heat-treatable aluminium alloys is hot cracking or solidification cracking. In an attempt to control hot cracking, these alloys are commonly welded using non-matching filler materials like Al-Si and Al-Mg alloys which are not heat-treatable. As a result, the weld metal does not respond to post-weld aging treatments and thus its strength properties far significantly below that of the base material. However, if the hot cracking propensity is controlled by some means which would allow the use of matching fillers during welding heat-treatable aluminium alloys, weld metal properties could be significantly improved. Refinement of the fusion zone grain size is a definite means to achieve this objective. Also, the formation of fine equiaxed grains, by itself can improve weld metal properties.

The possibility of refining the fusion zone grain size in a Russian origin Al-Li alloy 01441 using MAO and CP has been investigated [8]. The application of CP and MAO has resulted in substantial refinement of the fusion zone grain structure. Similar observations were also reported in another Al-Li alloy AA 2090 using MAO and CP [28].

7.3. Ferritic stainless steels [29-32]

Ferritic stainless steels have an economic advantage over austenitic stainless steels and there are certain applications, which require nickel free stainless steels. One such application is $TiCl_4$ reduction with magnesium, wherein Mg has a tendency to leach away Ni from the austenitic stainless steel and thus degrading mechanical properties. A major problem in the welding of ferritic stainless steel is grain growth in the welds, which leads to degradation in mechanical properties. The problem of grain growth was addressed by (a) Using pulsed gas tungsten arc welding process, which included optimization of pulse frequency, heat input and modifying the weld chemistry by the addition of alloying elements such as copper and titanium. Frequencies of 5 and 6 Hz have been found to be effective in grain refinement. Refinement in grain size has led to an improvement in the tensile strength and ductility in general. Addition of 2 %O_2 to the shielding gas resulted in a grain refinement at the fusion zone. Addition of copper and titanium led to an improved strength over that of the base alloy (b) Electron beam oscillation technique: Studied for the first time to refine the fusion zone grain size of ferritic stainless steel welds by using beam oscillation technique The different beam oscillation patterns are made transverse to the weld. The beam shapes studied are sinusoidal, square, elliptical, circular, triangular and ramp. The effects of the shape of the oscillating beam on the microstructure and tensile properties of the welds have been investigated. The results are compared with conventional electron beam welds. Elliptical beam oscillation resulted in maximum grain refinement.

8. Titanium Aluminides [33-34]

Titanium aluminides have the density advantage over nickel base super alloys. Additionally titanium aluminides exhibit superior creep strength], although, they lack room temperature ductility, which is a prerequisite for any engineering application calling for safe design and ease of fabrication. Ductility improvements have been achieved by the addition of β stabilizing elements namely niobium (Nb). Addition of Nb between 7.5 –12 at% results in $\alpha_2 + \beta$ microstructure. A further increase in Nb addition leads to orthorhombic (O) phase formation. The presence of 'O'phase imparts additional benefits with respect to creep properties. Majority of welding studies on α_2 titanium aluminides are confined to $\alpha_2 + \beta$ alloys. These studies have indicated that cooling rates have to be kept below about 10^0C/s or have to be greater than 100^0 C/s to achieve ductile welds. The observation was that at slow cooling rates an $\alpha_2 + \beta$ structure is observed while at cooling rates 100 ^{0}C/s retained β structure is obtained. The presence of β phase leads to ductile welds. Intermediate cooling rates give fine acicular microstructures consisting of $\alpha_2 + \alpha_2^|$. These microstructures lack ductility. Limited post weld heat treatment (PWHT) studies indicated that PWHT temperature as high as 980^0C is required to improve the properties. Welding studies on $\alpha_2 + O+ \beta$ class of alloys are not reported till date. An $\alpha_2 + O+ \beta$ titanium aluminide with Ti-24 at% Al-15.5 at% Nb has been investigated for its weldability characteristics. Conventional gas tungsten arc (GTA) welds and electron beam (EB) welds exhibited columnar fusion zone grains. Pulsed current and arc oscillated welds exhibited predominantly equiaxed fusion zone grains. The microstructure of GTA welds and pulsed current GTA welds exhibited $\alpha_2 + $O phases while arc oscillated GTA welds and EB welds had $\alpha_2 + O+ \beta_0 /\beta_2$, however β_0 /β_2 is predominant in EB welds. EB welds which contained $\alpha_2 + \beta_0 /\beta_2$ microstructure exhibited high strength and ductility compared to GTA welds. The observed microstructural trends are explained on the basis of possible weld thermal cycle and convective currents in the weld pool [33]. Friction welding is known to result in extremely fine recrystalised grain size. This can solve the problem of solidification cracking encountered in conventional fusion welding as no fusion is involved added to its fine recrystallized microstructure. The fine recrstalized microstructure that is throughout to be obtained in $\alpha_2 + O+ \beta$ titanium aluminide friction welds resulted in improved properties [34].

9. Friction Welding of Materials

9.1. Friction Welding of Similar Materials [35-36]

Fusion welds of aluminium alloy base metal matrix composites with SiC, as reinforcements are prone to cracking when exposed to moisture due the decomposition of Al_4C_3 that is formed. Friction welding is an answer for such material welding. However, even friction welding leads to disintegration of reinforcing particles and their increase in volume fraction in the weld region. This leads to low ductility welds although strength increment is observed with an increase in the volume fraction of reinforcing particles in the parent material. To keep the particle disintegration under control it is necessary to work at low friction and forging pressures.

Conventional continuous friction welding has two phases and they are heating phase and forging phase. Frictional heat generated in heating phase due to a relative motion between rotating member and non rotating member pressed against the rotating member is made use to weld under pressure. Forging phase aids in further consolidation of the weld at a higher pressure than the friction pressure. The nature of the welding process is such, that material deformation is uni-directional. This leads characteristic directional material flow lines in the direction of flow. This type of microstructures exhibit low impact toughness if the notch is located in the direction of this flow pattern due to least resistance of the microstructure to crack propagation. This phenomenon has been observed in the welds of Al-Li alloy AA 8090, ferritic stainless steel and titanium aluminide etc. The effect is pronounced with an increase in forge pressure although the strength increases with an increase in the forge pressure. Thus, there is need to identify optimum pressure combinations.

9.2. Friction welding of dissimilar metals [37-41]

Conventional fusion welding of some of the dissimilar metal combinations is not feasible due to metallurgical incompatibility leading to the formation of brittle intermetallics, thermal mismatch, wide differences in melting point etc. Solid-state welding processes like friction provide an answer to weld such combinations. Some of the material combinations are un-weldable even by friction welding. There is a need to keep the interaction as low as possible in order to keep under check the extent intermetallic formation to weld difficult to weld material combination.

In the welding of low alloy steels containing medium carbon levels to austenitic stainless steels that contain low carbon extended heating phase leads to lowering of impact toughness. Post weld heat treatment further aggravates the problem. Incorporation of Ni as a diffusion barrier could solve the problem.

In the welding dissimilar pure metal combinations it has been observed that in systems, which have tendency to form intermetallics (Fe-Ti, Ti-Cu) and insoluble systems (Fe-Cu) extended interaction time results in low strength. In soluble systems like Cu-Ni and Fe-Ni extended heating phase results in improved strength. X-Ray diffraction studies revealed that in intermetallic forming systems extension of heating phase leads to the formation of brittle phases.

It was made possible to weld the un-weldable Ti-Ni dissimilar metal combination with aluminium as interlayer material. The thickness of the interlayer dictated the properties. Too thin an interlayer could not act an effective barrier for diffusion to avoid intermetallic formation and consequent hard and brittle joints. If the interlayer was thick the joints were soft and weak. Interlayer materials like stainless steel also were found to be suitable candidates.

9.3. AISI 304 to AA 6061-Effect of electroplated interlayer:

This dissimilar combination is un-weldable by conventional fusion welding due to the tendency for brittle intermetallic formation. Solid state welding processes like friction welding are reported to be employed in such situations. Direct welding of this combination resulted in brittle joints with 0^0 bend angle due formation of Fe_2Al_5. To alleviate this problem continuous drive friction welding was carried out by incorporating Cu, Ni and Ag interlayer to act as diffusion barrier. The interlayer was incorporated by electroplating. Welds with Cu and Ni interlayer were also brittle due to presence of $CuAl_2$ and $NiAl_3$. Silver acted as an effective diffusion barrier for iron avoiding the formation of Fe_2Al_5. Therefore welds with Ag interlayer were strong and ductile and could be bent to an angle of 100^0 [41].

9.4. Maraging steel (MDN 250) to low alloy steel (AISI 4340)

Maraging steels and Ni-Cr-Mo low alloy high strength steels are employed for applications such as rocket motor casings, landing gears, leaf springs etc. Maraging steels exhibit ultrahigh strength and high toughness while low alloy steels equivalent to AISI 4340 posses high strength. The strength-toughness combination of maraging steels is due to precipitation hardening at 480°C, low alloy steels are subjected to austenitizing followed by quenching and tempering to achieve the required strength and moderate toughness. Certain applications call for joining of these dissimilar combinations, preferably by welding. Fusion welding of the dissimilar combination leads to migration of C, Mn, and S from the low alloy steel side towards maraging steel through the weld. The weld zone and adjacent maraging steel do not therefore respond to post weld ageing treatment i.e., applied to maraging steel. In addition low alloy steel also does not respond to ageing treatment. Similarly maraging steel also does not respond to heat treatment i.e., applicable to quench hardenable steels.

Solid state welding processes like, friction welding is a possible solution to weld such dissimilar metal combinations. Since the welding is performed in solid state, extensive migration of elements does not take place. In addition, as the materials are subjected to hot working, the weld zone exhibit fine grain size. The other advantages of solid state welding are that the welds are free from segregation; porosity and liquation cracking that are common in conventional fusion welding. Keeping the foregoing in view continuous friction welding studies on dissimilar metal combination of maraging steel (MDN 250) and low alloy steel (AISI 4340) has been taken up. For comparison similar combinations are also studied. The study consists of mechanical property evaluation such as tensile, impact and hardness coupled with microstructural examination. Influence of post weld heat treatments also forms part of the study. The post weld heat treatments investigated are quenching and tempering treatment i.e., employed for low alloy steel, post weld ageing at 480^0C and post weld solution treating and ageing which is adopted for maraging steel.

Maraging steel welds exhibited lower notch tensile strength and toughness compared to as received solution treated parent metal, due to aligned flow lines pattern. Post weld ageing, post weld solutionizing and ageing further reduced the impact toughness, due to precipitation hardening. However, quenching and tempering treatment similar to that employed for the low alloy steel, which does not involve precipitation hardening experienced marginal reduction in toughness (205J) from the as-welded condition (230J). In plain tensile testing, all the failures of similar and dissimilar welds are located outside the weld region. AISI 4340 welds exhibited low strength and toughness compared to parent metal. The strength could be improved by quenching and tempering treatment. No post weld heat treatment could improve the tough toughness. Dissimilar welds exhibited low strength, low notch tensile strength and low toughness compared to respective parent metal. Post weld heat treatment resulted in further reduction in toughness compared to that in the as-welded condition. The low toughness of AISI 4340 welds and dissimilar metal welds can be attributed to aligned sulphide inclusions. The observed mechanical properties are correlated with microstructure and fractures features [42]

9.5. Titanium to stainless steel

Continuous drive friction welding studies have been carried out on commercial pure titanium (CP Ti) to stainless steels (AISI 316) dissimilar metal combination. These studies involved the effect of friction welding parameters such as burn-off and friction force on hardness and tensile properties. The welds have been characterized employing optical microscopy, electron probe micro analysis (EPMA) and scanning electron microscopy (SEM) to explain the trends in mechanical properties. It has been observed that the maximum strength of the friction welds that is achievable is equivalent to that of CP Ti. However, all the welds exhibit no macroscopic deformation and hence no ductility. The interface regions of the welds contain intermetallics like $FeTi$, Fe_2Ti, Cr_2Ti, $NiTi$, $NiTi_2$ and sigma phases in addition to β Ti. The hardness and strength are observed to vary in a narrow band at varying burn-off as well as friction force. However, a definite trend showing the influence of burn-off as well as friction force is noted. Lowest burn off and friction force yielded maximum strength and low hardness.

This particular weld contained higher volume fraction of β Ti and lower volume fraction of intermetallics. In addition, this weld did not contain the intermetallic Fe_2Ti while, all other welds contain low volume fraction of β Ti and higher percentage of intermetallics including Fe_2Ti. The poor ductility of the welds is attributed to the presence of intermetallics. From the strength point of view, it is opined that the presence of Fe_2Ti results in lower strength. In general, Ti is observed to diffuse more towards stainless steel (SS). However, in the peripheral regions Fe and other elements in SS are found to diffuse more towards Ti. It may be added that the diffusion of elements towards Ti side is apparent as this phenomena is due to mechanical transport of SS towards Ti, which is observed in all these cases. Post weld heat treatments (PWHT), although reported to improve bend ductility without compromising strength, in our study, we observe a strength reduction after PWHT with no improvement in ductility [43]

Acknowledgements

The author is grateful to DRDO for funding the research activities and is thankful for all his colleagues and superiors for extending invaluable support enabling him to carry out research activities.

References

1. G. Madhusudhan Reddy, A.K. Mukhopadhyay and A. S. Rao, *Influence of scandium on the weldability of 7010 Al-alloy*, Science and Technology of Welding and Joining, Vol. 10, No. 4, 2005, p. 432.

2. A.K. Mukhopadhyay, G.M. Reddy, Influence of silver addition on the microstructure and properties of AA 7010 alloy sheets, 7th International Aluminium alloy conference, 2002, UK

3. S.R. K. Rao, G. Madhusudhan Reddy, P.S. Rao, M. Kamaraj and K. Prasad Rao, *Improving the mechanical properties of 2219 aluminium alloy GTA welds by Scandium addition*, Science and Technology of Welding and Joining, Vol. 10, No. 4, 2005, p. 418.

4. S.R. Koteshwara Rao, G. Madhusudhan Reddy, K. Srinivasa Rao, M. Kamaraj and K. Prasad Rao, *Grain refinement through arc manipulation techniques in Al-Cu alloy GTA Welds*, Materials Science and Engg., Vol. A, No. 404, 2005, p. 227.

5. S.R. Koteshwara Rao, G. Madhusudhan Reddy, K. Srinivasa Rao, M. Kamaraj and K. Prasad Rao, *Reasons for superior mechanical and corrosion properties of 2219 aluminium alloy electron beam welds*, Materials Characterization, Vol. 55, No. 4-5, November 2005, p. 345.

6. G. Madhusudhan Reddy, Amol A. Gokhale and K. Prasad Rao, *Optimization of pulse frequency in pulsed current gas tungsten arc welding of aluminium-lithium alloy weld*, Material Science and Technology, Vol. 14, 1998, p. 61.

7. G. Madhusudhan Reddy, Amol A. Gokhale, K. Satya Prasad and K. Prasad Rao, On *the chill zone formation in aluminium-lithium alloy welds*, Science and Technology of Welding and Joining, Vol.3, No. 3, 1998, p. 208.

8. G. Madhusudhan Reddy, Amol A. Gokhale and K. Prasad Rao, *Weld microstructure refinement in a 1441 grade Al-Li alloy*, Journal of Material Science, Vol. 32, 1997, p. 4117.

9. G. Madhusudhan Reddy and A.A. Gokhale, *Gas tungsten arc welding of AA 8090 Al-Li alloy*, Trans. Indian Inst. Met. Vol.46, No.1, February 1993, p 21.

10. G. Madhusudhan Reddy and Amol A Gokhale, *Weld Solidification Cracking In Aluminium-Lithium Alloys*, Metals Materials and Processes, Vol. 19, No.1-4, 2007, p 307.

11. G. M.Reddy, P. Sammaiah, CVS Murthy and T. Mohandas, Influence of welding techniques *microstructure and mechanical properties of AA 6061 gas tungsten arc welds*, National Conference on Processing of Metals, January 31-February 01, 2002, PSG College of Technology, Coimbatore, India.

12. G. Madhusudhan Reddy, C. V. Srinivasa Murthy, N. Viswanathan and K.P. Rao, *Effects of electron beam oscillation techniques on solidification behaviour and stress rupture properties of Inconel 718 welds*, Science and Technology of Welding and Joining, Vol 12, No.2, 2007, p. 106.

13. K. Sivaprasad, S. Ganesh Sundara Raman, P. Mastanaiah and G. Madhusudhan Reddy, Influence of magnetic arc oscillation and current pulsing on microstructure and high temperature tensile strength of alloy 718 TIG weldments, Materials Science and Engineering, Vol. A, No. 428, 2006, p. 327.

14. K. Sivaprasad, S.G.S. Raman, C.V.S. Murthy and G.Madhusudhan Reddy, *Coupled effect of heat input and beam oscillation on mechanical properties of alloy 718 electron beam weldements*, Science and Technology of Welding and Joining, Vol. 11, No. 1, 2006, p.129.

15. G.D. Janaki Ram, A. V. Reddy, K.P. Rao, G. Madhusudhan Reddy, J.K.S. Sundar, *Microstructure and tensile properties of Inconel 718 pulsed Nd-YAG laser welds*, Journal of Material Processing Technology, No.167, 2005, p. 73.

16. G.D. Janaki Ram, A.V. Reddy, K. P. Rao and G. Madhusudhan Reddy, *Micro structural characterization of Inconel 718 gas tungsten arc welds, Practical Metallography*, Vol. 21, No.10, 2005, p. 1132.

17. G.D. Janaki Ram, A. V. Reddy, K. P. Rao and G. Madhusudhan Reddy, *Control of Laves phase in Inconel 718 GTA welds with current pulsing*, Science and Technology of Welding and Joining, , Vol. 9, No. 10, 2004, p. 390.

18. G. Madhusudhan Reddy, T. Mohandas and D.S. Sarma, *Cold cracking studied on low alloy steel weldments - Effect of filler metal composition*, Science and Technology of Welding and Joining, Vol. 8, No.6, 2003, p. 407.

19. G. Madhusudhan Reddy and T. Mohandas, *Cracking in high strength low alloy steel weldments – A case study*, DRDO Workshop on Advanced Manufacturing Technology, 30th April-1st May, 2004. Organised by DRDL, Hyderabad.

20. G. Madhusudhan Reddy and T. Mohandas, *Ballistic performance of quenched and tempered high strength low alloy steel weldements*, Indian Welding Society Journal, , Weld 2, Bead 4, September 2005, p. 19.

21. G Madhusudhan Reddy, T. Mohandas and K. Papukutty, *Enhancement of ballistic capabilities of soft welds through hardfacing*, Int. Journal of Impact Engineering, Vol. 22, 1999, p. 775.

22. T. Mohandas, G. Madhusudhan Reddy and B. Satish Kumar, *Heat affected zone softening in high strength low alloy steels*, Journal of Materials Processing Technology, Vol. 88, 1999,   p. 284.

23. G. Madhusudhan Reddy and T. Mohandas, *Effect of welding process on the ballistic performance of high strength low alloy steel weldments*, Journal of Material Processing Technology,   Vol. 74, 1998, p. 27.

24. G. Madhusudhan Reddy, *'Insitu' hard layer formation in soft welds for improved ballistic Performance*, Issue No.11, Jan-Jun., 2002, Dhatu Drishti, An In house publication of Defence Metallurgical Research Laboratory, Hyderabad,

25. S. Sundaresan, G.D. Janakiram and G. Madhusudhan Reddy, *Microstructural refinement of weld fusion zones in α-β titanium alloys using pulsed current welding*, Material Science and Engineering, Vol. A, No. 262, 1999, p. 88.

26. N.K. Babu, S.G.Sundraraman and G. Madhusudhan Reddy, *Influence of beam oscillation pattern on the structure and mechanical properties of Ti-6Al-4V electron beam weldments*, Science and Technology of Welding and Joining, Vol. 10, No.5, 2005, p. 583.

27. G. Madhusudhan Reddy Swati Biswas, T. Mohandas, *Effect Of Electron Beam Oscillation On The Microstructure, Residual Stresses And Mechanical Properties of Ti 6Al 4V Welds*, Metals Materials and Processes, Vol. 19, No.1-4, 2007, p.307.

28. G.D. Janaki ram, G. Madhusudhan Reddy and S. Sundareshan, *Effect of pulsed welding current on the solidification structure in Al-Li-Cu and Al-Zn-Mg alloy welds*, Practical Metallography, Vol. 37, No.5, 2000, p. 277.

29. G. Madhusudhan Reddy and Suresh D Meshram, *Grain refinement in ferritic stainless steel welds through magnetic arc oscillation and its effect on tensile property*, Indian Welding Journal, Vol. 39, No.3, 2006, p. 35.

30. G. Madhusudhan Reddy and T. Mohandas, *Explorative studies on grain refinement of ferritic stainless steel*, Journal of Material Science Letters, Vol. 20, 2001, p. 721.

31. T. Mohandas, G. Madhusudhan Reddy and Mohammad Naveed, *A comparative evaluation of gas tungsten and shielded metal arc welds of a ferritic stainless steel*, Journal of Materials Processing Technology, 1998.

32. G. Madhusudhan Reddy and T. Mohandas, *The effect of electron beam shape on microstructure and mechanical properties of ferritic stainless steel*, 41st National Metallurgists' Day, 51st Annual Technical meeting, Science City, Kolkata, 14-16 November 2003, p. 171.

33. G. Madhusudhan Reddy and T. Mohandas, *Observation in welding of* $\alpha_2 + O + \beta$ *titanium aluminide*, Science and Technology of Welding and Joining, Vol. 6, No.5, 2001, p. 300.

34. G. Madhusudhan Reddy and T. Mohandas, *Friction welding studies of* $\alpha_2 + O + \beta$ *titanium aluminides, WMA (2) to WMA (10)*, Symposium on Joining of Materials, Organized by Welding Research Institute , Indian Welding Society, Southern zone, , Tiruchirapalli, 22-23 July, 2004

35. G. Madhusudhan Reddy and T. Mohandas, *Friction welding studies on AA 2124 Al-SiCp metal matrix composites–Effect of particle size and volume fraction of reinforcement*, National Welding Seminar, December 2002, Kolkata.

36. G. Madhusudhan Reddy, T. Mohandas and P. Sobhanachalam, Influence of friction welding parameters on mechanical properties of AA 8090 Al-Li alloy, International Welding Symposium on Emerging Trends in Welding, Organised by Indian Welding Society, HiTEX Exhibition Centre, Hyderabad, India, 22-23 February, 2002, p. 147.

37. S.D. Meshram, T. Mohandas and G. Madhusudhan Reddy, Friction welding of dissimilar pure metals, Journal of Materials Processing Technology, Vol. 184, 2007, p. 330.

38. G. Madhusudhan Reddy and T. Mohandas, *Dissimilar metal friction welding of maraging steel to low alloy steel*, 59th Annual Technical Meeting, IIT Madras, Chennai, 14-16 November 2005, PC 5 (TM-335), p. 161

39. G. Madhusudhan Reddy and T. Mohandas, *Dissimilar metal friction welding of maraging steel to low alloy steel*, National Welding Seminar 2006, Chennai Trade Center, Chennai, 24-26 November 2006, p.9.

40. V.V. Satyanarayana, T. Mohandas and G. Madhusudhan Reddy, *Dissimilar metal friction welding of austenitic-ferritic stainless steels*, Journal of Materials Processing Technology, No. 160 (2), 2005, p. 128.

41. G. Madhusudhan Reddy, T. Mohandas, and A. Sambasiva Rao, *Role of electroplated interlayer in continuous rive friction welding of AA 6061 to AISI 304 dissimilar metals*, Science and Technology of Welding and Joining, Vol. 13, No.7, 2008, p. 619.

42. G. Madhusudhan Reddy and T. Mohandas, *Dissimilar metal friction welding of maraging steel to low alloy steel*, National Welding Seminar 2006, Chennai Trade Center, Chennai, 24-26 November 2006, p. 9.

43. G. Madhusudhan Reddy and T. Mohandas, *Continuous drive friction welding of titanium to stainless steel dissimilar metal combination*, International Welding Conference 2007, Chennai Trade Center, Chennai.

EDUCATION, ENVIRONMENT AND MANAGEMENT

Repositioning Metallurgical Engineering Education in India to Meet Future Challenges

Supriya Das Gupta

Chairman and Managing Director
M.N. Dastur & Company (p) Ltd.
Consulting Engineers, Kolkata

1. The Changing Global Scenario

21st century is an era of numerous global challenges – the need to save our earth from impending disaster due to climate change and the greenhouse effect; growing demands for clean and alternative forms of affordable energy, reliable water supply, infrastructure for booming population, to name but a few. The engineering community, including the metallurgists and materials scientists, will be called upon to rise to the occasion and help mankind in successfully meeting these changallegs.

A more recent phenomenon is that of so called global melt down, and the associated challenges being faced by the world economy, resulting from short-term dips in demand on various fronts. This phenomenon has also had it s effect on the steel industry. As a consequence, steelmakers are now confronted with the need to enhance effiency, reduce costs and improve profitability, in order to ensure sustenance under adverse market conditions. However, it is expected that this growth path in the near future. In any case, the steel industry industry has successfully faced such situations in the past, thanks to the ingenuity and adaptability of the metallurgists.

It is generally acccpted that scientific and engineering knowledge doubles every 10 years or so. This geometric growth rate has been reflected in the accelerating rate of introduction and adoption of new technologies. The comfortable notion that a person learns all that he or she needs to know in a four-year engineering program is just not true and never was. Even the "fundamentals" are undergoing changes/modifications, as new processes and technologies are becoming available.

To keep pace with this fast changing scenario, engineers will have to accept the responsibility for their own continual re-educations, and engineering colleges will need to prepare engineers capable fo doing so by teaching them how to learn and how to think accordingly. Engineering colleges will also need to adopt appropriate organizational structures that will allow continuous adaptation to satisfy the professional needs of the engineering workforce, that are changing at an increasing rate.

2. The Metallurgical Heritage of India

The history of civilization is in many ways linked to the use of metals in antiquity. The commonly used metals in antiquity include gold, silver, copper, iron, tin, lead, zinc, and mercury. Although modern metallurgy has seen an exponential growth since the Industrial Revolution, it is interesting that many modern concepts in metallurgy have their seeds in ancient practices that pre-date the Industrial Revolution.

Many people may not be aware of the Indian contributions to process metallurgy in an ancient times. India's metallurgical achievements have been amongst the most important steps in the process of the large-scale production of some metals, such as Steel and Zinc.

The history of Indian metallurgy starts before 2500 BC. Early gold and silver ornaments dating back to around 3000 BC have been found from Indus Valley sites such as Mohenjodaro. Another significant development was the production of Wootz Steel, and its export for making the famous Damascus sword. The famous Iron Pillar, which has been in existence for over 1,500 years, is another example of the rich metallurgical heritage of India.

In the field of non-ferrous metallurgy, the earliest firm evidence of the production of metallic Zinc is from India. Investigations have shown that zinc technology was developed in India since 4th century Bc, and it culminated in large scale production of zinc in Zawar, Rajasthan in 13th Century AD.

3. Trends in Metallurgy & Materials Science

The skills and training to be acquired by a metallurgist through formalized, structured education, and often via specialized industrial experience, have to follow the trends in technological development. Accordingly, it would be revelant here to review briefly the recent technological advances in metals and materials in order to have an idea of the vast dimensions and directions of the changes required in metallurgical engineering education in the country .

The demands on materials to perform under complex service conditions, necessitates rapid advances in the design and development of materials. Development of new concepts starting at the electronic structure of materials and their modeling require good insights of structure, synthesis and performance. The ability of tailor the structure of materials is essential to realize quantum improvements in the performance of present day materials. Designing materials through modeling and simulation can substantially bring down cost and time involved in developing new materials and processes. Besides, the ability to synthesize such frontier materials is also essential to achieve technological breakthrough. A few typical examples of research and development al work, resulting in the redefining of frontiers in the design of materials are discussed below.

Continuous research is being carried out in the area of development of ultra low carbon steels with zirconium or niobium as micro alloying additions to meet the requirements of the nuclear power stations, which are expected to play a major role in meeting the huge power requirements of the country in the years to come. In the automobile sector, there is a continuing endeavour to produce steels with lower alloy additions and with proper combinations of strength and ductility. Similarly, high-strength alloy with desirable combination of properties are being developed for successful applications in the aerospace industry. Innovative steel processing concepts are being developed for achieving desired nanostructures, required for a new generation for ultra-high strength, weldable structural steels with exceptional toughness, and ductile fracture behavior at very low temperature.

Materials engineering, with a foresight into the future, has been rapidly attracting the attention of everyone. Nano-technology is developing at a vey past pace. In fact the number of publications having the word "nano" in the last tow decades has seen an exponential rise! The applications of metals and materials in the biomedical fields have been there for quite some time (e.g. use of titanium in equipments working in biofluids etc.)

In the area of steel production, improved melting and degassing techniques, continuous casting, controlled rolling, thermo-mechanical processing etc. have resulted in enhanced product quality, lower energy consumption, greater mitiation of environmental pollution and reductions in production cost. New ultra – low carbon, interstitial-free steels with improved strength and formability have been developed by microalloy additions of niobium, vanadium and titanium. There is greater emphasis now on near net –shape castings and direct rolling.

Numerous other developments in the field of metallurgy and materials science are in various stages of exploitation. Some of these are:

- Development of a better class of superalloys through advances in processing techniques such as directional solidification, single crystal solidification, ect.
- Increasing use of electron bean and plasma refining for better melting.
- Development of special alloys to better withstand corrosive environments.
- Special rapidly solidified alloys for aerospace allocations.

A new "miracle material" emerging on the horizon is graphence, which is basically a single-atom-thick sheet of graphite. Research has shown that grapheme has greater ability to conduct electricity at room temperature than any other known material, making it a likely candidate to replace silicon as a semi-conductor material.

4. Role of Metallurgical Institutions

Metallurgical education in India started at BHU way back in 1923. It was later in 1939, after a gap of some sixteen years, that the Calcutta University introduced a course in Metallurgy for undergraduate at the Bengal Engineering College, Shibpur. Much later, in the pre-independence years, a third institution – the Indian Institute of Science, Bangalore – commenced imparting Metallurgical education. After independence, Metallurgical education came to be part of the curricula of the newly established Indian Institute of Technology, the regional Engineering Colleges (presently called National Institutes of Technology) and other Universities in the country.

It has been projected that per capita consumptions of steel, Aluminum, Copper, Zinc, Lead, and Nickel could drastically increase in the course of the next 10-15 years. To meet the predicted increase in domestic consumption levels, the country's metal industry will have to grow and develop as a globally competitive sector. This will necessitate the ready availability of a pool of competent professionals possessing the required knowledge base, together with appropriate technical and leadership skills.

As Metallurgical development has become more complex and involved, today's metallurgist is different from that a decade age; and one can expect a similar change over the next decade. How do the engineering institutions keep pace with the changing needs and continue to turn out competent professionals with the required technical and leadership skills? This is a continuous self adjusting system, sometimes voluntary and sometimes forced, but always changing.

A debate on the importance and future of engineering education in a time of changing social needs, rapidly changing technology and production processes, within the framework of a globalized economy had already been initiated by a number of eminent metallurgists during 1990s. It has been suggested that the curriculum needs continuous renewal to serve the profession effectively. Our goal must be an economy where metallurgical development continues to lead and support innovative technologies. Strong emphasis has also been given on inclusion of environment, energy, development of advanced materials and basic mineral processing as a part of under graduate curriculum. This could also be achieved through more industry-academia interaction to create continual educational development for metallurgical graduates to suit the requirements of the metallurgical industry.

5. Concluding Remarks

With the national on the threshold of a renaissance in the metals industry, the metallurgical community to face with a colossal challenge. The growing attention to materials science and development of newer materials has placed new demands on conventional metallurgical fields such as extraction metallurgy and metallurgical processes, and physical metallurgy. Research and development in these areas will enable the Indian Industry to develop and thrive in the context of large scale increases in the production of conventional metals, and the growing demand for advanced metallurgical products in the expanding automotive, nuclear and aerospace industries.

There can be no doubt that basic research and scientific knowledge underpins development in technology. As the Indian economy prospers, scientists and engineers – especially academics, who train the next generation of students – will largely be responsible for sustaining the technological development. This, in turn, will necessitate constant review and revamping of the curricula in metallurgical institutions, and raising educational and research standards in areas revelant to the Indian Industry. Nothing is more fundamental to the future of our nation's metallurgical industry than attracting talented young men and women to the pursuit of an engineering degree and providing them with an education adapted for the 21st century.

Acknowledgement

I am thankful to Prof. P. Ramachandra Rao, Conference Chairman, Dr. J. Vilava Kumar, Conference convenor and the members of the Core committee of this International Conference, for inviting me to participate in the deliberations and deliver the Keynote Address. It is, indeed, an honour and a privilege to be present in such a gathering of luminaries in the field of Metallurgy and Materials Technology.

Futuristic View of the Role of the Metallurgical Engineering Professional in the Steel Industry

Bhaskar Yalamanchili

Director of Corporate Quality, Gerdau Ameristeel, Florida, USA
E.mail: byalamanchili@gerdauameristeel.com

Abstract: *The paper describes the author's experience as a Metallurgical Engineering Professional in the last 30 years on the production of long products from the recycled steel scrap-electric furnace-continuous casting process. During this period, metallurgical engineering professionals have led the development work that was necessary for improving process and product quality for more demanding products from rebar, structural, merchant, special bar and rod quality. All of these developments and improvements are now used world wide due to the global aspect of the steel industry.*

Recently, two new futuristic initiatives are developing; health and environment and recycled content of the finished steel product. The first initiative is a world wide effort to protect health and the environment by limiting the hazardous substances in materials including steel. Health and environment initiatives include the European Council (EC) Restriction on Hazardous Substances (RoHS), and the EC Regulation on the Restriction, Evaluation, and Authorization of Chemicals (REACH).

The second strong imitative is to increase recycling and other forms of recovery of materials to protect the environment. This is being led by the EC End-of-Life for Vehicles Directive that sets as goals for re-use and recycled content for automobiles. In the US, these efforts are led by the US Green Building Council's Leadership in Energy and Environmental Design (LEED) where goals are set for the recycling of materials including steel.

The author has led these efforts in North America for Gerdau Ameristeel, and will describe Gerdau Ameristeel's directives to protect health and the environment.

1. Introduction to Gerdau Ameristeel

The Gerdau Group is the world's 14th largest steelmaker and the largest producer of long steel products in the Americas. It has 272 industrial and commercial facilities, five joint ventures and two associated companies. Gerdau operates in Brazil, Argentina, Chile, Colombia, Peru, Uruguay, Mexico, Dominican Republic, Venezuela, the United States, Canada, Spain and India. Currently, Gerdau has an installed capacity of 22.6 million metric tons of steel per year and supply steel for civil construction, industry and agriculture.

The Gerdau Group started to operate in 1901, establishing Pontas de Paris Nail Factory in Porto Alegre, Brazil. The company continuously invests in the training of more than 35,000 employees and in the development of communities.

1.1 Introduction to the future of Metallurgical Engineering professional in the Steel Industry

The subject of this paper is to describe the role of steel industry professional in the in the production of long products around the world including bars, structural beams and shapes, rod and wire. Many of the developments in the steel industry are conceived and nurtured by Steel Industry Professionals and are adopted world-wide by globalization to achieve the efficiency and cost competitiveness that is necessary to be competitive in the global economy

The production of long products is concentrated in the electric furnace, continuous casting and hot-rolling process. Each of these processes has had major developments that have led to their improved production efficiency and over-all reduction in production costs. The total combination of these processes has led to rise of the mini-steel plant that utilizes these efficient processes to convert local raw materials usually in the form of steel scrap into steel products that are used locally. Frequently the mini steel mill location will be based upon the availability of the raw materials and a local market for finished products.

Professionals of all types have developed the processes upon which the mini-steel mill is based. Developments have occurred in process and equipment improvement, energy supply and utilization, raw materials and many other fields that are all important to the over-all process. The development work has led to the production of more demanding steel products. Initially most mini-mills produced rebar and then developed the know-how for producing more demanding products including merchant bars, structural products, low and high carbon wire rod and special quality bars. This paper will concentrate on the role that the Metallurgical Engineering professional has contributed to the over all success of the mini-mill. The following few paragraphs describe some of the general developments in steel melting, refining, casting and hot rolling followed by specific development work by the author.

2. General Development

2.1 Steel Melting

The electric furnace steel melting furnace has been developed into an extremely efficient method of melting steel scrap and carrying out preliminary chemistry adjustments primarily including phosphorous removal and nitrogen reduction. The process has been improved so that steel can be produced from blending types of recycled scrap that in most applications is the equal to steel melted from virgin iron units. Specifically, the electric furnace process has been improved by the following important processes:

2.2 Phosphorous Removal

Burnt high calcium lime and dolomitic lime base materials are added to the furnace early in the melting process to form a fluid basic slag that is effective in the removal of phosphorous at low temperatures. Pneumatic injection of carbon with the addition of oxygen assures an active carbon-oxygen boil that enhances the slag/metal reaction by increasing the slag-metal interfacial contact that is necessary for the transfer of phosphorous from metal to the slag.

2.3 Nitrogen Reduction

The reduction of nitrogen in the finished steel is important because of its adverse affect on strain aging in cold worked wire. Nitrogen is reduced in the electric furnace by an active carbon-oxygen boil from the addition of carbon to the charge materials or from injection or the addition of iron units high in carbon content including pig iron and direct reduced iron. Oxygen injected into the molten bath reacts with the carbon and results in a strong carbon-oxygen boil from the gaseous carbon-monoxide reaction product. Nitrogen dissolved in the liquid steel is removed by being absorbed into the carbon-monoxide bubbles formed in the liquid steel. The generation of the CO gas causes the formation of a foamy slag that covers the liquid steel and prevents the absorption of nitrogen from the nitrogen in the air. Process development in this field has resulted in steels with nitrogen content approaching that of steel made from virgin iron sources including pig iron and DRI.

2.4 Residual Elements

The control of residuals in the finished steel has become very precise by computerized blending of scrap types based on the residual content. For especially low residual content requirements, pig iron and DRI are often blended with the scrap mix.

2.5 Steel Refining

In most modern mini-mills, liquid steel is transferred from the melting furnace to the Ladle Metallurgy Furnace (LMF) for refining. In the LMF process, steel chemistry, temperature and deoxidation are all done in a controlled way so that steel of the precise chemistry and temperature is supplied to the continuous caster, the next step in the process.

2.6 Reduction of Sulfur, Dissolved Oxygen and Non-Metallic Inclusions

Sulfur and dissolved oxygen can be reduced to particularly low values by slag-metal reactions that are effective due to the formation of basic-reducing slag's and active stirring with inert argon gas or in some cases by electro-magnetic stirring. Inclusions formed in the LMF deoxidation products are absorbed by the slag to produce steels with low inclusion content.

2.7 Reduction of Hydrogen

Vacuum degassing of liquid steel has permitted the production of steels with very low hydrogen (1-2 PPM) content that is particularly important in high carbon and highly alloyed steel for demanding rail, carbon and alloy steel applications. Various processes have been developed to remove hydrogen separately or in combination with ladle refining processes.

2.8 Continuous Casting

The continuous casting process converts the liquid steel to cast shapes suitable for hot rolling without the need for heavy duty breakdown rolling mills previously required for ingot cast steel. Although the concept is straight forward, casting of suitable product without defects and with uniform properties required the development work by the Metallurgical Engineering Professional. Problems that have been solved have included the control of porosity, cracks and shape distortion in the cast product. Chemical uniformity has also been achieved by application of principals of nucleation and growth of crystalline structure during solidification.

2.9 Hot-rolling

Although much of the development in hot rolling is based upon increased efficiency achieved through equipment improvements, the Metallurgical Engineering Professional has made remarkable improvements in the mechanical properties of long products. These developments have included the vast field of micro alloying and controlled cooling processes to produce steels with higher strength, greater ductility and formability without the need for further heat treatment.

2.10 Specific developments

The remaining portions of this presentation will give insight into the professional's input in the steel industry based upon my specific experience for long products, and presented previously at the Wire Association's 2008 Mordica Award Paper (1).

3.Technical Aspects

3.1 Resistivity Measurements in Telephone Wire

During my first internship job in India, at Mukand Iron and Steel Company, I was asked to develop a low carbon grade of steel to meet the restricted resistivity requirements for telephone wire. With the help of Dr. Rau and my colleagues we produced several different chemistries for testing trials of both rod and customers' wire. Finally, we were successful in meeting all requirements specified by the leading International Standards governing this grade of wire.

Figure 1 Telephone Wire.

3.2 Use of Oxygen Probe to Assist Deoxidation Practice

While producing steel by the newly installed continuous casting process at Atlantic steel, we encountered a serious problem with under deoxidation that resulted in pinholes and blow holes in the billets as shown in Figure 1. With the help of Electro-

Nite we were able to measure the dissolved oxygen and thus optimize our deoxidation practice so as to avoid these defects.

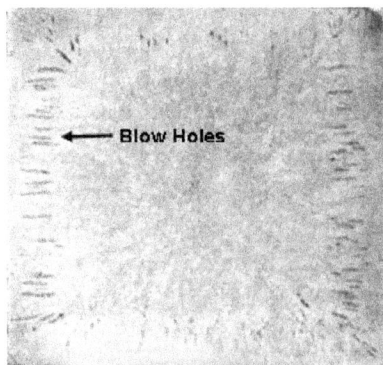

Under-Deoxidized Steel   Fully-Deoxidized Steel

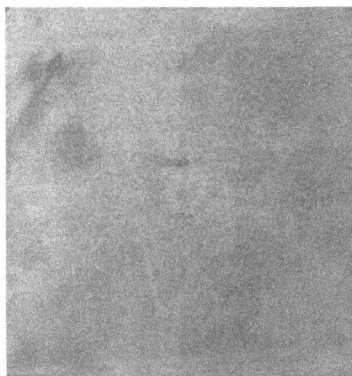

Figure 2 Macro-photographs of Continuous Cast Billet Cross Section

3.3 Deoxidation Process

In the LMF furnace, the dissolved oxygen content of the liquid steel is reduced by the following reaction:

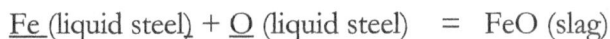

$$\underline{Fe}\ (liquid\ steel) + \underline{O}\ (liquid\ steel)\quad =\quad FeO\ (slag)$$

The FeO content of the slag is reduced by the addition of strong reducers such as silicon and calcium, which causes the following reaction to go to the right to establish equilibrium.

$$FeO\ (slag) + Ca = CaO\ (slag) + Fe\ (liquid\ steel)$$

Oxygen levels in the steel can be reduced to lower PPM values by the continual stirring and additions of reducing agents to the slag such as calcium and silicon. Measurement of the oxygen content of the steel allowed an optimum range to be determined for optimum casting conditions.

3.4 The Oxygen Probe

The oxygen probe uses a solid stabilized zirconia electrolyte to measure the electrical potential of a standard cell against the dissolved oxygen in steel as the other side of the cell. Electrical potential measurements in combination with liquid steel temperature can then be used to measure the oxygen activity of the steel.

3.5 Porosity in Cast Steel

For liquid steel, the following reaction occurs between carbon and oxygen:

$$C \text{ (steel)} + O \text{ (steel)} = CO \text{ (gas)}$$

The equilibrium value for this reaction is shown in Figure 3 for a temperature of 1600C. Oxygen content must be much below the equilibrium values to avoid the formation of porosity holes in the as cast steel because as the steel cools, the equilibrium shifts to lower solubility of carbon and oxygen resulting in CO gas formation. Evolved CO gas becomes entrapped in the as cast steel forming porosity holes. These holes if exposed to the surface during the reheating prior to hot-rolling will oxidize and will not weld up in rolling.

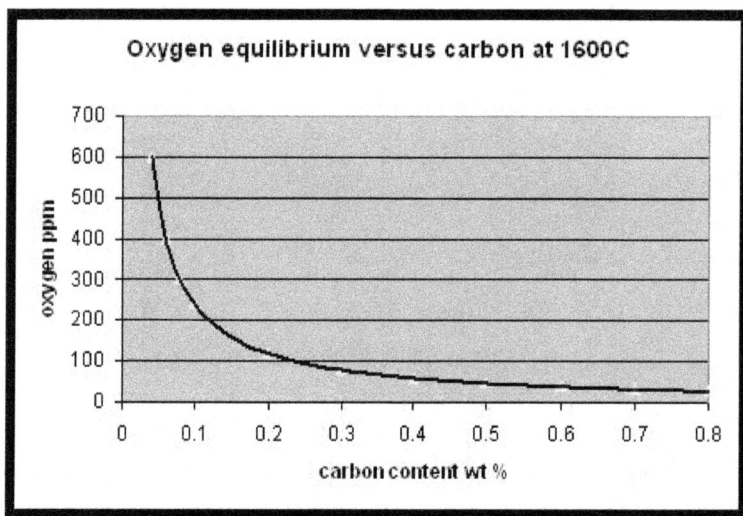

Figure 3 Carbon Oxygen equilibrium diagram for liquid steel at 1600 C.

This work, coupled with the help of Georgia Institute of Technology facilities, enabled me complete the Thesis requirement for my MS in Metallurgy degree. This thesis formed the basis for the measurement of dissolved oxygen in liquid steel to assure complete deoxidation that is still used today.

3.6 Boron in Low Carbon (Patented)

One of the most important contributions to the production of low tensile ductile wire was the treatment with boron to reduce tensile and increase ductility. An example of the need for better ductility was the breakage that occurred in low carbon wire is shown in Figure 4.

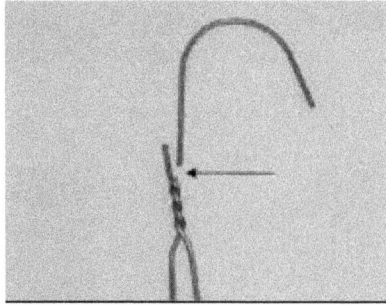

Figure 4 Wire breakage due to low ductility and formability.

The problem was related to the excessive work hardening in the wire drawing process due to the nitrogen content of the steel. This led to the idea of having a controlled amount of Boron to Nitrogen ratio in order to reduce the work hardening effect. Our team at the Beaumont Mill of Gerdau Ameristeel designed chemistry that reduced the work hardening during drawing as well as the subsequent aging. We faced some objection from the melt shop management at that time who were reluctant to use Boron due to its cost. Our team, however, was able to convince management of the value and get permission to produce just one heat. The results from that one heat were extremely interesting! Our Plant manager at that time agreed to proceed with making more heats in order to satisfy our quality goals. This eventually led to receipt of the patent "Process for producing low carbon steel for cold drawing".

The reduction of tensile gain obtained during drawing that resulted from this practice is substantial as shown in Figure 5.

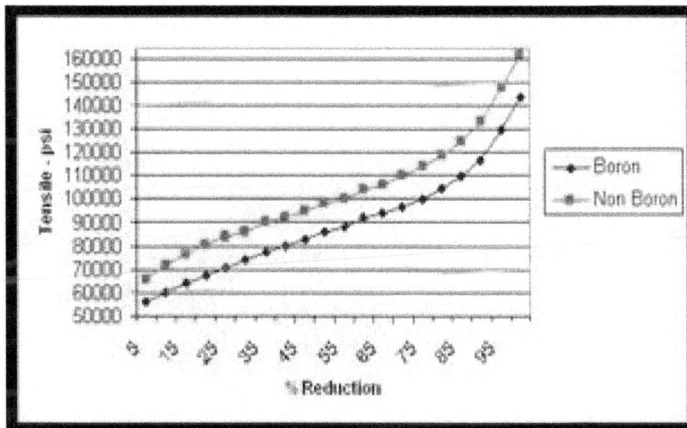

Figure 5 Tensile vs. % Wire Draft Reduction for Boron and Non-Boron Steel Wire.

Understanding the fundamental metallurgical principals enabled us to solve this problem. The nitrogen solute atoms cause increased resistance during wire drawing by pinning dislocations as shown in Figure 6.

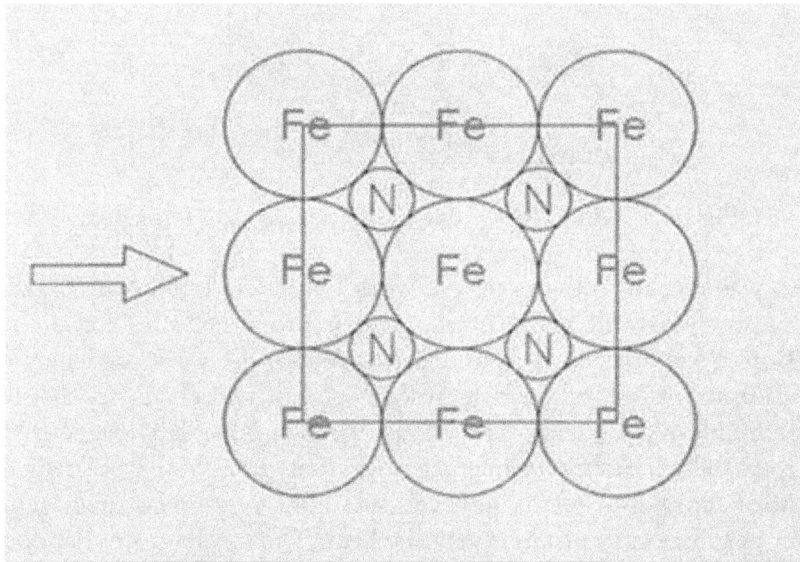

Figure 6 Resistance to Plastic Deformation of Metals (Presence of Nitrogen).

This adds to the amount of strain hardening by increasing the dislocation density that results from tangling (Figure 7). Boron combines with the nitrogen to form BN inclusions (Figure 8), thus removing its dislocation pinning effect. The resultant improvement in ductility from boron addition has become well recognized, and is now a normal practice in our industry.

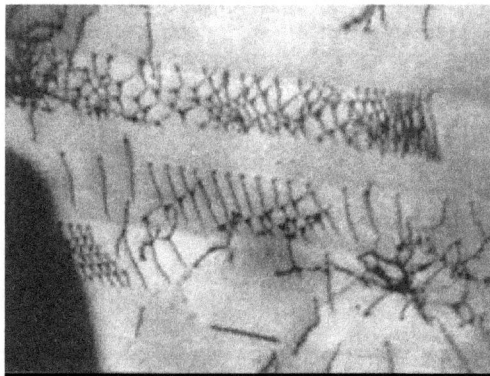

Figure 7 Dislocation Tangling.

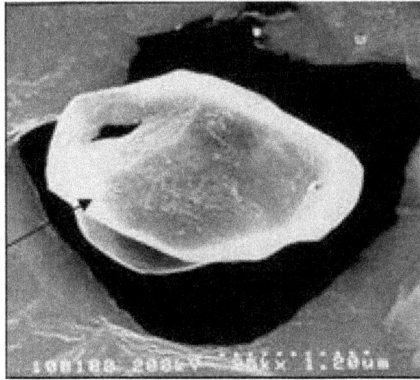

Figure 8 BN Inclusion.

3.7 Chevron Cracking Analysis

Chevron cracking in high carbon wire drawing is an all too common problem for wire drawers. In our experience it was mainly due to the segregation of elements at the center of the continuously cast high carbon billets. Our groups worked on this problem for many years to improve billet casting practices that were designed to minimize segregation. Much work was also done with the controlled cooling practices in the rod mill in order to minimize formation of the centerline hard, non-deformable, microstructures.

Since there are many causes for this phenomenon, we finally developed a Cause Analysis Chart as shown in Appendix 1, to guide us through the identification of the various causes for chevron cracking. Utilizing this technique, along with identifying some of the metallurgical phenomena involved, we were able to solve one of our customer's unique problems relating to chevron cracking. Publication of this approach which highlighted the steelmaking practices used to control nitrogen, ultimately led us to winning the WAI Silver Certificate Award for best ferrous paper in 2004.

3.8 Rust Guidelines

It is easy for anyone to say "give me a coil with no rust or very minimal rust". Our group thought this to be very subjective, and that we needed a more scientific measurement of the degree of rusting required. This would facilitate communication between customer and suppliers by having a "common language" for setting specifications. With this concept we developed a Rust Guide of photographs showing various rust levels on the rod coils, as shown in Appendix 2. This tool is now being utilized regularly by many of our customers. Through publication, it is now used by practically by many wire drawers and rod producers.

433

3.9 Mechanical Damage

After making good quality steel wire rod, if proper care is not taken, it is very easy to damage its surface as shown in Figure 9. It is also very common to produce surface Martensite in high carbon grades when a coil is scraped mechanically, especially in the winter time as shown in Figure 10. So, extreme care is needed to protect the surface in order to avoid costly mistakes and increasing the cost of quality. Vigorous training and explanations have been able to solve / reduce these quality problems.

Figure 9 As Received Rod mechanical Damage.

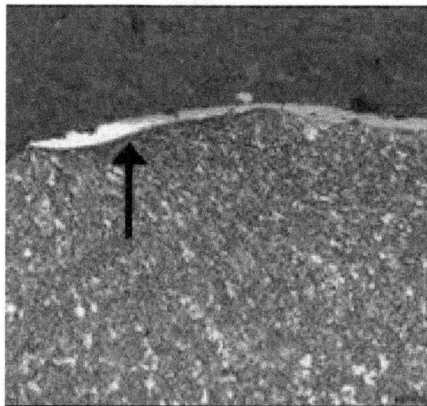

Figure 10 Cross Section with Surface Martensite.

3.10 Methods to Reduce Tensile Gain Rate

Tensile gain rate is reduced by optimizing chemistry, rod mill and wire drawing practices as shown in Appendix 3.

3.11 Cable Stay Wire

We have supplied Prestressed Concrete Strand wire rod to produce wire for Charleston Cable Stay Bridge in USA as shown in Figure 11. Internal and external metallurgical requirements are very stringent to satisfy demanding quality requirements.

Figure 11 Charleston Cable Stay Bridge.

4. Human Aspects

4.1 Mentoring

Over the years we have had several rewarding experiences in mentoring. I am particularly proud to have been able to succeed in having two department members work for several years during their spare time in order to receive their degrees some 25 years after high school graduation. They are now working for us at much higher level jobs. Planning for the future by mentoring associates and professionals new to the industry is a powerful way to contribute to the success of both your organization and the individuals involved.

4.2 Contributing to Professional Associations

Several of the examples given above were taken from our "regular work" that eventually served as the basis for papers that were published in the *Wire Journal*. We also work with several Universities and their students to provide them the opportunity to do industrial based projects that fulfill the requirements for their master's and doctoral theses. Two of these projects have resulted in Wire Association International papers that we co-authored with the students. While preparing the work for publication does add significant work, I can truthfully say that this has greatly contributed to the learning and professionalism of all involved. This, in turn, prepared us all to meet even greater challenges for our Company.

4.3 Continuous Improvement is Necessary

All Technical Professionals should look for ways to lead their organizations to the achievement of continuous quality improvement.

4.4 Guidelines for success

The following guidelines are offered for professional growth in the steel industry:

- **Working as a Team**
  - We have all heard it said, in very many ways, and from very many sources that we should work "as a team". Based on my experience, I emphatically add my voice to this chorus! Being part of a group that competently, objectively, works together definitely leads synergistically to the achievement of better results than any individual could ever do alone.

- **Problem Solving and Innovative Thinking**
  - Long term success requires this kind of thinking so that process improvements can be made in order to achieve the necessary quality requirements at least cost and maximum yield.

- **Achieving Product Quality Should Not Be a Needless Expense**
  I learned this through one of our strongly supportive customers. When a customer is pushing for extreme higher quality levels at your expense, you should challenge that customer to prove that it is actually necessary to have such stringent quality requirements for that particular end use. If the customer wants X product, give him the best quality that you can in that product, but not a Y product which is unnecessary and costs you more

- **Upper Management Support is a Must**
  - It is necessary that you win the support of Upper Management in order to achieve your quality goals. This requires credibility, objectivity, and an ongoing list of achievements.

- **Mentoring is Valuable**
  - o Be alert for opportunities to become actively involved in supporting the career development of others.

- **Participation in Industry and Professional Organizations**
  - o This is another valuable means to enhance the careers of all who become actively involved.

- **Ongoing Education**
  - o Always seek to improve your knowledge through continuing education and industry sponsored seminars.

- **Advancement of Technology**
  - o Recognize and implement opportunities for the advancement of technology in your work environment.

5. Environment and Health

Environment and health issues are important to everyone in the world and plant pollution as depicted in Figure 12 is no longer acceptable. New initiatives are taking place all over the world to produce steel with the least impact on the environment as possible.

Figure 12 Pollution that is no longer acceptable.

5.1 Environment

Recycling of steel scrap is now of paramount importance in the environmental effort around the world. In the USA, recycling of materials is part of the green building projects that are based in part upon the percentage of recycled content. Steel made from recycled steel scrap in the electric furnace or other melting unit is certainly the best method for achieving this goal.

5.2 Health

Initiatives are being developed and promulgated around the world to protect health through limiting the presence of hazardous chemicals in materials including steel, and secondly promoting the recycling of materials to reduce environmental impact.

5.3 Restriction of Hazardous Substances

The first initiative that is being broadly followed throughout the world is the European Directive of the Restriction of Hazardous Substances (ROHS) (3). This directive sets limits for certain organic substances and for the metals industry: lead, mercury, cadmium, and hexavalent chromium. In steel there are no organic compounds and all metals are in the zero valence state, leaving lead, mercury and cadmium of concern. Our approach at Gerdau Ameristeel was to conduct a chemical analysis of products produced by each of our plants for these elements to determine compliance. Results of these analyses performed by atomic absorption showed compliance at all of Gerdau Ameristeel plants.

5.4 Registration, Evaluation and Authorization of Chemicals (REACH)

Directives are developing and being phased in over time from the EC for the control of hazardous chemicals that are imported into the European Union. This Directive referred to as REACH or Registration, Evaluation, and Authorization of Chemicals (4) applies directly to chemical substances and also to articles that might contain hazardous chemicals. Steel is classified as an article and therefore falls into the REACH Directive. Although steel contains no hazardous chemicals as compounds, it does contain or could contain hazardous metals including lead, mercury and cadmium. These elements will need to be reported under the REACH Directive for articles of steel imported into the EU.

5.5 Environmental Impact of Steel

Steel industry environmental issues are very important to the entire world because of the size and scope of the steel industry. According the International Iron and Steel Institute (5) a global steel industry approach is essential to preserving the environment because:

- Steel is an essential material for engineering and construction
- 2007 world steel production was 1.3 billion tons
- Future growth world wide is expected to be 3-5 % per year
- Steel production accounts for 3-4 % of world's greenhouse gases
- Steel production is energy intensive
- Approximately 1.9 ton of CO_2 is produced per ton of steel
- Steel is produced now in three process routes;
  - Blast furnace/basic oxygen furnace – most CO_2 per ton of steel
  - Electric arc furnace/ Direct reduced iron –medium CO_2 per ton of steel
  - Electric arc furnace/recycled steel scrap least CO_2 per ton of steel
- New technology is needed to be developed with less generation of CO_2
  - Requires joint global effort through government support
  - Requires new professionals in the field of process metallurgy to develop new process based on new methods of steel production
  - Requires major R&D expense
    - Requires break through as
    - Processes generating less CO_2
      - Hydrogen reduction
    - Carbon capture

The other approach to reducing green-house gases is through the efficiencies in the use of steel as a material and to design for ease in recycling for a to life cycle approach. Efficiencies must be achieved in the:
- Automotive design for the total life cycle including recycling

- Housing and Building design for energy efficiency and the use of recycled materials

6. Automotive Design for total life cycle

The total life cycle for automotive vehicles has been addressed throughout the world and one good example is the European Directive on the End-of-life for vehicles. The EC Directive on the end-of-life vehicles (6) is a basic outline of how the recycling and re-use of materials can be planned into the design of vehicles. It includes statements on the control or elimination of hazardous substances to prevent release into the environment and to facilitate recycling. In particular the use of lead, mercury, cadmium and hexavalent chromium are limited or prohibited in automotive materials. Similar restrictions have been set for hazardous materials in non-metals, particularly plastics.

The European Directive has set minimum levels for the percentages of recyclable materials in vehicles. These limits are:

- Beginning in January. 2006, the minimum recyclable material shall be 80 % by weight, and for vehicle manufactured prior to 1980 the minimum level is 75 %.
- By January 1, 2015, for end of life vehicles, the reuse and recovery shall be increased to a minimum 85 % by an average weight per vehicle.

To facilitate compliance with directives such as the vehicle-end-of-live EC Directive, Automobile groups have been formed to develop data bases on materials and their possibility for containment of hazardous materials. One very large organization is the International Materials data system (IMDS) (7) that through a central computer system allows automotive companies store and access data on materials used in all vehicles. Each company can privately access data that is necessary to determine hazardous material data content for a given material. As the components of a vehicle are assembled the total content if any of hazardous materials can be accumulated. This system that is world wide in scope is up and running at the EDS in Dallas, Texas.

6.1 Green Housing and Buildings

The design of housing and buildings for energy conservation and environmental preservation is well underway in many countries. In the USA this program is being championed by the US Green Building Council's LEED program. LEED stands for Leadership in Energy and Design. The LEED program has credits for design features, but for the steel industry the significant portion of the LEED program is that LEED issues credits for the recycled content in steel materials, and therefore for steel made from recycled steel scrap. Another feature of the LEED program is that it requires local content be used in the manufacture of materials. For Steel it is recommended that raw materials be gathered locally and by that it is meant within a 500 mile radius. LEED is an example of programs developing around the world that promote the use of recycled materials. It will be up to the Metallurgical Engineering Professionals to produce suitable steel products from recycled steel scrap.

7. Conclusions

As outlined in the body of this presentation, all Quality professionals have to look for ways to lead their organizations to achieve continuous quality, health and environmental improvement. One way to think and improve continuously is through the standard life-cycle graph as shown in Figure 13. You need to think always that you are on the left side "Good" of this curve and need to continually strive for "Great". If you think you are already "Great, and ignore improvement, the next step soon will be the "Grave" of obsolescence. So, always think at the bottom and keep aiming for the top seeking excellence and you will not only achieve your goals but also will enable you to make valuable and necessary contributions to the advancement of the steel industry.

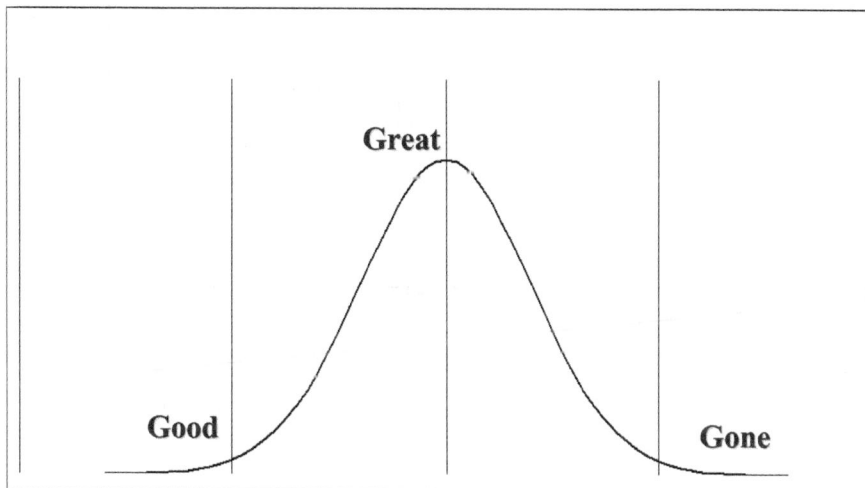

Figure 13 Standard Life cycle graph.

Appendix 1 Chevron Cracking Cause Analysis Chart

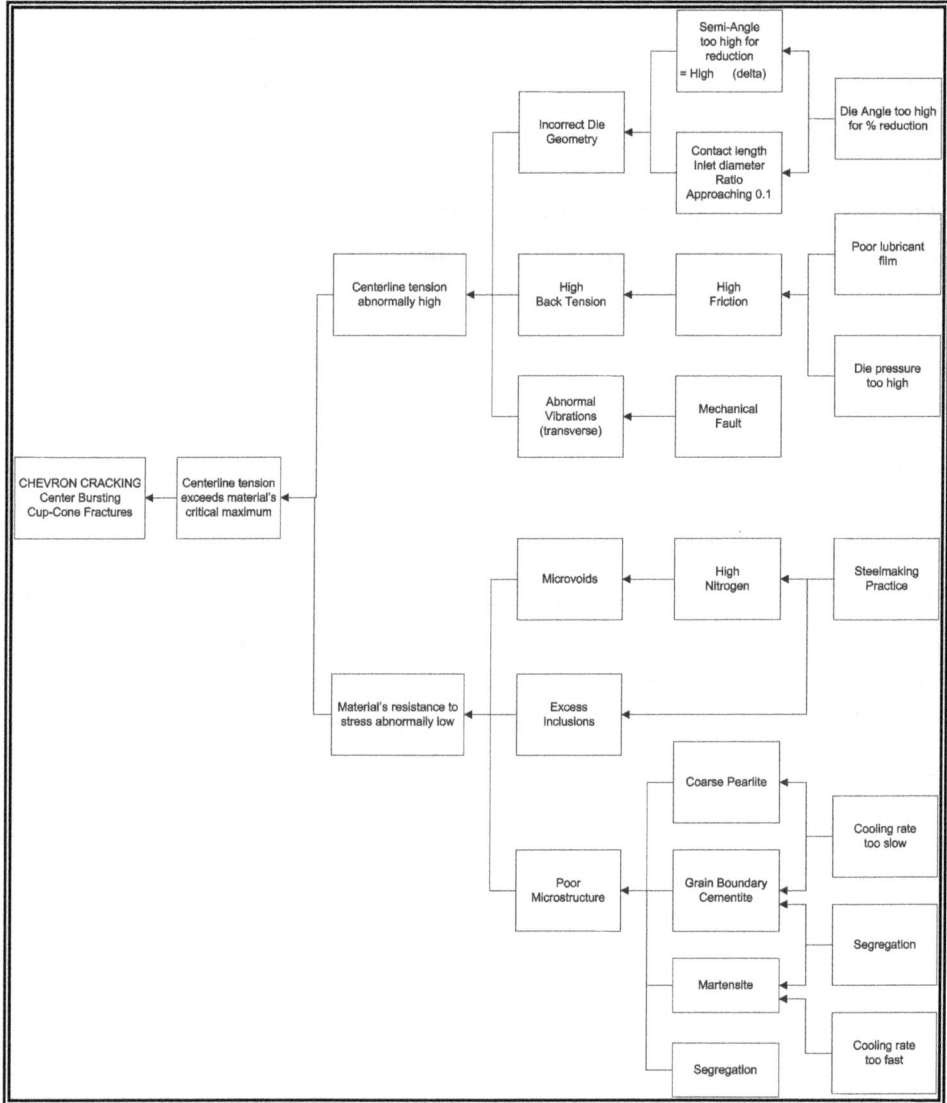

Appendix 2 Rod coil Rust Guide

ROD COIL RUST GUIDE			
Rust Level	Example Photo	Rust	Average Pit Depth Thousandth in. (Microns)
1		None	0 (0)
2		Very Light	0.22 (6)
3		Light	0.32 (8)
4		20%	0.47 (12)
5		50%	0.67 (17)
6		75%	1.00 (26)
7		90%	1.50 (38)
8		100%	2.21 (56)

Appendix 3 Methods to Reduce Tensile Gain Rate

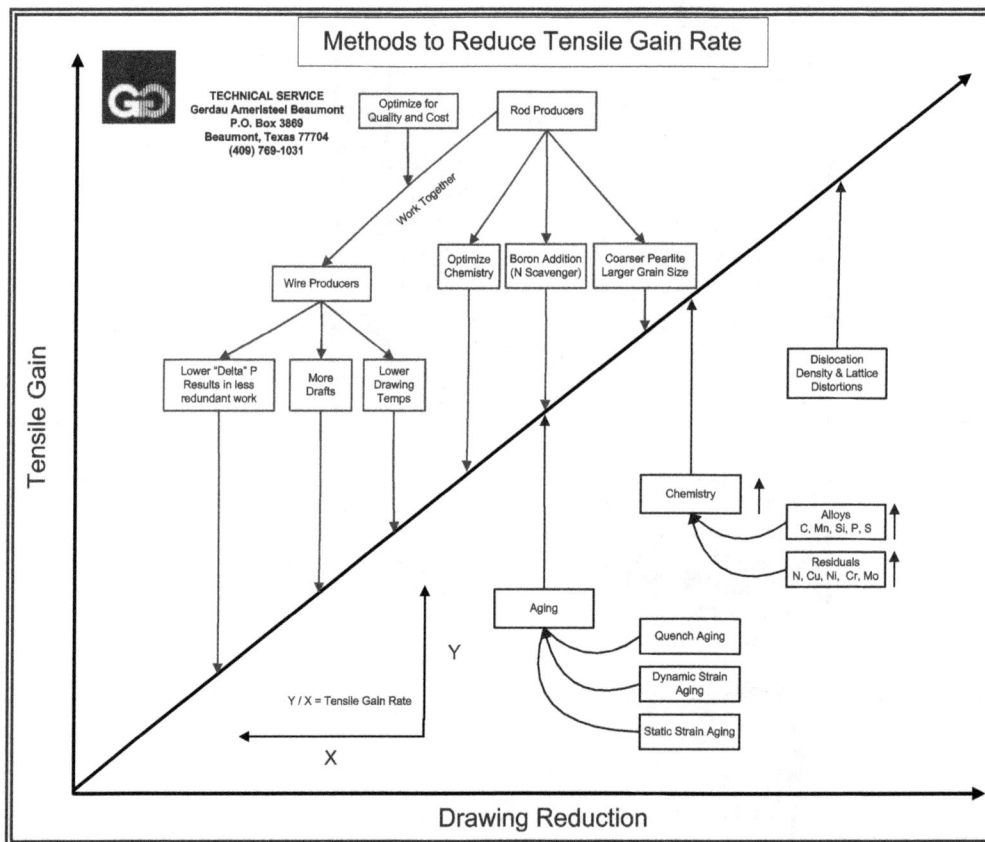

Methods to Reduce Tensile Gain Rate

Acknowledgements

I would like to thank organizers of this symposium especially Dr Viplava Kumar and his staff to give me an opportunity to present at this prestigious International Conference. I also like to thank my company Gerdau Ameristeel allowing me to present this paper.

444

References

1. Bhaskar Yalamanchili, *My Professional Career and Experiences in the Wire Industry*, Wire Association, Mordica Lecture, Wire Association Show, June 9, 2008

2. US Green Building Council's LEED program, U.S. Green Building Council (USGBC) 1800 Massachusetts Ave. NW, suite 300 Washington, DC 20036 (www.usgbc.org)

3. *Restriction of the use of certain hazardous substances in electrical and electronic equipment (RoHS,)* Directive 2002/95/EC of the European Parliament and of the Council of 27 January 2003.

4. *Registration, evaluation, Authorization and Restriction of Chemicals (REACH),* Regulation (EC) No 1907/2006 of the European parliament and of the Council of 18 December, 2006.

5. Steels Commitment to Climate Change, Ian Christmas, Secretary General, IISI, AISI, Phoenix, May 6, 2008

6. *End - of life vehicles,* Directive 2000/53/EC of the European Parliament and of the Council of 18, September 2000

7. International Materials Data System (IMDS) administered by EDS, Plano, TX, USA

Materials Technology for Water Filter for Bottom of the Pyramid (BOP) Consumer

VCS Prasad

Advisor, Business Systems and Cybernetics Center

Tata Consultancy Services, Hyderabad

Abstract : *Most of the science and technology efforts in the world currently are aimed at serving the top of the economic pyramid community – in other words , largely to the developed and to the top rung of the developing world which constitute only a mere 20% of the world population. The rest constitute the bottom serve this community are somewhat different than those that work for the high end top of the pyramid consumer. of the pyramid (BOP) whose average income / day is of the order of two to three dollars a day. Because of their poor affordability and lack of access to information and knowledge, they are unable to participate in the development process the developed world is witnessing. The purpose of this paper is to draw attention to the possibilities of serving the BOP consumer and sensitize materials scientists towards the opportunities in this direction. The principles of innovation to How some of these principles are relevant to serve the BOP consumer are brought out by taking the example of low cost water filter technology based on ceramic materials. The science and technology aspects in the development of this filter and need for appropriate business models to make such a work successful are briefly discussed.*

1. BOP Technologies and Markets

This conference is on metallurgy and materials science. This means sharing knowledge related to recent research in this field. Most of this research and in fact, perhaps all of it is aimed at improving the properties of the materials which are meant for use by the developed world or to some extent the developing world. In fact most of the development is aimed at the people who are at the top and the middle of the economic pyramid constituting nearly the top 1/5 of the world population in terms of the purchasing power. The rest of them constitute nearly $4/5^{th}$ of the population (in the range of 4-5 billion people) who are in the income groups of $ 2-3 / day. They constitute the bottom of the economic pyramid (the BOP). Obviously they do not have the purchasing power to participate with dignity in the overall development that is going on around the world. In short, the development is not inclusive. This problem was addressed admirably by C.K Prahlad (2005) in his path breaking book 'The Fortunes at the Bottom of the Pyramid (BOP)'. Since then the interest on developing technologies and business models to serve the BOP markets has risen sharply over the last 3 years. This is because of the huge untapped markets which are estimated to be a few trillion dollars if only the poor also get an opportunity to earn and spend thereby inclusively contributing in the development process. More than 80 % of the population in India could be in the BOP category. However, to serve this market which is extremely challenging, the principles of innovation ought to be different and Prof. Prahlad has come up with 12 principles of innovation as guidelines for managing this challenge.

These are:

1. Focus on price performance of the product and service .Serving BOP markets is not just about lower price but creating a new price – performance envelope.

2. Innovation requires hybrid solutions. BOP consumer problems cannot be solved with old technologies. Most scalable, price-performance-enhancing solutions need advanced and emerging technologies that are creatively blended with the existing and rapidly evolving infrastructures.

3. As BOP markets are large, solutions that are developed must be scalable and transportable across countries, cultures and languages.

4. The developed markets are accustomed to resource wastage. All innovations must focus on conserving resources: eliminate, reduce and recycle.

5. Product development must start from a deep understanding of functionality, not just form.

6. Process innovations are just as critical in BOP markets as product innovations.

7. Deskilling work is critical. Most BOP markets are poor in skills. The design of products must take into account the skill levels, poor infrastructure and difficulty of access in remote areas.

8. Education of customers on product usage is a key. Innovations in educating a semi-literate group on the use of a new product can pose interesting challenges.

9. The product must work in hostile environments. Products must also be developed to accommodate the low quality of the infrastructure, like electricity.

10. Research on the interfaces is critical, given the nature of the consumer population. The heterogeneity of the consumer base in terms of language, culture and skill levels and prior familiarity with the function/feature is a challenge.

11. Innovations must reach the consumer. Both the highly dispersed rural market and a highly dense urban market at the BOP represent an opportunity to innovate in methods of distribution. Designing methods for accessing the poor at low cost is critical.

12. The feature and function evolution in BOP markets can be very rapid. Product development must focus on the broad architecture of the system — the platform — so the new features can be easily incorporated. BOP markets allow us to challenge existing paradigms.

Some of these principles are used in the development of a low cost water filter for use by the rural poor, the typical BOP consumer. An entrepreneurship model is presented to aid develop rural entrepreneurship for this product.

The problem of safe drinking water:

"According to the UN, 1.3 billion people lack access to *'safe'* drinking water, and with global consumption of water currently doubling every 20 years, 48 countries are expected to face chronic water shortages by 2025. This means higher prices for clean water, and as a result, higher rates of diarrhea and other illnesses; at any one time, it is estimated that half of the world's hospital beds are occupied by patients suffering from water-borne diseases " –(Abstracted from Health Care delivery : www next billion.net John Paul , 2005)

In India there are about 6 lakh villages and a good percentage of them do not have access to safe drinking water. Although this problem has received lot of attention both by the International agencies (such as WHO and DFID) as well as by the Govt of India and NGOs , a cost effective and sustainable solution for providing safe drinking water to the poor is still elusive.

1.1 Use of Relevant Technologies:

Ceramics, by their very nature are materials that the BOP consumer is generally familiar with, and to some extent also knows how to work with them. Example, the clay based materials are used by rural artisans for making pots and other related earthen ware items. What BOP additionally advocates is the *'Development through enterprise'* with an emphasis on business models driven by a profit motive that meets the needs of the *'underserved communities'* in emerging economies. The strategy is: targeting the huge market with low margins, developing innovative means of reaching the scattered markets, and train the local people for contributing to the supply chain. Serving BOP markets is not just about lower price but creating a new price-performance envelope.

In this paper, how ceramic materials technology can be used in developing low cost filters with an aim of addressing not only the safe drinking water problem for the poor, but also the problem of *'Development through enterprise'* is demonstrated. Proper use of the BOP innovation principles can make such an initiative sustainable. Earlier there were projects in Kenya where NGOs were taught how to make small batch water filters on their own. But the project never took off till a micro enterprise component was added to it to make it sustainable. Martin Fisher , the project director admits that one of the most fundamental lessons he learnt while interacting with the poor is that *the 'poorest people in the world are also among the most entrepreneurial .They do not want handouts they want opportunities'* (Ref: *'The chicken Metric'* by E.A. Lambert – internet site) . Likewise there are number of examples of low cost materials for use by the BOP consumer, like the use of straw as a building material because of its toughness and durability.

2. The proposed water filter technology for the BOP markets:

The traditional ceramic technology is very old and is known to human kind from several millennia. It widely practiced by a few artisans in the rural areas which is the BOP market, the subject of this paper. Examples are the brick and pottery type of industries. Village artisans are able to make livelihood out of these traditional technologies. But the drawbacks are: product quality, business growth and sustainability. If the BOP innovation principles are used, these problems can be minimized while also serving a larger population of the rural poor. We shall illustrate these ideas through the water filter technology developed more than a decade back at the Tata Research and Development Center (TRDDC) at Pune, India, a division of the Tata Consultancy Services.

Significant research at IIT Kanpur and later in TRDDC led to the finding that commonly available rice husk ash (RHA) , a waste material in rural India can be used as a filtration medium for water filtration (Kapur , 1985). The presence of activated silica (80-85 %) in RHA facilitates the entrapment of pathogenic bacteria while activated carbon (~ 5 %) in the burnt husk helps in the removal of colour and odour. RHA can be obtained from different sources and these details along with these characteristics are given in Table 1. The material RHA is too fluffy and voluminous and cannot be directly handled for filtration purposes to get reproducible end results. It is to be made rigid in a cost effective manner using materials that are commonly available in the rural areas. This is achieved by using commonly available pebbles (on the road). The binding of the RHA to the pebbles is achieved through the use of another low cost material that is available in rural areas Viz. cement . The resulting mix can be fabricated into the shape of a "bed ". The concept of the 'bed' has come from the sand bed filtration technologies that are commonly used in the water works in towns and cities. However, while the traditional sand bed filters are of height 10 '-20', the pebble matrix filter (PM) that was developed has a bed height of only 3 '-6' and is able to trap the harmful E coli bacteria as well as suspended matter to near 99% (Sundaram et al, 2002)! Such miniaturization in the bed height (when compared with the sand filter beds) was possible because of the close attention to the choice of the size mix of the pebbles used (the concept of the use of different sizes to improve the packing factor of the bed) in trapping the suspended matter as well as the effectiveness of RHA in trapping the E coli bacteria

Table 1 Characterization of ash from different sources

Property	Rice husk ash source		
	Batch A	Batch B	Batch C
Size	-300 micron	-300 micron	-300 micron
Color	White	Gray	Black
Bulk density	0.4 gm/cc	0.3 gm/cc	0.3 gm/cc
Carbon content	4.47 %	4.7 %	13.35%
Moisture content	0.2 %	0.2 %	0.4 %

Ash A : Ash obtained from brick kiln

Ash B : Ash obtained form Tube in basket burner , a special burner designed for efficient burning of ash.

Ash C: Ash from industrial boilers where rice husk is used as a fuel.

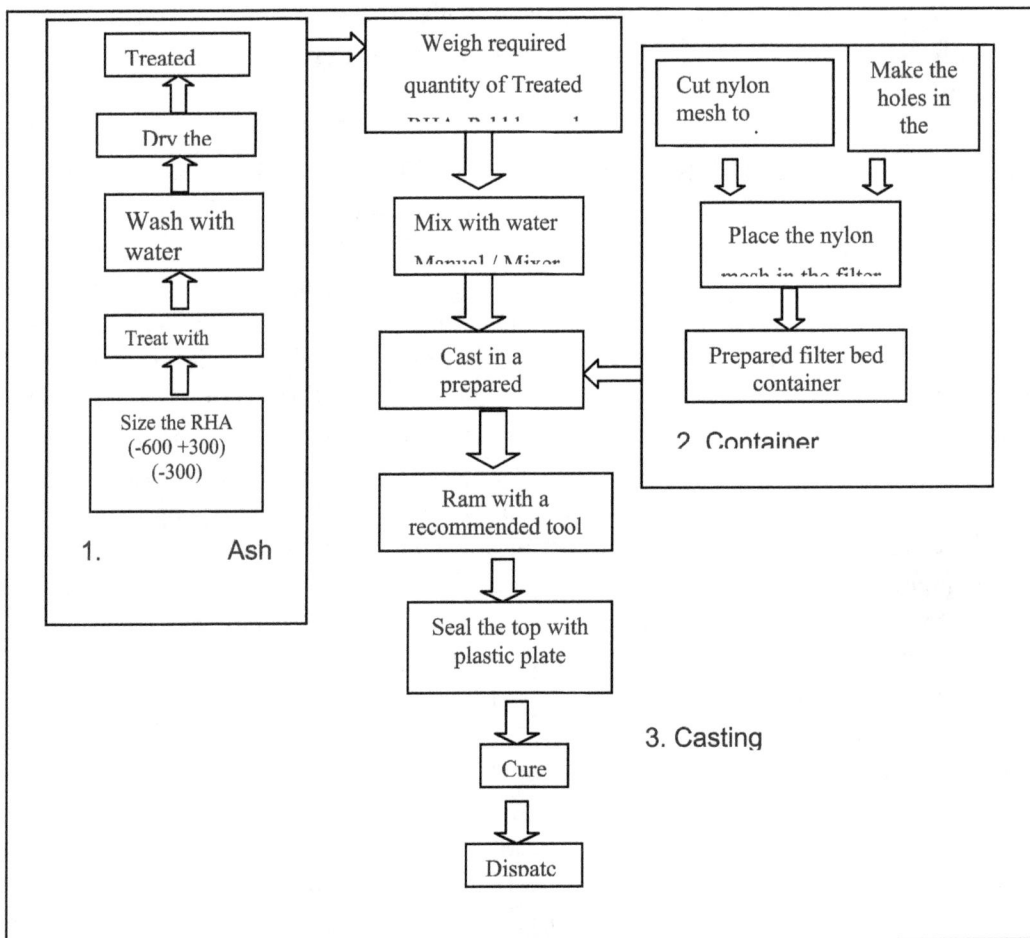

Figure 1 Filter fabrication process.

A flow chart for the fabrication of this filter is given in Figure 1. The process has three sub processes *viz*, ash treatment, container preparation and casting (fabrication) process. The details of the filter fabrication process and design are published earlier (Prasad, 2000 and Mehar & Prasad, 2004). Briefly, the process of making the filter element involves thorough mixing of RHA (~14%), pebbles (~85%) and cement (~1%) in the required proportion along with water to the consistency of a concrete mix. The mix was poured into a container (preferably plastic) with holes at the bottom and then rammed with specially designed tool (placed on a tray to have a hard surface for ramming). After ramming, the top end of the bed is also covered with a plastic plate. The list of tools/ equipment used for the filter fabrication is given in Table 2.

Table 2 List of tools used in filter fabrication

Item	Quantity
Tube in basket burner (2 x3)	1
Sieves with 1, 3, 5 and 10 mm. mesh (3'x 3')	1 each
Sieve 420 micron (37 mesh, Size 12 inch)	1
Weighing Machine (10 kg capacity)	1
Tubs (100 liter)	2
Bucket (25 liter)	2
Ramming Tool (2.5 kg)	1
Mixing tool (Tapi)/ mechanical mixer	1
Trays (1' x 2')	2

The setting characteristics of the cement impart the necessary strength to the filter element. The element is replaceable and can sit tightly into an outer tapered plastic container. The use of a rubber gasket around the element (cooker gasket) makes the assembly leak proof (Fig 2). It is to be noted that all materials are available in the rural areas and the equipment used doesn't need any power source to operate. Therefore, fabrication can also be done in power dark areas of the country side. The output characteristics measured are: bacterial trapping (1995) turbidity, pH and filtration rate.

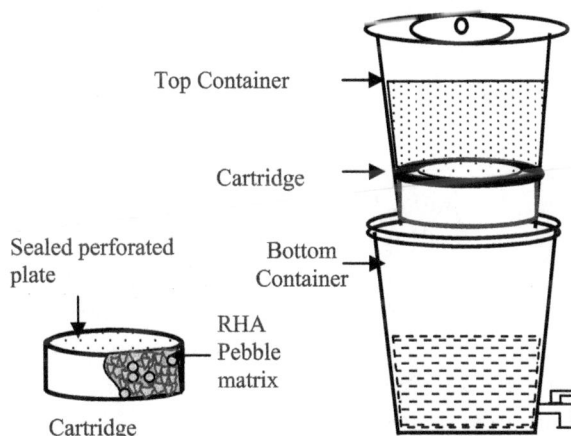

451

Fig. 2 Assembly of the filter and the element

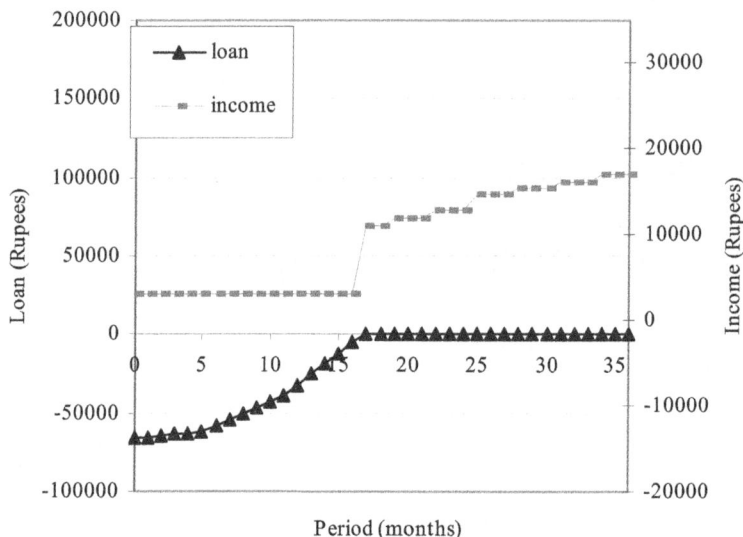

Fig. 3 Financial model for a micro enterprise.

Table 3 Performance of filters made by an elementary school educated operator after half a day's training (batch of ten filters)

Parameters	Filtered water										Raw water
Filter number	1	2	3	4	5	6	7	8	9	10	
Bacterial counts (CFU/100ml)	120	180	200	140	120	100	140	180	200	240	2800
% bacterial trapping	96	94	93	94	96	97	95	94	93	91	

3. Elemental trapping

There can be harmful elements such as Fluorides and Arsenic in the ground water that is used for filtration. Example: In India, 20 states are affected with high fluoride level (more than 1 ppm) in ground water. Prolonged drinking of high fluoride contaminated water (more than 1.5 ppm) causes skeletal and dental fluorosis. There are many existing technologies (such as activated alumina) for the removal of fluoride from water but they are costly, need maintenance and therefore not very effective. The main issue with fluoride removal is its cost, since concentrations of fluoride is anywhere from 5-7 ppm to as high as 40 ppm in some places. Vivek etal (2006) have developed a Rice Husk Ash (RHA) based technology to remove fluoride ion from the fluoride contaminated water with the coating of aluminum hydroxide on RHA. The very high surface area of RHA

(80-90 m2/gm) acts as a good substrate for coating of aluminum hydroxide. The active aluminum hydroxide on the RHA surface adsorbs the fluoride ions from the fluoride contaminated water. This technology is simple, economical and convenient to implement in rural settings. No leaching of fluoride from the fluoride adsorbed coated RHA was noticed. The estimated treatment cost is 15 – 20 paisa per liter of water. Its adsorption capacity is 3-3.5 mg/gm of coated RHA as against 1-2 mg/ g for the activated alumina filters. While the initial field tests carried out in the Guntur region. are promising , further work is required in terms of reproducibility and improving the life.

3.1 Understanding functionality and the price performance envelope

Since the materials used in the filter fabrication are of low grade and are not synthesized to specific characteristics (excepting cement which constitutes a mere 0.85% by weight of the total materials used), it is hard to control the variation. Therefore, the trapping efficiency of the filters when made in large numbers is found to vary from 71 to 99% for filter to filter. The influent bacterial levels (E-coli) of water in some of the villages (around Pune) tried is found to vary from 1000 CFU (Colony Forming Units) /100ml to 3,000 CFU/100 ml based on the source of water. This means at the lower trapping level, the filtered water will have a bacterial content of approximately 300 CFU/100 to 870CFU/100 ml. Such water is still unsafe for drinking purposes particularly for children whose immunity level is not as high as for rural adults.

Therefore, to assure minimum safety of drinking water particularly for children, a higher mean and lower variation for bacterial trapping efficiency is required for all the filters fabricated. This is achieved through the use of the well known DMAIC approach (Define, Measure, Analyze, Improve and Control) (Pande, 2000) that is generally used for variation reduction in six sigma processes. A controlled and robust process gives confidence in the output characteristics even if the bacterial testing is not routinely done in the village environment. The DMAIC approach helped in arriving at a specification of 95+ 4 % for the filters fabricated in the village environment (Prasad and Vivek 2005). It was possible to fabricate the filters to this specification by an elementary school educated operator with just half day training. Table 2 shows that the variation within ten filters is within the above specification.

4. Field studies related to rural health:

An out of specification product is likely to give water borne diseases because the bacterial level of filtered water crosses the immunity level of the villagers. The water borne diseases can result in increased medical expenses of the rural families.

Pilot studies were carried out in a carefully chosen village (Pusane) near Pune, India to understand the overall improvements due to the use of the water filter. This village had consistently reported over the years water borne diseases like Diarrhea and Cholera due to unsafe drinking water. Filters were installed in almost all the homes in this village (186 families). The quality of filtered water was tested at regular intervals over a period of one year and was found to be within specification limits. During this year, the reduction in water borne diseases as reported by the doctors is 90%.

On an average, the medical expenses per family per annum in this village (based on discussion with villagers) are ~ Rs 2000 (~US $ 40). Out of this total expense, 80% is assumed to be due to unsafe drinking water (based on discussion with local doctors). Therefore, the average expenses per family due to water borne diseases is (80% of 2000) ie; Rs. 1600 (~US $ 32). Hence the estimated saving in medical cost per annum due to reduction in water borne diseases in the village is 1600 X 186 = Rs. 297600 (~US $5952).

The above figure is not strictly representative. The savings can vary from region to region and season to season and a systematic analysis related towards such a data is not available. However, UNDP (United Nations Development Program) studies indicate that the medical expenses form a substantial portion of the average income of the villager (India development forum, 1997).

More than ten thousand filters were used in different parts of rural India using the technology outlined above.

4.1 A win-win proposition for the rural consumer and rural entrepreneur

The cost of the filter if fabricated in rural areas is in the range of Rs 150–500 depending upon the container costs. For an earthenware container, it is low and for stainless steel, it is high. For plastic, it is intermediate. The cost of a replaceable filter bed is Rs 30, which forms the periodic maintenance cost for the filter. A simple economic model, which is implementable in rural India, is worked out as shown in Appendix A2. With an investment of Rs. 8000), the village entrepreneur can fabricate 50 filters/month and earn Rs. 2000/pm. Higher income is possible with more investments and sale. For instance, with an investment of Rs. 15,000 and a loan of Rs 61,000, the entrepreneur will be able to make a few hundred filters/month using larger size equipment. If he makes 100 filters / month and is able to sell all of them, he can earn Rs. 15,000/month after repaying the loan during a 17-month period. The model thus indicates the potential for steady decent income for livelihood generation. The numbers in the model are illustrative and refinements may be required. For example, one is not sure whether there will be a year-wise increase or decrease of sale of filters. The sale may reach a peak after a couple of years, and then decrease. Nevertheless, the model shows the potential of generating steady income for the entrepreneur. Depending upon the village density and the population density in the village, the size of the micro enterprise can be planned in different locations. For relatively large size enterprises, additional jobs in marketing and distribution are required.

The BOP consumer has to make a one-time investment in the first year of Rs150–500 for the purchase of the filter. Even if the filter bed is replaced 4 times in a year (studies have shown that the life of the filter bed is 3–6 months based on the influent turbidity levels), his maintenance costs is Rs. 120/annum Therefore, the total cost of using the filter in the first year varies from Rs 270–620 depending upon the type of filter. For subsequent years, the maintenance cost will be Rs. 120 /annum whereas the medical expenses per family is of the order of Rs 1600/annum. Therefore, there is significant savings to the extent of at least Rs1100) per annum if the filtered water is used particularly in places where there is high incidence of water-borne diseases. Thus, there is a win–win situation for the common villager, as well as the village entrepreneur. The bonus is improved health and well-being, resulting in improved quality of life.

5. Financial model for small scale filter fabrication

The average population in an Indian village is approximately 3,000 people (Prahalad, 2005). Let us consider that a cluster of 10 villages exists within a 10 km radius, to proliferate water filters from the filters fabrication site. Assume the average number of members in a family in a village is 5. Therefore, the average number of families in a village will be 600 and the total number of families in 10 villages will be 6000, in the filter proliferation area.

If a villager can invest Rs. 15,000 and raise a loan of Rs 61,000, he will be able to make 100 filters per month. If we target 20 % of the total families in the cluster considered above, the target numbers work out to 1200, which can be met with the above investment. The period we have considered in this case was 3 years.

Assumptions
1. Period of operation considered : 3 years
2. An entrepreneur is able to manufacture 100 filters per month and sell them at the same rate i.e.100 filters/month for the 1st year.
3. The filter demand will increase over time, i.e., 120 filters per month for the 2nd year and 140 filters/month for the 3rd year.

4. The entrepreneur's sustained income comes through the filter element replacement. The replacement time for the filter elements is 4 months.

5. The fabrication cost of the filter based on material and labour costs: Rs. 250

6. The selling price of the plastic filter: Rs.300

7. The fabrication cost of the filter element: Rs. 15

8. The selling price of the filter element: Rs.30

9. The transportation cost for the filter and filter element is Rs. 5 per filter and Rs 2 per filter element respectively

10. Interest rate considered: 6.5 % per annum (Pearce, 1992)

Based on these assumptions, a financial model is built, the results of which are shown in the Figure 3. The pay-back period (PBP) is 17 months. Around Rs.3000/- per month, as income for the entrepreneur during PBP is also accounted for his livelihood. After repaying, the loan income would be Rs ~12,000 / month ($ ~240 per month). This would be sustained or grows (Rs. ~15000–18000 per month) depending on his income on replacement of the filter elements. Fig 3 illustrates this model.

The above illustration is based on some ball park figures which were found reasonable when the illustration was developed. Bulk of the filter cost is in the container cost. But container costs can be brought down to nearly 50% by bulk purchases.

6. The role of Information Technology

Internet is playing a great role in Information diffusion which is critical for the BOP type initiatives to realize the potential of the large and somewhat scattered BOP markets. But internet access is not possible in power dark villages. TCS has been working with a leading NGO to develop a system to help rural communities, even those that do not have direct access to not only organize, use and manage information but most importantly to be able to share it among them quickly and easily. TCS is also working on an innovative Adult Literacy program (ALP). This program brought out software which enables an uneducated adult to learn to read easily. Improved literacy in the villages improves the scope for innovations and entrepreneurship amongst the BOP community, which help improve the success rate of the BOP initiatives. Example, it was noticed that in the Guntur area when the ALP and the water filter training programs were conducted at the same place , the readiness to take up the water filter project for entrepreneurship was evident. Although this could be a stray incident, it is generally known that higher the knowledge levels, better the chances of receptivity to new innovations (example ; Lyytinen and Rose , 2004) .

7. Summary

It is shown that how a simple ceramic based filter technology can be used to develop rural entrepreneurship. Some of the BOP innovation principles are used in this work. Refinements in the business model are required to ensure sustainability. Use of the filters in a complete village near Pune reduced the water borne diseases significantly and also saved the costs of medicines to the villagers .It is suggested that the outline of the approach suggested in this paper can be used for any simple ceramic based materials technology which can be suitably fabricated in rural India. It is further suggested that more research and development work is needed in materials technologies to address the BOP needs while the opportunities seem to be immense.

While the results presented in this paper were from the data obtained nearly ten years back , further research during this period has resulted in major improvements in terms of improving the specification of the filter to near zero variation in characteristics so that it is very roust for use in all environments . Further, technologies were also improved and made robust for elemental trapping. Accordingly improved business models are also getting evolved to effectively serve the BOP segment.

Acknowledgement: to Mr. Vivek Ganvir, Scientist TRDDC, Pune for readily providing information on his work on fluoride trapping , Dr. Vikram Jamwal and Prof. Nori for review of the manuscript.

References

1. Ganvir V, Das K and Kapur P.C *A filter media for removal of fluoride ion contamination from water, a device and a method for purifying fluoride ion contaminated water*, Indian Patent No. 207190

2. Kapur P C, *Production of bio-silica from the combustion of rice husk in TiB burner*, Powder Tech 44, 1985, p 63

3. .Lyytinen,K and Rose, G.M (2004) "Explaining radical innovation in system development organization" *Winter*, vol 4, p 1

4. Meher K K and Prasad V C S, *A rice husk ash based domestic water filter*, Indian Patent No 193692, 2004

5. Membrane filter technology for Coliform group *Standard Methods for the examination of water and wastewater*, Sixteen-edition, 1985, p 887

6. Pande, Peter S, *Six sigma way. How GE, Motorola and other top companies are honing their performance*, McGraw–Hill N.Y, 2000

7. Prahalad C.K, *The fortune at the bottom of the pyramid*, Wharton School publishing, University of Pennsylvania, Printed in India by Saurabh Printers, 2005

8. Prasad V C S, *Low cost domestic water filter:* The case for a process-based approach for the development of a rural technology product, Water SA Vol. 28 No, 2 , 2002, p 139

9. Prasad, VCS and Ganvir .V, (2005) *Study of the Principles of innovation for the BOP consumer – the case of rural water filter,* Int. J. of Innovation and Technology Management , Vol. 2 (4), 2005, p 349

10. Sundaram S K, Meher K K, Kapur P C, *A rice husk ash based domestic water filter,* Indian patent no. 187147 (2002)

Modern Management Philosophy & Practices at Global Ispat Koksana Industrija

Guntupalli Jagannadham,

Managing Director, GIKIL, Lukavac, Bosnia-Herzegovina

1. Introduction

The authors are indeed extremely happy to be present at this conference and share their experience with and learn from all the participants.Our company is located in Eastern part of Europe, in - Bosnia and Herzegovina. Country rich with natural recourses, forests, beautiful landscape, art, music and culture. Its key business is metallurgical coke, by-products of coke and energy from coke gas. In the same site we also have chemical plants for manufacturing of Mineral fertiliser and Anhydrite Maleic acid. We also provide services like processing of potable water for civil, industrial use and Engineering services for the companies around us. Overall plant area is 1sq.kilo meter.

Our company started operating from 1952 as a major Coke and chemical producer of former Yugoslavia. In 1962 to 1991 plant underwent a major various expansions, addition of plants to fulfill increasing demand of Coke and chemicals. The war that took place in 1992 halted our operations indefinitely. The supply chain network that existed before the war was totally disrupted and it was impossible to resume production for 12 long years. A ray of hope arose with the proposal of Global steel Holding Limited (GSHL), multinational company showing interest in taking over the company. The takeover materialized in November 2003 with a contract being signed between Government of Bosnia & Herzegovina and GSHL. The great challenge of reviving this closed unit and also restarting the plant and machinery that has remained idle for 12 years faced GSHL. With everyone's support and encouragement the plant resumed operations in July 2004. Revival and merger brought-in new life into the company which also witnessed flow of modern management philosophies. As part of this, Total Productive Maintenance (TPM) methodology was implemented from day one of plant commissioning.

In 2006, the anhydride maleic acid plant underwent a major technological change. The cumulative effect of such developments enabled a turnaround in our business by the middle of 2007. The year 2008 has also been also been a quite eventful year. It has been year of accomplishments. We could sustain the sales growth and profitability besides 100% utilization of plant capacities. Also, we are glad to share here, that in 2008 we became first company producing metallurgical coke to achieve TPM excellence award outside Japan. Our historical development is shown in Fig.1. The present organizational structure is shown in Fig.2.

Our main products are metallurgical coke with an annual capacity of 682,000 T; anhydride maleic acid of 9600 Tpa; Mineral fertiliser 125,000 Tpa; and electrical energy (16.5 MWh). Of the total power produced we consume 9.5MWh and the rest is sold to our National grid. The growth in production of various items over the last few years is shown in Fig.3. This production is made possible by a dedicated team of 1082 employees. Less than 200 of these constitute administrative staff. A reduction, from the level of 1247 employees in the year 2005, was made possible thanks to multiskilling, training and succession planning to take advantage of natural retirements. Company activities are guided through statement of Policies and Objectives (Fig.4). We have defined policies for all major functions of company including Quality Objectives (Fig.5). The policies are displayed and communicated to percolate the culture among all the employees. The policies arise from an assessment and estimation of the internal and external challenges before the company (Fig.6). Investments are continuously carried out after detailed analysis of Return on investments. Loss reduction is also a parallel activity. We use Focused improvement methodology to systematically resolve losses in the manufacturing. These activities have yielded : Increase in OPE by 50 %; Total costs reduction by 36% and Cumulative effect on Sales which is increased by 117%.

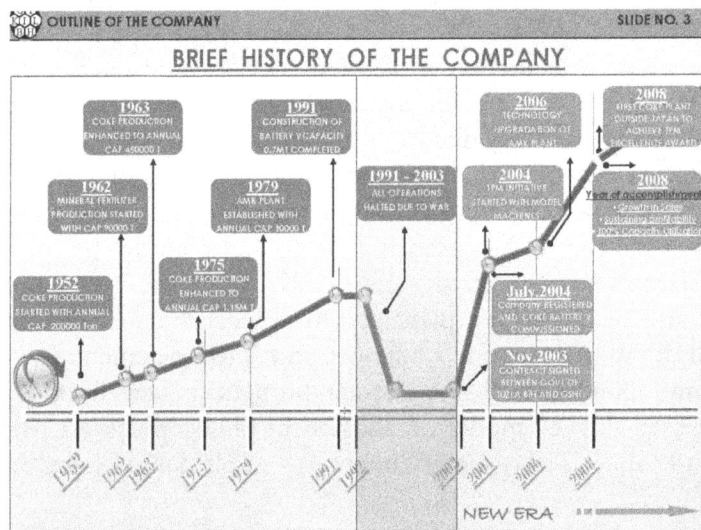

Proactive management is followed on new projects, making `First time right' and `Vertical start up'as objectives. For example, manual packing was converted to automated packing- stretching – palletizing line. This is to cater European markets with new packing protocols. Our new packaging process won the prestigious `Best packing quality' award. Even during the current economic tsunami and its effects thereafter, proactive management has helped us take specific preventive measures in order to reduce potential losses. An analysis of expected scenario under favorable and non favorable conditions led to the identification of a brief list of initiatives taken up to prevent effects of global recession on our operations. Some of these initiatives are: Proactively exploring new markets by adapting to newer quality expectations; Reduction of Fixed costs by redefining operating norms; reducing administrative expenditures and reduction of manpower by voluntary retirements and multi skilling; Reduction of variable cost through reducing logistic charges and laying off all temporary manpower.

At the operational level, planned maintenance systems have made the plant more reliable for increased production demands. We make use of comprehensive approach of zero breakdowns. Mapping is made for vital machines and their critical components.

Well knitted balance between operators and maintenance experts have made machines operate smoothly (Fig.8).

People are our key strength. It was the effort of every employee in bringing life into machines that remained standstill for 12 years. This is a clear example of the strong will and dedication of our employees. To further enhance skills and multi tasking ability of our staff we have developed a systematic education and training process. We have set up a technical training centre within our site to provide technical know how.

Recognition of success is done regularly by top management visits and rewards.

Our Production process is constantly improved in select areas like:

1. Product quality,
2. Balancing quality and cost
3. Searching for new coal sources and testing feasibility.
4. Refining existing SOPs and ensuring the compliance.

All these have resulted in increasing our customer base and secured future for plant. Vital operating norms are monitored through Key Performance Indices under productivity, quality, cost, delivery, safety and employee morale . We have corporate defined Key Performance Indices (KPIs) as well as plant -wise KPIs (Fig.9). About a total of182 indices are being monitored regularly. These are tracked and communicated through display boards and controlled by corrective actions. These have resulted in the optimization of production resources ; raw materials , utilities, consumables and product yields. Quality control is attained through maintaining 4M conditions, such as Man, Machine, Material and Methods. Critical quality points in each plant are mapped and made quality maintenance plan. As a result quality variations have reduced by nearly 90% from the base year (Fig.10). Support services like offices are made efficient by adopting 5S techniques and visual controls. Supply chain of coal is monitored through visual board and daily scenario is displayed. These tools have led to eliminate huge losses in its logistics. Also constant monitoring and control of administrative costs have led cost reduction by 50%. Safety, Health and Environment systems are established.

We monitor and resolve unsafe conditions and unsafe acts in order to prevent potential hazards. Regular safety drills and health checkups are carried out. Steady reduction of accidents has taken place and we aim for zero-injury plant.

We provide a glimpse of the improvements achieved in key results areas which through implementation of modern management practices. On manpower productivity, a cumulative effect of increased volumes and decrease in manpower has led to 200% raise in the coke production and 374% increase in the productivity of the energy plant. Similarly AMK plant has become 500% productive. Production of mineral fertiliser increased by 470% of that in the base year 2005. Quality has improved significantly moving towards zero defect level. A number of loss reduction activities, big or small, have led to 3.9 million savings. Breakdowns have reduced by 71%. Machine availability has increased by 68% and maintenance costs were optimised to 60% of those in the base year. Customer service and satisfaction is monitored by delivery compliance that is Right product at Right time. Delivery compliance has improved to 95%. A number of in-tangible changes have taken place. To name few we have seen: Cleaner work environment ; Neat and clean machines, Team spirit and harmony at work and Win-win interaction between management and employees.

Marching further, we want to be a

-       World class company ,
-       Value enhancement to our stake holders,
-       And We would like to become Preferred supplier by World class customers.

While concluding my presentation, I would like to express my acknowledgements to all those who participated throughout the journey and going to be party with future prospects.

First our sincere gratitude to our chairman Mr. Pramod Mittal who has been showing extreme confidence on our team,

> Our Managing Director Mr.Guntupalli Jagannadham
> Expertise and support from institutions like JIPM and Deliotte,
> And all our functional heads and their teams in making a `Real difference'

METALLURGICAL HERITAGE

The Enigmas of Bidri

Paul T. Craddock

Visiting Faculty, Department of Materials and Metallurgical Engineering
Indian Institute of Technology, Kanpur, INDIA
Ex-Dept. of Conservation, Documentation and Science
The British Museum, London, UNITED KINGDOM

Abstract : *The production of bidri is a traditional craft of India that continues to be practised to this day. Bidri wares are cast from an alloy containing about 95% of zinc and 5% of copper. The castings are then inlaid with silver, or more rarely brass or gold before being treated to produce a rich black patina that forms a perfect background for the shiny metal inlay.*

The origins of bidri are very uncertain, India was producing zinc around a thousand years ago much earlier than anywhere else in the world, and thus it is natural to link zinc and bidri together. Zinc was being produced in India at Zawar in the Aravallli Hills of Rajasthan, about 30 km south of Udaipur and production was flourishing there at the time that bidri is said to have been invented. However, Rajasthan has traditionally been a staunchly Hindu part of India, yet bidri making has always been associated with Moslem craftsmen in the Moslem parts of India in the south and east, thousands of km distant from the putative source of the zinc. Furthermore, lead isotope analysis on some early bidri wares strongly suggest that the zinc did not come from Zawar, yet there are no other sources in India that were exploited in those times. The bidri makers themselves claim that the technique was introduced from Iran which has neither bidri wares nor a tradition of early zinc production. Finally, although the processes of producing the patina are well known, the nature of the patina remains a mystery, several scientific studies having failed to discover why the surface is black!

1. Introduction

The production of bidri is a traditional craft of India that continues to be practised to this day Stronge (1985 & 1993); Lal (1991); Mehta (1960, pp.28-30) (Figs.1 & 2). Bidri wares are cast from an alloy containing about 95% of zinc and 5% of copper. The castings are then inlaid with silver, or more rarely brass or gold before being treated to produce a rich black patina that forms a perfect background for the shiny metal inlay (Untracht 1968, pp.138-49; Gowd ed. 1964, p.p.3-21; Craddock 1993). Bidri has and is being used for a wide range of artefacts, but vessels, boxes and plates for various uses continue to dominate the range. These include ewers (Fig.3), vases of various shapes and the bases of *huqqa* (water pipes), spittoons together with boxes for betel nut and *pan*, or for pens and brushes (Fig. 4). Bidri has also been used for furniture fittings and even for whole panels, such things still made for clients in the Middle East, where bidri wares have always been popular. The traditional items continue to be made but have now been joined by smaller items aimed specifically at the tourist market. Although museums around the world contain displays of bidri and it is made in a number of centres in India there are still many unanswered questions over its history, the sources of the materials and even the nature of the patina.

The origins of bidri are very uncertain, India was producing zinc around a thousand years ago much earlier than anywhere else in the world, and thus it is natural to link zinc and bidri together. Zinc was being produced in India at Zawar in the Aravallli Hills of Rajasthan, about 30 km south of Udaipur (Fig. 2) and production was flourishing there at the time that bidri is said to have been invented (Fig. 5) (Craddock 1995, pp.309-17; Craddock et al 1998). However, Rajasthan has traditionally been a staunchly Hindu part of India, yet bidri making has always been associated with Moslem craftsmen in the Moslem parts of India in the south and east, thousands of km distant from the putative source of the zinc. Furthermore, lead isotope analysis on some early bidri wares strongly suggest that the zinc did not come from Zawar, yet there are no other sources in India that were exploited in those times. The bidri makers themselves claim that the technique was introduced from Iran which has neither bidri wares nor a tradition of early zinc production. Finally, although the processes of producing the patina are well known, the nature of the patina remains a mystery, several scientific studies having failed to discover why the surface is black!

Fig. 1 Craftsmen at the Yaqoob Bros. Bidri Works at Hyderabad. All stages of the manufacture are carried with the various specialists working side by side. This private establishment is quite large, many others have only 2 or 3 craftsmen and apprentices, especially in Bidar itself. (Photo. P.T. Craddock, 1991)

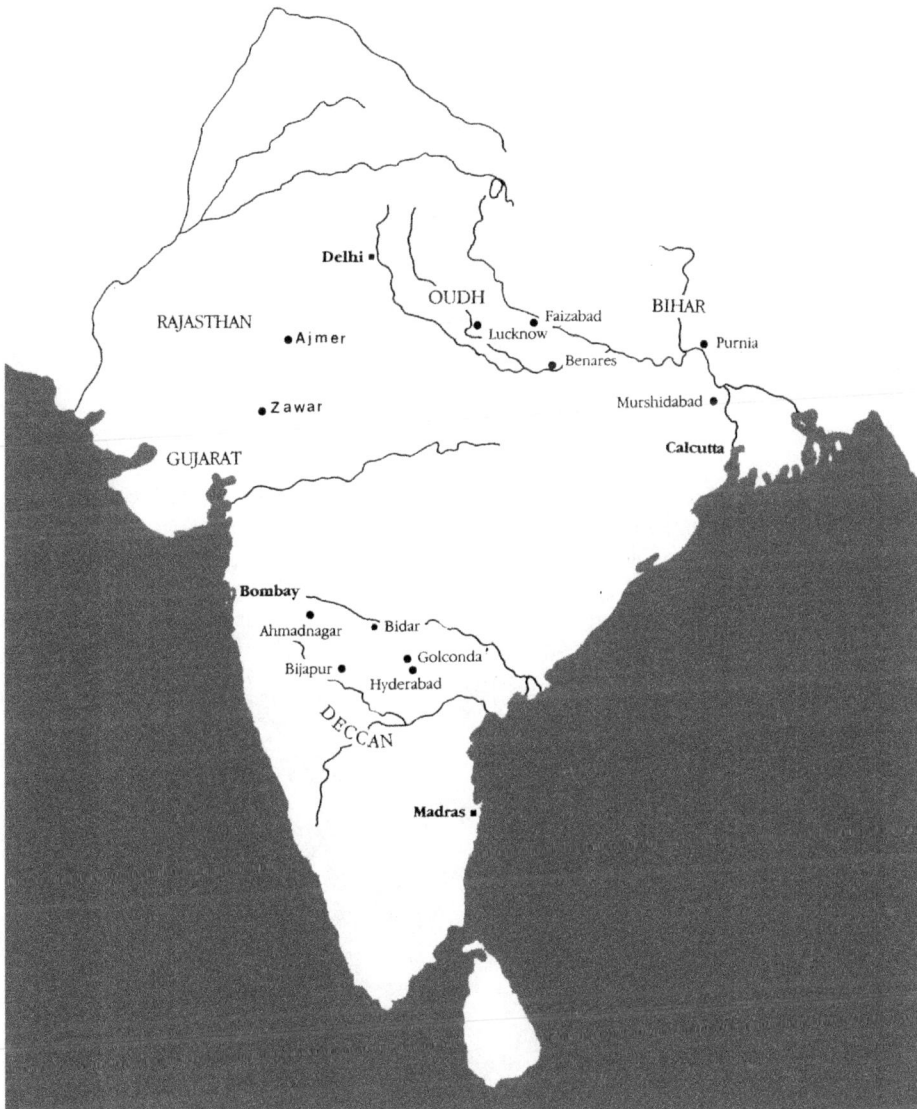

Fig. 2 Zawar, the early zinc mine, and some traditional bidri production centres (adapted from Lal 1990).

Fig. 3 Mid 17th century bidri ewer inlaid with silver and brass made in the Deccan, probably at Bidar (V & A Reg.1479-1904. Stronge 1985 Cat.2). (Photo. By courtesy of the Board of Trustees of the Victoria & Albert Museum)

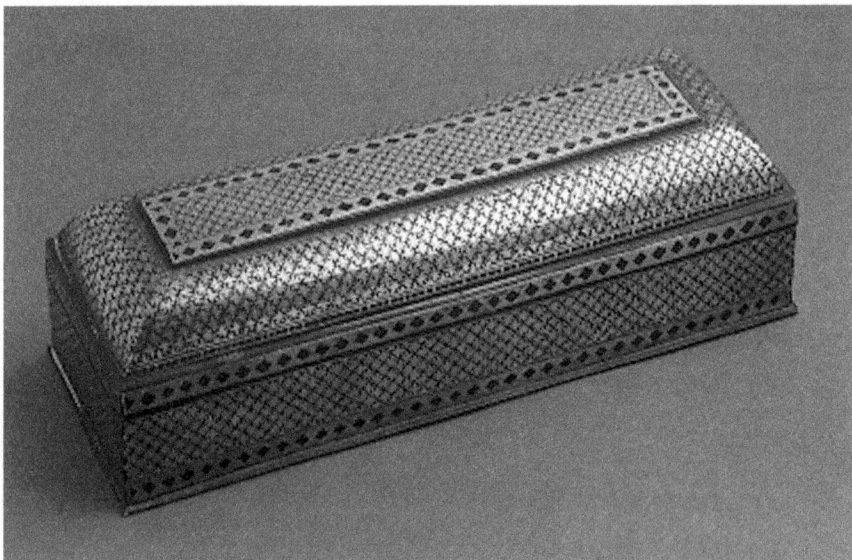

Fig. 4 Pen box inlaid with stars of silver wire. The style is tradition but was made at the Mumtaz / Gulistan Bidri Co-operatives in Hyderabad in 1986. (Photo. J. Heffron)

Fig. 5 Bank of furnaces at Zawar being loaded with retorts for the smelting and distillation of zinc as they would have appeared on an evening sometime in the 14th or 15th centuries AD (B.R. Craddock)

2. History of the Process

Bidri is historically associated with the city of Bidar, from which it derives its name, the one-time capital of the Bahmani realms in the Deccan of central India (Fig. 2) (Yazdhani 1947). Legends relate that the secrets of bidri making came to Bidar in the 15th century, apparently from Iran, via Ajmer in Rajasthan. However, the earliest extant historical record of zinc production at Bidar dates from as late as 1759, although taken from a document of the 1720's. This states 'In this *subah* (Bidar) the fine and rare bidri vessels are made....The craftsmen of this place make them with such delicacy that even a painter could not imagine them.'. Thus bidri making was clearly well established in Bidar by the early 18th century and the earliest surviving pieces probably belong to the late 16th century.

The shapes and decoration of many of the earliest pieces are indeed Iranian in style, but it seems unlikely that the technique could have originated in Iran which did not have access to zinc. Yazdhani (1947, p.20) makes the interesting comparison between bidri and the patination technique of *kuftgari*, that is, steel inlaid with gold and silver and then given a dark grey patina (Untracht 1968, 153-8), which was widely practiced in the Middle East as well as in India.

The connection with Ajmer is more significant, as Zawar was a dependency of the *subah* of Ajmer in the 16th century. Possibly craftsmen originally from Iran developed the new technique in Rajasthan where zinc would have been available before moving south to Bidar. Certainly the thriving Bahmani dynasty attracted a great number of craftsmen to its court in the 15th and 16th centuries, and so some elements of the legends may be substantially correct. Production seems to have been confined to the Deccan until Bidar's incorporation into the Moghul Empire under Shah Jahan in 1656, there after production began additionally in centres such as Purnea and Murshidibad in present day Bihar, and in Lucknow, in present day Uttah Pradesh, all in the Ganges valley (Fig. 2). There is no evidence that bidri was ever made in any of the Rajput states of Rajasthan, and this rather surprising fact is supported by the lead isotope analyses of a selection of early pieces.

3. Production

The alloy is predominantly of zinc with just a few percent of copper, which is however essential (see below). There seems to have been a good deal of local variation in composition as recorded by Lal (1990, p.4) and Mehta (1960, p.29) at various regional centres across India, but the amounts of copper are typically around 5%, varying in reported analyses between 1.5% to 12% (La Niece and Martin 1987; Lal 1990) (Fig. 6). The All India Handicrafts Board in conjunction with the India Standards Bureau state that the copper content should be between 6 and 7%. The alloy often has some lead, the small amounts of a few percent probably come with the zinc metal, but some pieces have between 10 and 20%, which must be deliberate adulterations and some also have a few percent of tin which is a little surprising. Neither lead nor tin are necessary for the development of the patina. Lal (1990, p.4) records that some of the directions for bidri making in 19th century Lucknow state that about 20% of powdered steel should be added to the zinc and copper. The iron would dissolve slowly in the molten zinc at temperatures around 500°C, more readily at higher temperatures, to form a series of complex intermetallic compounds (Smith 1918, pp.181-2; Morgan 1985, pp.159-60). The iron in these intermetallics could be expected to oxidise to form hematite or goethite during the patination treatment giving the bidri surface a more ruddy appearance. It would also be expected to make the patina more unstable owing to the greater activity of the compounds. However, high iron contents have never been encountered in any of the analysed pieces, and it is likely that this alloy was not used extensively, if at all.

Victoria & Albert Museum acquisition nos.	Description	%				
		Zinc	Copper	Lead	Tin	Iron
2539-1883 I.S.	*Huqqa*	91.2	3.2	2.9	1.2	0.2
1578-1904	*Huqqa*	98.0	3.3	1.4	0.1	0.2
45-1905	*Huqqa*	88.2	3.0	2.9	0.7	0.5
I.S.46-1977	Weight	92.6	4.7	1.4	5.7	0.1
1479-1904	Ewer	89.3	4.8	0.8	<0.1	0.8
I.S.10-1973	Bowl	92.1	3.6	1.4	0.3	0.1
02942 (I.S.)	*Huqqa*	84.6	3.3	1.0	0.1	0.5
I.S.131-1958	Pan box	95.0	3.1	1.5	0.1	0.1
I.S.181-1965	*Huqqa*	76.1	10.1	8.2	11.4	0.6
I.S.31-1976	Bottle	98.6	2.6	1.0	0.4	1.2
857-1874	*Huqqa*	92.1	3.6	0.8	0.6	0.4
I.S.17-1970	Pan box	81.3	2.0	2.0	<0.1	0.5
02949	Bottle	83.8	2.5	6.6	<0.1	0.2
02949	Bottle	93.2	2.4	1.1	0.1	0.2
855-1874	*Huqqa*	99.2	2.6	0.4	0.2	0.6
02941 (I.S.)	Basin	91.2	2.8	2.1	0.1	0.2
I.S.4-1977	*Huqqa*	83.6	3.7	0.8	<0.1	0.6
I.S.19-1978	*Huqqa*	85.3	4.6	3.0	1.3	<0.1
120-1886	Bottle	86.7	5.3	1.9	0.2	0.1
I.M.224-1921	Bottle	91.7	3.6	0.7	0.7	0.7
2066-1883	Box	79.6	2.7	19.9	0.1	0.5
1402-1903	*Huqqa*	95.3	2.7	1.1	3.0	0.5
I.S.11-1973	Vessel	97.6	3.4	1.8	<0.1	0.1
I.S.39-1976	*Huqqa*	89.2	3.4	0.5	0.7	0.7
I.S.19-1980	*Huqqa*	80.6	3.6	7.3	6.4	0.9
856-1874	*Huqqa*	77.4	4.4	0.9	0.4	0.1

(Analysis by Graham Martin)

Fig. 6 Composition of some bidriware in the Victoria and Albert Museum. The analyses were performed by atomic absorption Spectrometry. (From Stronge 1993)

Fig. 7 Direct evidence of the zinc trade with China. Chinese zinc ingots (BM MLA Reg. 1997 2-2, 1 & 2) recovered from the East India ship *Diana*, which sank in the Mallacca Straits off Malaysia in 1816 *en route* to Madras and Calcutta. (Photo. A. Milton / British Museum)

In the early days of the industry surely Zawar would have been the source of the zinc, but the very limited programme of lead isotope analysis on pieces dated stylistically to the 18th century showed that Zawar zinc was not represented. This fits in with the historical evidence that by this time quantities of zinc were being shipped to the south and east ports of India from China (Fig. 7), and more specifically Ball (1886) specifically stated that much of the zinc used to make bidri came by sea from China. In the 1980's at the Mumtaz and Gulistan Bidri Co-operatives in Hyderabad used old zinc lithography plates and scrap copper cable as the source of much of the metal they used.

The artefacts are all sand castings (Fig. 8) and the earlier pieces also seem to have been made in this way, (although Yazdhani (1947, p.20) in a brief but detailed account describing the production of bidri wares at Bidar in the mid 20th century, stated that the metal was cast in clay moulds). Complex pieces were, and are still, cast separately and joined by lead tin solders with the necessary low melting points. Historically sand casting is not well attested in India, especially for artistic productions for which lost wax casting was and is preferred. The technique was used in the Middle East in the medieval period (La Niece 2003), which may provide some support for the idea of a connection with the Islamic craftsmen of Iran.

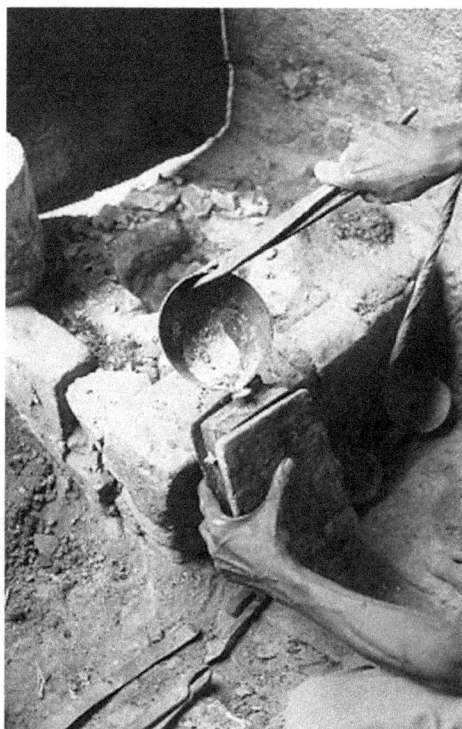

Fig. 8 Pouring the bidri alloy into a sand mould (Photo. P.T. Craddock, Yaqoob Bros., Hyderabad, 1991)

The zinc is melted in crucibles at temperatures of around 500°C and the small amounts of copper added and dissolved in the molten zinc, a maximum of approximately 2.7% of copper will enter into solid solution, the remainder will form intermetallic compounds with the zinc (Morgan 1985, pp.158-60). The fires are usually of charcoal, sometimes supplemented by partially combusted coal collected from railway tracks in the recent past whilst India's rail network was still largely steam driven.

After casting the artefact is cleaned by scrapping, filing and polished with the buffing wheel. At this stage the surface is dull and leaden in colour and prior to working the designs the surface is first given a temporary black patination to make the incised design more visible by rubbing it with a damp cloth bag containing copper sulphate, 'blue vitriol', crystals (NB copper sulphate was sometimes listed as one of the ingredients of the main patinating agent but this is unnecessary and seems likely to be a conflation of the two patinating stages). The metal, although quite brittle is soft and easy to carve. First the designs are scribed into the surface using compasses and steel points (Fig. 9). Then the lines are punched out using a small tracer (Figs. 10 & 11) and the broader areas for inlay are chiselled out with the edges undercut to hold the inlays. The inlay metals, either wires for the lines or thin sheet for the plates, are then hammered into place (Fig. 12). When the inlay is completed the completed piece is again carefully filed and polished, nowadays invariably on a buffing wheel, ready for patination. This was and is done with a clay poultice rich in common salt (sodium chloride), saltpetre (sodium nitrate) and sal ammoniac (ammonium chloride). This special clay is still collected from 'under old walls' in Bidar. Anyone who has travelled in India will appreciate that walls in towns serve as public urinals and thus the soil beneath them are likely to be well saturated in the chemicals mentioned above. Gairola (1956) analysed Bidar earth and found it to contain 14% by weight of water soluble materials, made up of nitrates and chlorides but no sulphates or, more surprisingly, ammoniac salts, such as urea. Ammonium chloride is routinely added to the soil. At the time of the author's last visit in 1991, the bidri makers of Hyderabad claimed that they still used soil collected in Bidar, about 80 km away, for their patination work, selecting soil that tasted salty. The soil is finely ground and hot water added to make a warm wet paste which is applied to the surface of the bidri which itself is gently warmed before application. The deep black patina develops in seconds, although the inlays are not affected at all. After this the soil is washed off and the black surface given a final polish with a little coconut oil.

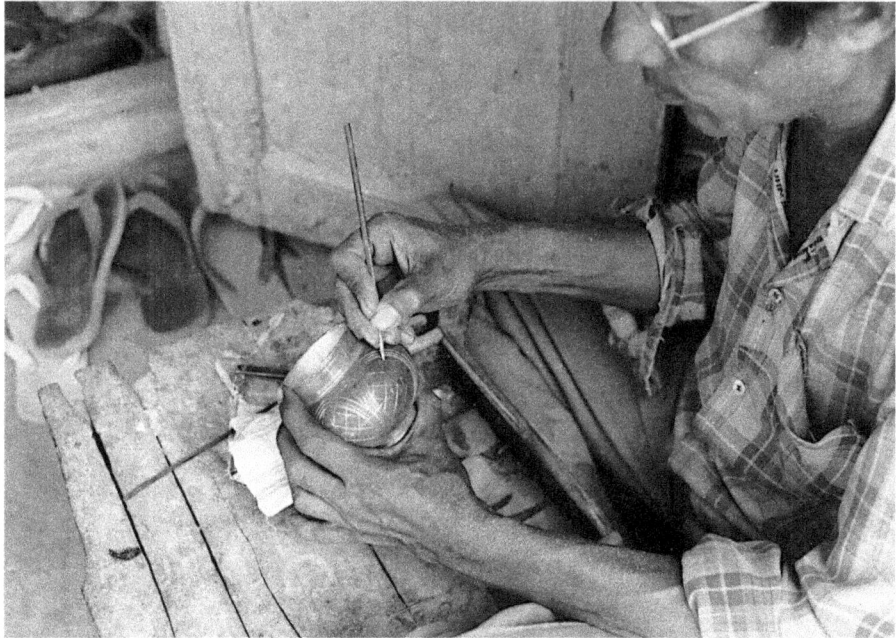

Fig. 9 Scribing the design to be engraved on the surface of the bidri casting temporarily blackened with copper sulphate. (Photo. P.T. Craddock, Yaqoob Bros., Hyderabad, 1991)

Fig. 10 Chasing a line using a small chisel or chaser. The work piece is imbedded in bitumen (Photo. P.T. Craddock, Yaqoob Bros., Hyderabad, 1991)

Fig. 11 Bidri making in 18th century Bidar, (Detail from the border of a map of the *subah* of Bidar prepared for Colonel Gentil Faizabad in 1770. India Office Library Add.Or. 4039 folio 13)

4. Imitations and Inspirations

idri wares have been imitated quite widely in India and in the Middle East.   At Moradabad in Uttar Pradesh vessels of thin copper or brass sheet that had been tinned were hammered to create large sunken areas just leaving a few areas raised.   The depressions were then filled with lac or some other black mastic which mimicked the bidri and the tinned areas standing proud mimicked the silver inlay (Mehta 1960, p.35). Both in India and in the Islamic world black-patinated iron inlaid with silver, a form of *kuftgari*, is used to imitate bidri.

5. The Patina

The nature of the bidri patina has been preliminarily investigated by La Niece and Martin (1987).   The patina on a replicate piece made in exactly the same way as recorded in India was sectioned and found to be about 10 microns in thickness. Energy dispersive X-ray fluorescence analysis of the patinated surfaces of this and old pieces from India revealed zinc, copper and chlorine. X-ray diffraction of the patina of 12 of pieces found zinc hydroxide chloride hydrate (JCPDS 7-155), $Zn_5(OH)_8Cl_2.H_2O$, known as simonkoileite, and zinc oxide (JCPDS 5-644), ZnO, in all the samples together with silver chloride, AgCl (presumably from the inlays), and cuprite, Cu_2O, in some (Fig. 13).

Fig. 12 Hammering a silver wire into an engraved channel on the lid of a bidri box. (Photo. P.T. Craddock, Yaqoob Bros., Hyderabad, 1991)

Line spacing (d in Å)	Intensity I/I_0	Probable identity				
7.9	100	$Zn_5(OH)_8Cl_2$				
5.4	18	,,				
3.98*	14	,,				
3.58	17	,,				
3.17	17	,,				
2.94	8	,,				
2.86	17	,,	$+ZnO$			
2.72	14	,,				
2.67	23	,,				
2.59	14	,,	$+ZnO$			
2.47	16	,,	$+ZnO$	$+Zn$	$+Cu_2O$	
2.37	25	,,				
2.31	3			Zn		
2.15	9	$Zn_5(OH)_8Cl_2$			$+Cu_2O$	
2.09	37			Zn		
2.02*	12	$Zn_5(OH)_8Cl_2$				
1.96	6	,,				
1.9	5	,,	$+ZnO$			
1.82	3	,,				
1.77*	7	,,				
1.69*	10	,,		$+Zn$		
1.63	3		ZnO			
1.58	14	$Zn_5(OH)_8Cl_2$				
1.55	7	,,				
1.51*	5	,,			$+Cu_2O$	
1.47	5	$Zn_5(OH)_8Cl_2$ $+ZnO$				
1.42*	3	,,				
1.37*	3	,,	$+ZnO$			
1.34	5			Zn		
1.32	4			Zn		
1.29	4				Cu_2O	

*Diffuse or wide line

Fig. 13 X-ray diffraction survey of samples taken from the patina on base of a *huqqa* base, dated 1802 AD, now in the Victoria and Albert Museum, Reg. 401-1876. (from la Niece and Martin 1987)

476

None of the common zinc salts are black, and the only common black copper oxide is tenorite, CuO, which was not detected, and as it usually forms under reducing conditions, would not be expected anyway. Replicate pieces of a bidri alloy containing 3% of copper and 0.5% of lead were made and treated with a warm aqueous solution of potassium nitrate and ammonium chloride, causing a deep black, well-adhering patina to form almost instantly. The same treatment on a clean sheet of pure zinc produced just a pale grey patina of zinc chloride and oxide, thus demonstrating that the copper is essential for the formation of the black patina. La Niece and Martin postulated that the zinc was attacked and rapidly dissolved by the ammonium chloride effectively exposing very finely divided elemental copper, which in turn was oxidised by the nitrates to form cuprite. This seems eminently reasonable, the only objections being that cuprite was not detected in all the samples and is usually red or brown. However oxides which have formed very rapidly and are very finely dispersed can appear black and are often to all intents and purposes amorphous. Thus, for example, the well-known Japanese *shakudo* alloys of copper with a very little gold and silver, readily patinate to produce superb deep purple-black surfaces in which only cuprite can be detected by X-ray diffraction (Notis 1988; Murakami et al 1988; Craddock and Giumlia Mair 1993).

6. Decline and Recovery of the Industry

The bidri industry largely ceased in northern India after the first War of Independence in the mid 19th century, and was in sharp decline in the south. Bidri was displayed and greatly admired in the Great Exhibition at the Crystal Palace in London in 1851, and enthusiastically taken up by the Arts And Crafts movement in England, the small Indian workshops appealing to the movement's concept of the artist-craftsman, and more practically, the London store, Liberty's began to stock bidri (Stronge 1985, p.28: 1993). However this interest was not enough and by the early 20th century the industry was almost extinct in India. It was revived in the then state of Hyderabad in which both the cities of Hyderabad and Bidar then lay, by the Nizam's government. After independence in 1948 Hyderabad became capital of the state of Andhra Pradesh but Bidar became part of the state of Karnataka. Feeling isolated from their main support and markets most of the remaining bidri makers in Bidar emigrated to Hyderabad, although some remained and the industry was certainly undergoing something of a revival in the 1990's in that state with additional production centres opening in Aurangabad. State support continued in Hyderabad and at the time of the 1961 census (Gowd 1964) it was estimated that there were 8 established workshops in Hyderabad, such as that of the Yacoob Brothers in the Gunfoundry district of the City (Fig. 1), together with the Mumtaz and Gulistan manufacturing and training co-operatives (Lal 1993, p.16). At the time of the author's visits in the 1980's these institutions were flourishing (Fig. 14) with about 125 employed in Hyderabad and 35 in Bidar. As well as selling directly from their workshops the products are also sold through the State Craft Emporia (Fig. 15), with an ever widening range of artefacts. The traditional vessels and boxes are probably still the staple products but have been joined by a wide range of cheap 'airport art' tourist souvenirs, through a range of better quality new but traditionally inspired designs aimed at India's burgeoning middle classes (Figs. 16 & 17), to special individual commissions, many still coming from the Middle East.

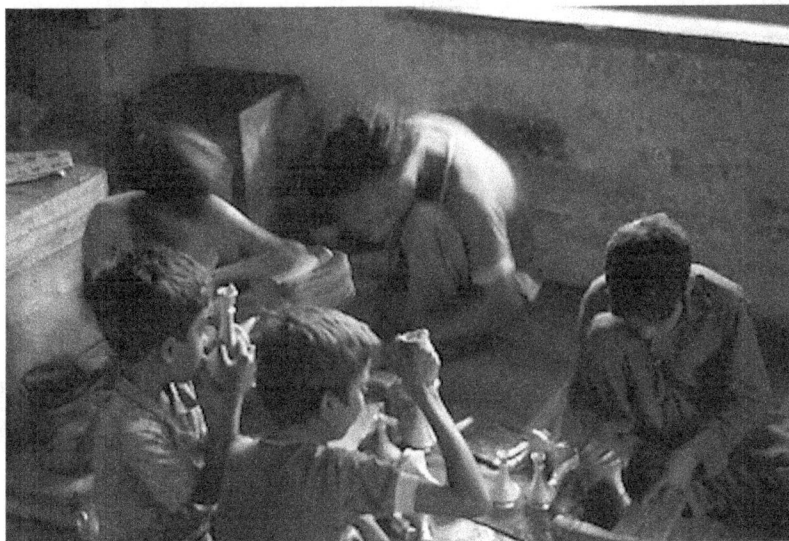

Fig. 14 Young apprentices learning the craft of bidri making at the Mumtaz / Gulistan cooperative, Hyderabad. (Photo. P.T. Craddock, 1991)

Fig. 15 Andhra Pradesh State Craft Emporium for bidri sales in Hyderabad. Promotion of bidri at these outlets has helped revive the traditional craft (Photo. P.T. Craddock 1986)

Fig. 16 Modern Bidri designs: Large plate with typical floral design inlaid with silver, made by Yaqoob Bros., Hyderabad 1991 (British Museum Reg. OA 1992, 7-31, 1) (Photo. A. Milton / British Museum)

Fig. 17 Modern Bidri designs: Plaque inlaid with silver and brass made by MD Shafiuddin of Bidar, 1991 (Photo. A. Milton / British Museum)

Thus at the end of the 20th century the bidri trade was thriving, and for the first time in centuries, was using zinc from Zawar.

References

1. Ball, V., *zinc and zinc ores*, Scientific Proceedings of the Royal Dublin Society, Vol. **5**, No.5, 1886, p.321.

2. Craddock, P.T., *Contemporary production of bidri ware in Hyderabad and Bidar*, Appendix to Stronge, 1993, p.142.

3. Craddock, P.T., *Early Metal Mining and Production*, Edinburgh University Press, Edinburgh, 1995.

4. Craddock, P.T. and Giumlia-Mair, A. R. *Hśmn-km, Corinthian bronze, shakudo: black-patinated bronze in the ancient world*, Metal Plating and Patination, S. La Niece and P.T. Craddock (eds.), Butterworth Heinemann, Oxford, 1993. p.101.

5. Craddock, P.T., Freestone, I.C., Middleton A., Gurjar, L.K. and Willies, L., *Zinc in India, in 2,000 Years of Zinc and Brass*, P.T. Craddock (ed.). British Museum Occasional Paper 50, London, 1998. p. 27.

6. Gairola, T.R., *Bidri Ware*, Ancient India, Vol. 12, 1956, p.116.

7. Gowd, K.V.N., (ed.) *Selected crafts of Andhra Pradesh*, Census of India 1961 Vol. II, pt.VII-A3, Government of India, Delhi, 1964.

8. Lal, K., *National Museum Collection: Bidri Ware*, National Museum, New Delhi, 1990.

9. La Niece, S., *Medieval Islamic Metal Technology, in Scientific Research in the Field of Asian Art*, P. Jett (ed.), Archetype Books, London, 2003. p. 90.

10. La Niece, S. and Martin, G., *The technical examination of bidri ware*, Studies in Conservation Vol. 32, No. 3, 1987, p.97.

11. Morgan, S.W.K., *Zinc and its alloys and compounds*, Ellis Horwood Co., Chichester, Sussex, 1985.

12. Murakami, R., Niiyama, S. and Kitada, M., *Characterization of the black surface on a copper alloy coloured by a traditional Japanese surface treatment*, Proceedings of the IIC Congress in Kyoto, N.R. Bromelle (ed.), IIC, London, 1988, p.133.

13. Mehta, R.J., *The Handicrafts and Industrial Arts of India*, D.B. Taraporevala and Sons, Bombay, 1960.

14. Notis, M.R., *The Japanese alloy shakudo: its history and its patination*, The Beginnings of the Use of Metals and Alloys, R. Maddin (ed.), MIT Press, Cambridge, Ma. 1988, p.315.

15. Smith, E.A., *The Zinc Industry*, Longmans Green and Co, London, 1918

16. Stronge, S., *Bidri Ware*, Victoria and Albert Museum, London, 1985.

17. Stronge, S., *Bidri ware of India*, in La Niece and Craddock (eds.), 1993, p.135.

18. Untracht, O., *Metal Techniques for Craftsmen*, Doubleday, New York, 1968

19. Yazdani, G., *Bidar: Its History and Monuments*, Published under the authority of the Nizam. Hyderabad, 1947

Novel Phosphoric Irons for Concrete Reinforcement based on Studies on the Delhi Iron Pillar

R. Balasubramaniam

Department of Materials and Metallurgical Engineering

Indian Institute of Technology, Kanpur, India

Abstract: *The Gupta-period 1600-year old Delhi Iron Pillar has attracted the attention of metallurgists and corrosion scientists for its excellent resistance to corrosion. The talk will first summarize new insights, relating to its history, astronomical significance, engineering details and manufacturing methodology, based on the researches of the author. The corrosion resistance of the Pillar will be discussed in detail, with a brief review of earlier theories of its excellent corrosion resistance. The results of a detailed characterization of the Delhi Iron Pillar's rust clearly established that the major constituents of the scale were crystalline iron hydrogen phosphate hydrate, goethite, lepidocrocite, amorphous delta-FeOOH and magnetite. The process of protective rust formation on Delhi pillar iron will be understood and compared with that of mild steel and weathering steel, based on the rust analysis. Some notable characteristics of the protective passive film on the Delhi Iron Pillar will be summarized. Based on the known beneficial effect of phosphorus on corrosion of the Delhi Iron Pillar, the second section of the presentation will be devoted to understanding the possible use of phosphoric irons in a modern application, namely for concrete embedment. Three phosphoric irons (Fe-0.11P-0.028C, Fe-0.32P-0.028C and Fe-0.49P-0.028C, wt.%) were ingot melted (85 kg ingots) and forged between 1050 and 1100 degree Celsius in the dual phase region (of the iron-phosphorus phase diagram) to minimize grain boundary segregation of phosphorus. The importance of this heat treatment in inducing ductility will be explained, touching upon mechanical testing and associated studies of fracture surfaces. The phosphorus distribution in the microstructure will be understood by detailed microstructural characterization. The results of potentiodynamic polarization, linear polarization and electrochemical impedance spectroscopy (EIS) studies in two different simulated concrete environments (a) saturated calcium hydroxide of pH 12.5 and (b) 0.3M NaHCO3 + 0.1M Na2CO3 solution of pH 9 will be presented and discussed. The electrochemical behavior will be contrasted with that of two commercial reinforcement steels of composition Fe-0.18%C-0.8% Mn (mild steel) and Fe-0.18%C-0.34%Cu-0.09%P-0.8%Mn (equivalent to weathering steel). The threshold chloride contents for passive film breakdown were higher for phosphoric irons. The phosphoric irons were particularly superior in chloride containing high pH environments. The superior corrosion resistance of the phosphoric irons will be discussed in terms of phosphorus maintained in solid solution in the matrix and a bipolar passive film model. Future directions for research and development on phosphoric irons will be set forth in the concluding section with special emphasis on large scale manufacture of phosphoric irons.*

1. Introduction

The damage caused by corrosion constitutes a major drain on the useful resources of the country. It is particularly not quite evident, since a major portion of the costs are indirect in nature. In view of this cost of corrosion, corrosion engineering is a topic of great concern for any nation, seeking to industrialize and develop on a large scale. This very much applies to the case of India, where the economy has been growing at significant rates over the last several years. At this juncture, we also note that the related subjects of teaching of corrosion and training of skilled manpower are also very important vital components of an economy that is striving for rapid industrialization.

Considering the importance of corrosion to the country, the present communication will focus attention on how the study of the past can result in useful engineering materials for the future, with particular emphasis on materials engineering for resisting corrosion. The philosophy of corrosion prevention and control will be first highlighted with special emphasis on materials selection aspects. The communication will then highlight the development and utility of novel phosphoric irons that have been developed recently at IIT Kanpur, particularly their exceptional resistance to corrosion in concrete environments. The paper will therefore offer insights into how the study of the past (i.e. the Delhi Iron Pillar [1-3]) has resulted in the development of these novel phosphoric irons. The science of corrosion prevention of these phosphoric irons will be finally emphasized, at which point in time, the philosophy of materials engineering for corrosion prevention will be revisited to critically analyze and assess the optimum strategy in the selection of engineering materials to prevent corrosion.

2. Philosophy of Corrosion Prevention and Control

Corrosion plays a very important role in three stages in the life of an industrial part. Therefore, considerations of corrosion come into play in the entire life of an engineering part, in particular, or engineered object, in general. The three "stages" in the life of an engineering component can be viewed as the design stage, production stage and finally the application stage [4]. In all these stages, corrosion needs to be taken into account. The reasons and requirement will be different for each stage. In the design stage, the concerns are regarding the design of geometric shape and size of the product such that corrosion is minimal, the choice of materials, the operating environment the part will be subjected to (and the possible planned and unplanned changes that can occur to this environment), and, finally, whether protective coatings, inhibitors or electrical protection methods will be used or not. Once the product is designed it has to be manufactured as per the specifications. The points that need to be noted during the manufacturing stage such that corrosion does not become a problem later on include accurate reproduction of design, the use of specified materials, the use of correct heat treatments, the use of correct fabrication methods, and the correct application of protective coatings. After the product has been manufactured, it will finally find service with the customer. The customer has to take certain precautions to ensure that corrosion does not become a problem. These include the use of correct materials during any replacement, monitoring of the environment, maintenance of protective coatings and maintenance of electrical protection. It is important that all the steps mentioned above are carefully adhered in order to ensure good product life for the product. It is important to stress that in case there is **failure to maintain any of these above points at any stage**, then this **will result in premature failure of the product.**

Before addressing the specific issue of materials engineering in the process of corrosion prevention and control, it may be educative to understand the various methods of prevention of corrosion, namely the philosophy of corrosion protection. We can talk either in terms of corrosion control or corrosion prevention. The *mantra* of minimizing economic losses is clearly "prevention is better than control." This will be the ultimate aim of any corrosion engineering. However, in several cases, it would just not be possible to completely prevent corrosion and in such cases, it is important to implement corrosion control methods. Corrosion control seeks to regulate corrosion and keep it within acceptable or predictable limits.

What are the means available to the engineer? It may be enlightening to state a verse from Buddha in the *Dhammapada* and relate it to corrosion. Verse number 191 in the chapter titled *Buddhavaggo* says [5]:

> *dukham, dukkhasamuppadam, dukkhassa ca atikkamam*
> *ariyam c'atthangikam maggam, dukkupasamagaminam*

Suffering, the origin of suffering, the cessation of suffering, and the noble eightfold path which leads to the cessation of suffering.

The above verse informs that the way to prevention of suffering is by the noble eightfold path consisting of right views, right aspirations, right speech, right actions, right living, right exertion, right recollection and right meditation.

In a similar manner, the *mantra* of corrosion prevention and control can be stated as: Corrosion, the origin of corrosion, the cessation of corrosion, and the eightfold path which leads to the cessation of corrosion.

The mantra of corrosion prevention and control informs about the eight methods by which corrosion can be prevented. These are by (a) selection of suitable material, (b) using protective coatings, (c) changing service conditions (like temperature, pressure, velocity of fluid, etc.), (d) changing the environment chemistry like pH, concentration, aeration, etc., (e) adding corrosion inhibitors, (f) shifting the potential by cathodic or anodic protection, (g) modifying the design to usher in changes, and (h) allowing the material to corrode at predictable rates and replacing the product at opportune times.

In the present communication, the emphasis will be on aspects of materials selection to prevent corrosion of engineering assets, with special attention focused on the corrosion resistance imparted by the formation of protective passive films on the surface. The inherent corrosion resistance of the material is thus emphasized rather than external factors like protective coatings, inhibitors, electrical protection methods, etc.

3. Corrosion Control by Materials Selection

There are several materials available in modern technology. Their right selection will help in mitigating corrosion problems. However, it is important to keep economics in mind while selecting materials for any particular application. For example, it is be most costly to use a corrosion-resistant material but the low maintenance and higher life of the material may offset the initial high investment. On the other hand, a cheap material would require frequent maintenance and also will have to be replaced more regularly such that the overall cost may become more than using a corrosion-resistant material. It is for this reason that several different considerations are taken into account for finding out the right choice based on economics. One example is the method of discounted cash flow (NACE standard RP-02-72).

The available engineering materials in terms of materials selection issues are detailed in several standard reference books and text books and there will not be any need to elaborate on this particular aspect in this communication. In case of metallic materials, the availability of a variety of steels, stainless steels, cast irons, Ni alloys, Al alloys, Cu alloys, Ti/Zr alloys, etc. have resulted in a fairly comprehensive understanding of the advantages and disadvantages of using these materials in particular corrosive environments. Needless to add, one of the critical deciding factors that will determine material selection is the cost of the material involved and the frequency and cost of maintenance of the selected material. In addition to metallic materials, there is also a range of non-metallic materials that are available for corrosion prevention and these include a wide range of ceramic and polymeric materials. In addition, there are special issues to consider while addressing corrosion in case of electronic materials. All these issues are well addressed in the literature.

At this point in time, it is important to address the crucial question as to what makes one material superior to another while it comes to selection of materials for corrosion resistance. Or to put this in another words, what would be the property that would be ideally required in an engineering material that is selected for corrosion resistance. The fundamental question is therefore what constitutes a good design of an engineering material. To answer this important question, the article will now change focus on a very important example from history and then relate how the knowledge gained from that material regarding corrosion resistance has helped in the development of modern material with very attractive corrosion resistance properties. The example that will be discussed is the legendary 1600-year old Delhi Iron Pillar [1-3] and the modern engineering material developed based on the knowledge gained from detailed studies of Pillar is phosphoric iron. A large scale application where materials selection plays a crucial role is in reinforcement of concrete. The usefulness of phosphoric irons in this specific application will be highlighted and as a consequence, the philosophy of designing an engineering material for maximum corrosion resistance will be understood.

4. Lessons from The Past: The Delhi Iron Pillar

One of the metallurgical wonders of the world is the Gupta period 1600 years old Iron Pillar located now at Delhi (see Figure 1). It is testimony to the high level of skill achieved by the ancient Indian iron smiths in the extraction and processing of iron [1-3]. This Pillar has attracted the attention of corrosion technologists because it has withstood corrosion for the last 1600 years. The fact that an iron structure could remain resistant to corrosion over such a long time is indeed incredible. The very low rate of corrosion of the material of the Pillar is evident when one looks carefully at the surface, for example the location of where the oldest Sanskrit inscription has been die struck on the Pillar (Figure 2). Notice how the die struck imprints of the letters are clearly delineated with minimal corrosion damage, such that is very difficult to believe that this structure is 1600 years old!

Figure 1 Delhi Iron Pillar at Qutub Minar complex in New Delhi.

Figure 2 Some die-struck characters of the oldest Sanskrit inscription showing how the
imprints are clearly delineated with minimal corrosion damage.

There are several theories proposed for the resistance to corrosion of the Pillar. They
will be briefly summarized here with a critical look. A full detailed discussion can be
found in the books of the author [2,3]. The theories can be divided into two broad
categories: environmental and material The proponents of the environment theory state
that the mild climate of Delhi is responsible for the corrosion resistance of the Delhi
Iron Pillar as the relative humidity at Delhi does not exceed 70% for significant periods
of time in the year, which therefore results in very mild corrosion of the pillar.
However, this is not entirely true as the contact period of the Pillar with water is
significant if one considers rainfall in addition to condensation due to low temperatures
(in winters). Mathematical modeling of the water residence time on the Pillar over its
entire lifetime has revealed that had the Pillar rusted even minimally, the rust layer on
the surface would have been far too significant than what is actually observed [6]. The
shining nature of the surface indicates that the surface layer is very thin indeed [7].
Therefore, this discounts the environment theory. A strong reason why this may not be
the factor is again revealed by archaeological evidences of the iron beams at Konarak
(Figure 3) which are relatively unaffected, even in the moist saline environment as
Konarak is located on the sea-coast. This provides the strongest clue that the material
of construction of these beams is resistant to chloride attack in addition to being
corrosion resistant in the atmosphere. This is a very important point and this will be
emphasized in the later discussion on the novel material, phosphoric iron, that were
developed based on the studies on the Delhi Iron Pillar.

Several investigators have stressed the importance of the material of construction as the
primary cause for its corrosion resistance. The ideas proposed in this regard are the
relatively pure composition of the iron used, presence of phosphorus, and absence of
deleterious MnS inclusions (due to absence of S and Mn in the iron), its slag enveloped
metal grain structure, and formation of protective passive film. Other theories to
explain the corrosion resistance are also to be found in the literature like the mass metal
effect, initial exposure to an alkaline and ammonical environment, residual stresses

resulting from the surface finishing (hammering) operation, freedom from sulfur contamination both in the metal and in the air, presence of layers of cinder (charcoal) in the metal thereby not allowing corrosion to proceed beyond the layer (cinder theory) and surface coatings provided to the Pillar after manufacture (treating the surface with steam and slag coating) and during use (coating with clarified butter). That the material of construction may be the important factor in determining the corrosion resistance of ancient Indian iron is attested by the presence of ancient massive iron objects located in areas where the relative humidity is high for significant periods in the year (for example, the iron beams in the Surya temple at Konarak in coastal Orissa and the iron pillar at Mookambika temple at top of the Kodachadri Hills on the western coast).

(a)

(b)

(c)

Figure 3 (a) Konarak temple showing (b) iron beams in the superstructure. (c) Iron beams in the yard on raised platform at Konarak. Note that these iron beams are relatively unaffected in the moist saline environment since Konarak is located on the sea-coast.

When one views critically the material theories, the following conclusions can be arrived at. The absence of MnS inclusions is not critical because there are entrapped slag inclusions which will act as initiating sites for corrosion. Recent in-situ micrographs obtained from the Iron Pillar [8] do not show "slag enveloped grain structure." This was a wrong interpretation of the microstructures obtained by Bardgett and Stanners [9] in early 1960s because the pearlite regions at the grain boundaries were wrongly called as slag and this started the confusion regarding this issue. Therefore, the "slag coated structure" theories can be conveniently discounted. The presence of phosphorus certainly is the critical issue and this is the common factor in the ancient Indian irons that are resistant to corrosion. Therefore, the most important factor from the material side is the relatively high P content of the Pillar. Further, it is now known, based on detailed studies conducted on the long-term rust of the Delhi Iron Pillar [10], the process by which phosphorus beneficially affects the corrosion resistance. It is due to the formation of a protective passive film [11]. The complete details of the protective passive film are also now available and the phosphates play a very important role in the protection. The possible method by which phosphates can affect the corrosion resistance is briefly summarized at the end of the article. The surface coating theories can all be discounted because a freshly-cut surface achieves the surface finish of the rest of the Pillar in a matter of three years, thereby proving that no artificial coatings were applied on the Pillar. The only other theory that is worth considering is the mass metal effect theory [12]. As per this theory, the massive weight of the Pillar allows for faster heating up of the Pillar and thereby lowering the residence time of moisture on the surface. However, it must be borne in mind that the larger heat capacity of the Pillar will also result in faster cooling of the Pillar and therefore the residence time is also increased due to this effect. It is not possible to quantify the relative importance of these two effects. However, the mass metal effect cannot be significant because mathematical modeling of the moisture residence time on the Delhi Iron Pillar has shown that the time of wetting due to rainfall is much more than the time of wetting due to the moisture condensation [6]. Therefore, the most significant factor dictating the superlative corrosion resistance of the Iron Pillar is the formation of protective passive film, which is aided by the presence of phosphorus in the Iron Pillar [11].

5. Material for The Future: Phosphoric Irons

The presence of phosphorus leads to the formation of a protective passive film on the surface, as discussed above, which provides the Delhi Iron Pillar its exceptional corrosion resistance properties. Therefore, it is anticipated that phosphoric irons will also show enhanced corrosion resistance.

How do we use the knowledge learnt from the study of the Iron Pillar for modern day technological applications? It is now worthwhile to explore phosphorus-containing iron in more detail. They will be generically called phosphoric irons in the rest of this communication. In normal steels used in modern day industry, the phosphorus is controlled to less than 0.05 weight % due to its deleterious effects on ductility, especially when phosphorus is present at the grain boundaries. However, the ancients normally used phosphorus contents that were almost hundred times this amount. Therefore, in order to understand how the ancients handled phosphorus in iron, detailed studies were conducted on archaeological Indian irons, both at low magnifications in the optical microscope and at very high magnification in the transmission electron microscope [13,14]. These studies revealed several important alloy design criteria to produce ductile phosphoric irons. First, precipitating austenite allotriomorphs by a high temperature anneal can restrict P segregation to the grain boundaries. Secondly, phosphorous segregation can be avoided by locating a small amount of carbon/carbide along the grain boundaries. Thirdly, precipitation of phosphide phase in the ferrite matrix, in a manner similar to age-hardenable Al alloys, would prevent P segregation to the grain boundaries apart from providing additional strength. This indicated that by suitable heat treatments and thermomechanical processing, one can easily achieve the required microstructure as per the alloy design proposed such that ductile phosphoric irons can be produced [15,16].

In order to provide answers to this question, a detailed study [17] was undertaken by Gadadhar Sahoo at IIT Kanpur, under the guidance of this author, to understand possible industrial applications of phosphoric irons. The first aim was to render the phosphoric irons ductile. The second aim was to locate a modern application wherein the corrosion resistance of phosphoric irons could be put to good use.

Ductility is a very important property from the point of practical application of any engineering material. How do we make phosphoric irons ductile? It is known that phosphorus segregation to the grain boundaries makes these locations weak and results in poor ductility. Therefore, the first aim of the work was to find out methods to keep phosphorus away from the grain boundary. It was first realized that phosphorus will not be present in regions where carbon is located in the iron matrix because phosphorus is a substitutional solute element whereas carbon is an interstitial solute element. Therefore, a small amount of carbon needs to be maintained in phosphoric irons. The next challenge is to locate these carbon atoms along the grain boundaries, thereby keeping the phosphorus atoms away from these locations. In order to achieve this, an intelligent use of the phase transformations in the iron-phosphorus system was applied.

The phase diagram of iron-phosphorus shows a $(\alpha + \gamma)$ dual phase region at high temperature (see Figure 4). If a phosphoric iron in the composition range 0.25 and 0.50 wt %P, is soaked in the two phase region (temperature between 1000°C and 1100°C), then austenite phase ($\square$) will precipitate on the grain boundaries of ferrite phase (α). This is well known in physical metallurgy. Austenite has a higher solubility for carbon

490

than phosphorus and therefore all the carbon is pushed to the grain boundary region while the phosphorus is removed from the grain boundary region. After a suitable soaking time at high temperature, the phosphoric irons can be air cooled to room temperature. The beneficial aspect of this treatment is that phosphorus, which was removed from the grain boundary regions, does not go back to these regions during air cooling because phosphorus requires time to diffuse to the grain boundary regions. In this manner, a high temperature soaking in the two phase region and subsequent air cooling should result in good ductility for phosphoric irons [18].

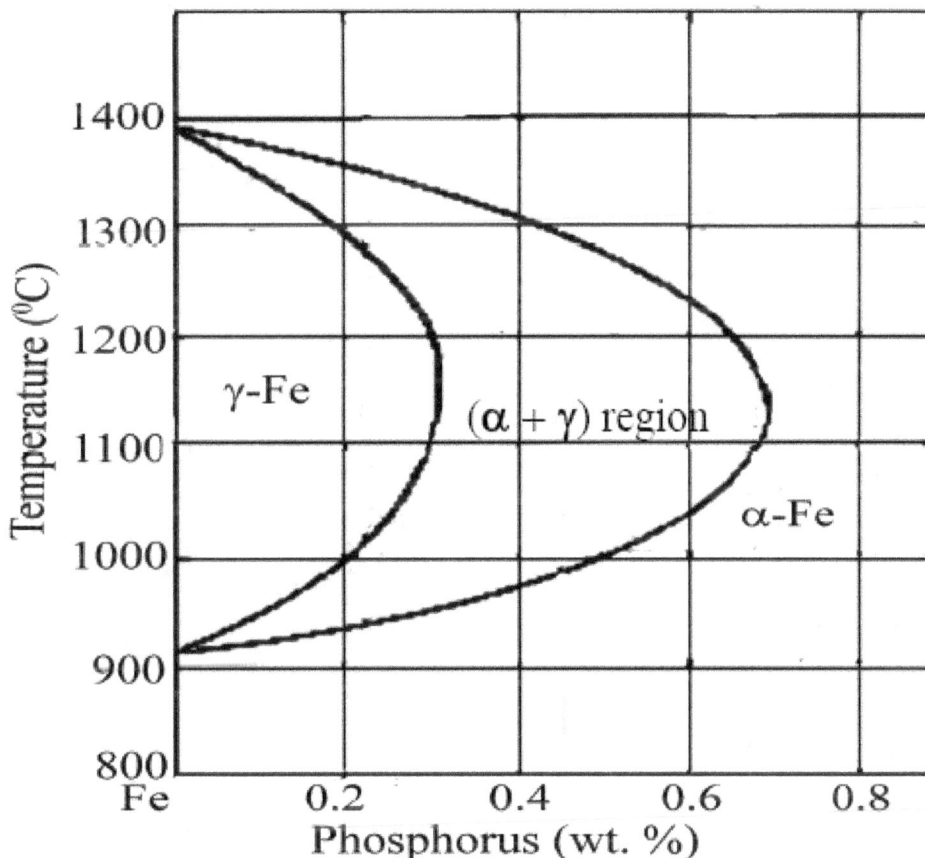

Figure 4 The high temperature gamma loop region of the Fe–P phase diagram.

In order to test this hypothesis, three phosphoric irons P1 (0.11 wt. % P), P2 (0.32 wt. % P) and P3 (0.49 wt. % P) were ingot cast and later forged to rods. In all these phosphoric irons, a small amount (0.02 wt. %) of carbon was maintained. Appropriate heat treatments were devised for these irons, as per the philosophy outlined above. Tensile testing of triplicate samples indicated good ducitlities for phosphoric irons, especially P_1 and P_2 (see Figure 5) [19]. In the same figure, the tensile test result of a commercial reinforcement bar (TISCON of Tata Steel) used for concrete reinforcement applications is also shown.

Good ductilities were obtained for phosphoric irons because phosphorus was kept away from the grain boundaries. This was confirmed by microscopy (see Figure 6). In this figure, the optical micrograph of P_2, that was soaked in the $(\alpha + \gamma)$ dual phase region is shown, after etching with Nital. The light contrast at the grain boundaries indicates locations of low phosphorus content. Interestingly, this structure is known as a "ghost" structure because although the entire structure is ferrite at room temperature, the location of prior austenite and prior ferrite is revealed by the contrast that obtains due to differences in phosphorus contents. Well, this is not important anyway here, but the idea of "ghosts" does also apply in metallurgy!

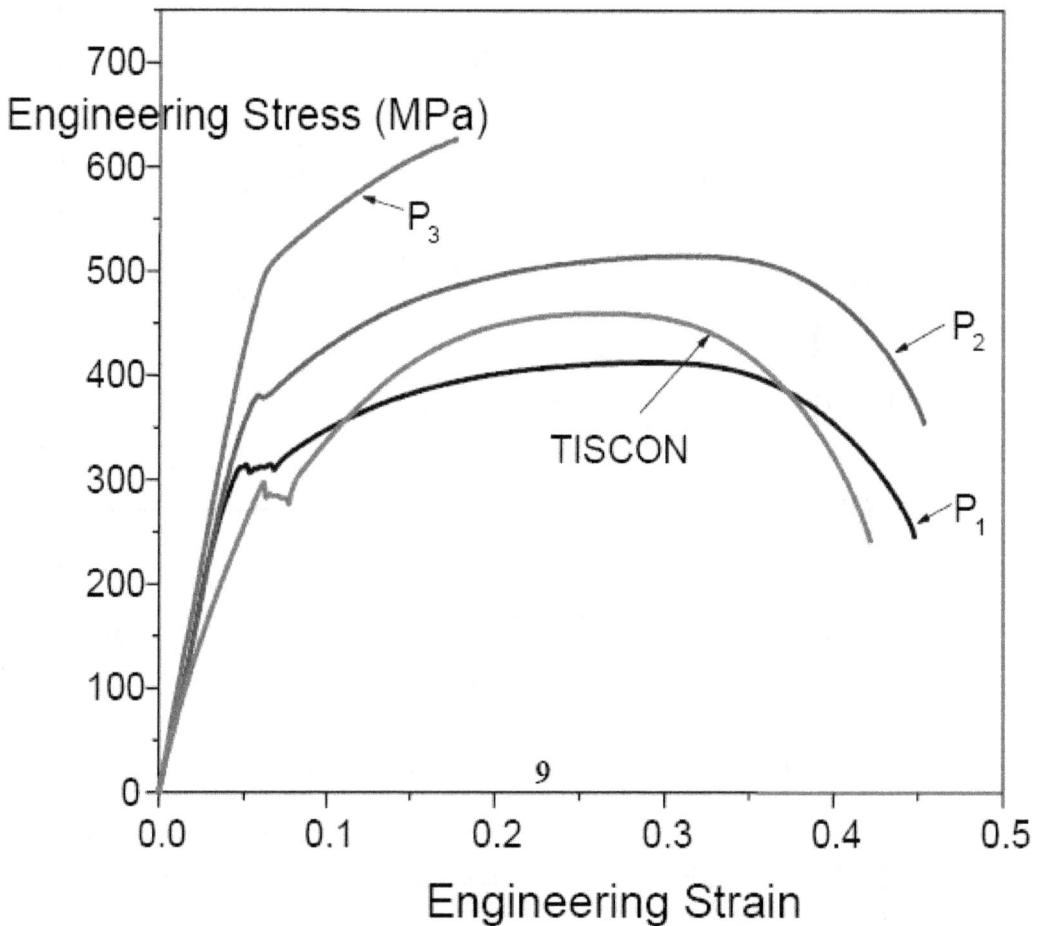

Figure 5 Typical engineering stress-strain curve of the specimens P1, P2, P3 and TISCON.

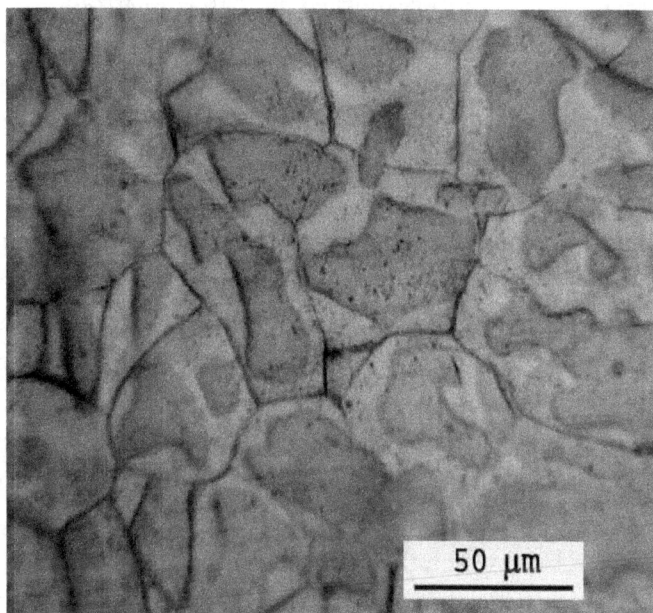

Figure 6 Optical micrograph of heat-treated specimen P$_2$ after etching with Nital.

The good room temperature ductility obtained in phosphoric irons will remain as long as the materials are not exposed to the temperature range of 200°C to 500°C for long times. When phosphorus containing steels are exposed in this temperature range, phosphorus diffuses and segregates to the grain boundaries, leading to their embrittlement. This is known as temper embrittlement. In case phosphoric irons have to be applied commercially in ambient temperature conditions, there would be absolutely no problems regarding their ductility.

The corrosion behavior of phosphoric irons has been evaluated in a wide variety of environments [20-24]. It was concluded that phosphoric irons possess necessary corrosion resistance in near neutral and alkaline conditions. A typical industrial application for which phosphoric irons may be applied is for reinforcing concrete. Therefore, the corrosion resistance of phosphoric irons was evaluated extensively in simulated concrete environments and compared with commercially available material from Tata Steel, Jamshedpur. The results of the detailed study indicate the superior corrosion resistance of phosphoric irons in concrete conditions. For example, the corrosion rate of TISCON was much higher compared to phosphoric irons. Interestingly, phosphoric irons exhibited good passivity in concrete environments. A wide variety of techniques was used to study corrosion in concrete conditions [21,22].

These included polarization methods (linear and Tafel polarization), potentiodynamic polarization and potentiostatic polarization, electrochemical impedance spectroscopy, salt fog immersion, complete immersion and atmospheric immersion experiments. Apart from samples exposed to aqueous solutions, cement-grouted samples were also tested to simulate actual application conditions [21].

It is known that chloride ions are damaging to surfaces because they generally tend to destabilize the protective passive films on the surface. Surprisingly, phosphoric irons showed excellent corrosion resistance even in their presence. This significant result indicates that phosphoric irons will provide good service in environments where chloride ions are present, like in case of structures near seacoast. As India possesses a large coast line, there is wide scope for using phosphoric irons in reinforced concrete structures near coastal areas, which generally call for higher material quality. This is the reason that the major steel makers in India, SAIL and TATA STEEL, market another quality of reinforcement bars, which they call CRS indicating "corrosion resistant steel." Generally, these steels contain a little amount of Cu and Cr as alloying additions and they are costly compared to normal reinforcement bars. Phosphoric irons can perform as good as, if not better, than these commercial CRS materials. An added advantage is that they will not be costly compared normal TMT bars.

In order to visually show the good corrosion behavior of phosphoric irons, the result of immersion testing in saturated $Ca(OH)_2$ solution containing 5 % NaCl is shown in Figure 7. The nature of surface after 125 days of immersion can be noted for three phosphoric irons and two commercial grades of reinforcement steels from Tata Steel (TISCON and CRS). The severe nature of corrosion in case of P_1, TISCON and CRS is quite clear. Phosphoric iron P_3 is the most resistant, followed by P_2 and then P_1. These visual images clearly prove the beneficial effect of phosphorus in conferring corrosion resistance, especially in simulated concrete pore solution containing a relatively high concentration of aggressive chloride ions. In fact, it took about 90 days for corrosion to initiate in the case of phosphoric irons, thus indicating the longer lives that can be anticipate in the case of phosphoric irons when compared with commercial rebar material.

It is important that we realize that ideas for developing phosphoric irons originated from the study of the Metallurgical Wonder of India – the Delhi Iron Pillar, which goes to show that "the best of the new is often the long forgotten past."

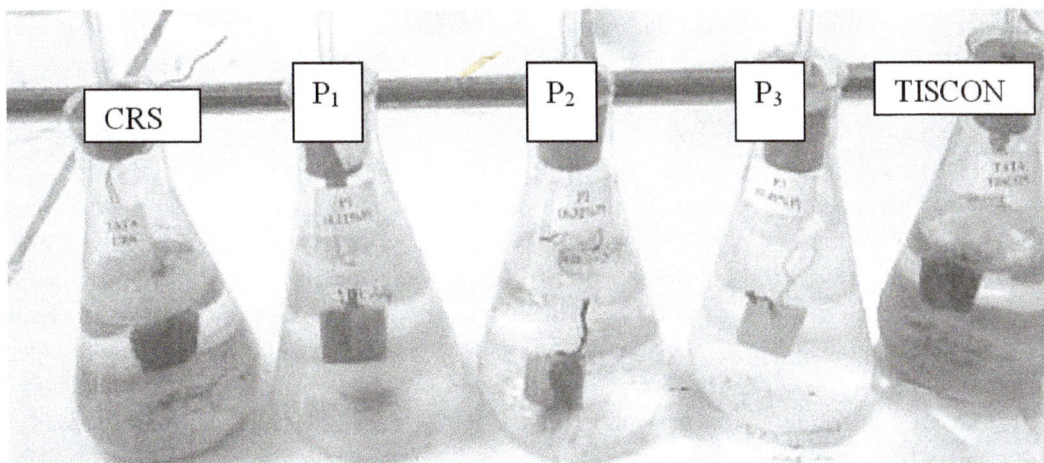

Figure 7 Immersion testing samples after 125 days of immersion in saturated Ca(OH)$_2$ solution containing 3.5 % NaCl.

6. Large Scale Manufacture of Phoshphoric Irons

The possible ways in which phosphoric irons can be manufactured and processed on a large scale will be briefly discussed. As regards the metal extraction stage, the iron produced from the blast furnace would be the primary material for producing phosphoric iron. This ensures that the large productivities that are achieved in the blast furnaces are maintained. The phosphorus impurity from the iron ore remains in the extracted metal because of the reducing conditions maintained in the blast furnace. Therefore, the selection of higher phosphorous containing ores for iron extraction can be envisaged. These high-P ores are deliberately not being mined at present and one beneficial effect that will accrue is that the large reserves of high phosphorus iron ores can be mined for iron extraction. In the next step, the carbon content of the extracted pig iron has to be reduced. Levels down to 0.05%C can be easily achieved in the popular oxygen converter without entailing significant additional costs and change of operational parameters. It must be realized that P will also be removed during the decarburization process in the converter and this can be easily replenished by the well-established techniques of re-phosphorization (which is currently used for producing automotive grade steel). The ingots thus obtained need to be soaked at high temperature for further thermomechanical processing.

In this article, we are considering the production of reinforcing bars because a large tonnage of steel produced in a developing country like India is used for reinforcement of concrete in structural applications. The reinforcement bars have to possess the necessary strength, ductility and corrosion resistance. Modern-day reinforcement bars are strengthened by three methods, namely, microalloying with costly elements like Nb, Ti, etc., cold twisting (which used to be popular earlier), and by controlled cooling in the bar mill plus temp-core cooling process. The bar mill plus temp-core cooling process can produce bars in large tonnage with high productivity. In this process, the soaked ingots are first reduced in size to the shape of bars by means of rolling through a set of rolling strands. When the bar comes out of the final rolling strand, it moves through a quenching stand, wherein surface quenching results in martensitic structure in the outer region of the bar. However, the higher temperature existing in the core, heats up the surface, as the bar proceeds away from the quenching stands, and tempers the surface martensite. In this way, this process produces a hard and tough surface layer with a ductile interior. This is one of the most popular methods for producing large quantities of reinforcing bar material. The complete bar mill plus temp-core cooling process has been explained here because the phosphoric irons can be processed by the same method utilizing the existing arrangements, with the major difference being the ingot soaking, bar quenching and further cooling arrangements have to be fine tuned to produce phosphoric iron with a tough surface and a strong interior, by achieving the three design criteria. The temperatures that are used in commercial temp-core process is amenable for processing phosphoric irons.

There are several advantages of utilizing phosphoric irons for reinforcement applications. The most important, of course, is that there is no need to change the existing iron and steel making technologies and thermomechanical processing technologies for manufacturing phosphoric iron bars on a large scale. If the presence of P can be tolerated, there will be no need to address dephosphorization of steel and phosphorous reversal phenomenon that occurs in the ladles. The addition of expensive microalloying elements can be avoided because P in solid solution can produce the necessary strengths in the bars. The solid solution strengthening effect of phosphorus is particularly remarkable, as the tensile test results of the phosphoric irons reveal (Figure 5). The final and most important advantage of phosphoric irons is their corrosion resistance, especially even in the presence of dangerous chloride ions.

7. Philosophy of Corrosion Prevention by Materials Selection

It will be possible to highlight the important issue involved in the selection and design of engineering materials for prevention of corrosion taking the same example of concrete reinforcement application. This is an important application because, in terms of quantity of material used, the engineering material that is used in the largest amount is reinforced concrete (RC) structures. Concrete is reinforced with reinforcement bars, which are commonly referred to as rebars and these offer additional strength to the structure. Reinforcement bars in normal RC construction are protected against corrosion by the passive film formed on these bars due to the high alkalinity of concrete pore solution (pH ~ 12). However, the presence of chloride ions in excess of a threshold value disrupts this passivating film, rendering the bars susceptible to corrosion [23] In fact, chloride-induced corrosion of reinforcement bars leads to deterioration of reinforced concrete structures and has been a major challenge to civil engineers around the world. As far as the source of chloride ions is concerned, they may be either present in the initial constituents of concrete (e.g., in cases when marine sand or aggregate is used in construction) or diffuse into hardened concrete as in the case of marine structures [24] or structures subjected to deicing salts [25]. The reported annual direct cost of chloride-induced bridge infrastructure corrosion to the U.S. economy was estimated at $8.3 billion per year, and indirect cost was reported many times higher than this [26].

Several methods are used for improved corrosion performance of rebars in concrete structures. One of the popular methods has been to use protective coatings on the surface. These include the use of galvanized and epoxy-coated rebars. For one, they will certainly be costlier options then simple uncoated rebars. While applications have been found, especially in the United States, Europe and Japan, their cost effectiveness and other technical issues still prevent widespread use of these methods. One of the major concerns, especially in India, has been the condition of the epoxy coating on rebars during pouring of concrete. In case there is any damage to the surface epoxy coating, severe corrosion occurs at these breaks and this may lead to intensive localized attack. Therefore, there is care required in selecting coated rebars in applications.

A simple method for preventing rebar corrosion is the use of inhibitors. Cathodic protection has also been suggested as a viable method to prevent corrosion of embedment in concrete. A costly method is the use of expensive stainless steel as rebars.

The above study on the development of phosphoric irons has shown that they fare much better than the commercial rebar materials in concrete environments. The superior corrosion resistance of phosphoric irons in concrete environments is due to the formation of a more protective scale on the surface, one that is capable of withstanding significant amounts of chloride attack. Let us explore what makes the surface passive film work in the case of phosphoric irons.

8. Protection offered by Phosphorus in Iron

Normally, the presence of phosphates in the environment is addressed, like an inhibitor. Phosphates and hydrogen phosphates, like calcium hydrogen phosphates ($CaHPO_4$), sodium phosphate (Na_3PO_4) and sodium hydrogen phosphate (Na_2HPO_4) are known to be anodic corrosion inhibitors in concrete environments. Phosphate species generally adsorb on steel surfaces and prevent them from corrosion. The process of protection is not simple and there are several intricacies that are now being understood. In particular, orthophosphates are considered to be very effective as corrosion inhibitors for iron and carbon steels in aqueous environments related to drinking water distribution systems [27]. Their non-toxic nature is of particular interest in this application. They are also very effectively used as inhibitors for steel reinforcements in concrete [28]. They are also added as zinc phosphate to anticorrosive paints [29]. They come under the category of non oxidizing anodic inhibitors, and are supposed to be efficient in the presence of oxygen. Their relatively low cost is a big advantage. They are relatively less toxic than the nitrite inhibitors used in concrete mixture and chromate inhibitors in painting systems.

The process of film formation on the steel surface in the presence of phosphates is being understood. The nature of film formed depends on several parameters like pH, oxygen presence and applied potential. One of the important parameters if pH [30]. Alkaline pH favors surface films that have an inner region of iron oxyhydroxides and an outer region of iron phosphates [31] while acidic pH favors the formation of iron phosphates [32]. The superior resistance is attributed to the presence of phosphates on the surface. In the particular case of localized corrosion, which is a far more dangerous phenomena, the processes are not very clear at present. Recent studies have indicated that the mechanism of prevention of localized corrosion by phosphates in carbon steels may be related to repassivation due to the formation of protective phosphate contains iron compounds inside the pits [33,34]

Several modern tools of electrochemical research like micro-Raman spectroscopy and scanning vibrating electrode technique (SVET) have been applied in the recent past to monitor *in situ* the pitting processes of carbon steel and gain understanding of the role of pH, dissolved iron and phosphate species concentrations, and dissolved oxygen concentration on localized corrosion processes. [35, 36]. In particular, the phosphate species interact strongly with the iron oxyhydroxide film by adsorbing on the surface and leading to precipitation of iron phosphates. The extremely low solubility of the phosphates has been a significant contributory factor. The importance of oxygen supply to the pits has also been emphasized.

These studies have clearly shown the beneficial aspects of phosphates on corrosion with phosphates in the environment added externally. In the case of phosphoric irons, it is anticipated that phosphates will form in-situ on the surface due to the local interaction of corrosion processes with the material. This is predicted by the Pourbaix diagram of the phosphorus-water system [37]. This has also been confirmed by understanding the evolution of phosphate concentrations in concrete pore solution in which phosphoric irons were immersed [22]. Since the phosphate ions come from the corrosion of the underlying material, it is easy to envisage that these phosphates will be also populated in the inner surface of the surface film. In this manner, the protection offered by a film rich in phosphates in the inner region will be superior to the protection offered by a film right in phosphates in the outer region only. The actual mechanism by which phosphates, which originate from the material (i.e. phosphoric irons), affect corrosion has not been studied and it needs to be taken up to gain further insights on the role of phosphorus and the role of protective passive film containing phosphates in imparting superior corrosion resistance to iron and steels.

The example discussed in great detail in this communication has shown that the best method of protection would be to develop a material that will grow a protective passive film on the surface. This basic philosophy underlying materials selection has been understood in this article with particular reference to phosphoric irons.

In conclusion, the best philosophy of materials selection would be to choose a material with alloying element that can form a passive film on the surface, one that is self healing.

Acknowledgements

The author would like to acknowledge the co-operation of the Archaeological Survey of India in his research on the Delhi Iron Pillar. The author also acknowledges the PhD thesis work of Dr. Gadadhar Sahoo, some of which were used in this article.

References

1. T.R. Anantharaman, *The Rustless Wonder – A Study of the Iron Pillar at Delhi*, Vigyan Prasar, New Delhi, 1996.

2. R. Balasubramaniam, *Delhi Iron Pillar - New Insights*, Indian Institute of Advanced Study, Shimla and Aryan Books International, New Delhi, 2002.

3. R. Balasubramaniam, *Story of the Delhi Iron Pillar*, Foundation Books, New Delhi, 2005.

4. K.R. Trethewey and J. Chamberlain, *Corrosion for Students of Science and Engineering*, Addison Wesley Longman, Second Edition, 1995.

5. S. Radhakrishnan, *The Dhammapada*, Oxford University Press, New Delhi, 1996, p.   122-123.

6. S. Halder, G.K. Gupta and R. Balasubramaniam, Current Science, Vol. 86, 2004, p.   559-566.

7. R. Balasubramaniam, Current Science, Vol., 82, 2002, p.1357.

8. B. Raj, P. Kalyanasundaram, T. Jayakumar, C. Babu Rao, B. Venkataraman, U. Kamachi Mudali, A. Joseph, A. Kumar and K.V. Rajkumar, Current Science, 88 , 2005, p. 1948.

9. W.E. Bardgett and J.F. Stanners, J. Iron and Steel Inst., Vol. 210, 1963, p. 3-10 and NML Technical J., Vol. 5, 1963, p. 24.

10. R. Balasubramaniam and A.V. Ramesh Kumar, Corrosion Science, Vol. 42, 2000, p. 2085.

11. R. Balasubramaniam, Corrosion Science, Vol. 42, 2000, p. 2103.

12. B. Sanyal and R. Preston, *Note on Delhi Pillar*, Chemical Research Laboratory, London, 1952.

13. R. Balasubramaniam, A.V. Ramesh Kumar and P. Dillmann, Current Science, Vol. 85, 2003, p. 1546.

14. Gouthama and R. Balasubramaniam, Bulletin of Materials Science, Vol. 26, 2003, p. 483.

15. R. Balasubramaniam, Current Science, Vol. 84, 2003, p. 126.

16. R. Balasubramaniam, Current Science, Vol. 85, 2003, p. 9.

17. Gadadhar Sahoo, *Corrosion of Novel Phosphoric Irons for Concrete Reinforcement Applications,* PhD. Thesis, IIT Kanpur, 2007.

18. Gadadhar Sahoo and R. Balasubramaniam, Metallurgical and Materials Transactions, Vol. A No. 38, 2007, 1692.

19. Gadadhar Sahoo and R. Balasubramaniam, Scripta Materialia, Vol. 56, 2007, p. 117.

20. Gadadhar Sahoo and R Balasubramaniam, J. ASTM Int., Vol 5, No 5, 2008, p. 2.

21. Sahoo Gadadhar, R. Balasubramaniam and S. Mishra, Corrosion Science, Vol. 63, 2007, p. 975.

22. Gadadhar Sahoo and R. Balasubramaniam, Corrosion Science, Vol. 50, 2008, p. 131.

23. C.L. Page and K.W.J. Treadaway, Nature, No. 297, 1982, p. 109.

24. O. Poupard, V.L. Hostis, S. Catinaud and L. Petre-Lazar, Cern. Concr. Res., Vol. 36, 2006, p. 504.

25. C.D. Johnston, Concr. Int., Vol. 16, No.8, 1994, p. 48.

26. S.D. Cramer, B.S. Covino, S.J. Bullanrd, G.R. Holcomb, J.H. Russell, M. Ziomek-moroz, Y.P. Voramo, J.T. Butler, F.J. Nelson and N.G. Thompson, ISIJ Int., Vol. 42, 2002, p. 1376.

27. C. Volk, E. Dundore, J. Schiermannn and M. Lechevallier, Water Res., Vol. 34, 2000, p. 1967.

28. J.-M.R. Génin, L. Dhouibi, Ph. Refait, M. Abdelmoula and E. Triki, Corrosion Science, Vol. 58, 2002, p. 467.

29. M. Zubielewicz, E. Kaminsja-Tarnawska and A. Kozlowska, Prog. Org. Coat., Vol. 53, 2005 p. 276.

30. Z. Szklarska-Smialowska and R.W. Staehle, J. Electrochem. Soc., Vol. 191, 1974, p. 1393.

31. I.V. Sieber, H. Hildebrand, S. Virtanen and P. Schmuki, Corros. Sci., Vol. 48 ,2006, p. 3472.

32. C.A. Borras, R. Romagnoli and R.O. Lezna, Electrochem. Acta, Vol. 45, 2000, p. 1717.

33. S. Simard, H. Ménard and L. Brossard, J. Appl. Electrochem., Vol. 28, 1998, p. 593.

34. S. Simard, M. Odziemkowski, D.E. Irish, L. Brossard and H. Ménard, J. Appl. Electrochem., Vol. 31, 2001, p. 913.

35. M. Reffass, R. Sabot, M. Jeannin, C. Berziou and Ph. Refait, Electrochim. Acta, Vol. 52, 2007, p. 7599.

36. M. Reffass, R. Sabot, C. Savall, M. Jeannin, J. Creus and Ph. Refait, Corros. Sci., Vol. 48, 2006, p. 709.

37. J. Van Muylder and M. Pourbaix, Atlas of Electrochemical Equilibria In Aqueous Solutions, 1st ed., Cebelcor, Pergamon Press, Brussels, 1966, p. 504.